Blackbeard

The Truth Revealed

Robert Jacob

Published by
DocUmeant Publishing
244 5th Avenue, Suite G-200
NY, NY 10001

646-233-4366

©2024 Robert Jacob. All rights reserved

Limit of Liability and Disclaimer of Warranty: The design, content, editorial accuracy, and views expressed or implied in this work are those of the author.

No part of this publication may be reproduced, stored in a retrieval system, or transmitted in any way by any means—electronic, mechanical, photocopy, recording, or otherwise—without the prior permission of the copyright holder, except as provided by USA copyright law.

For permission contact the publisher at Publisher@DocUmeantPublishing.com

Cover Art: An original painting by artist Sharon Glaze, facebook.com/sharon.h.glaze

Editor: Anne C. Jacob, Popin Edits, AnneCJacob.com

Cover Design: Patti Knowles, www.virtualgraphicartsdepartment.com

Sloop Illustration: Ginger Marks, DocumeantDesigns.com

Layout & Design: Ginger Marks, DocumeantDesigns.com

First Edition

3 4 5 6 7 8 9 10

Library of Congress Cataloging-in-Publication Data

Names: Jacob, Robert (Historian), author.

Title: Blackbeard : the truth revealed / Robert Jacob.

Description: First edition. | NY, NY : DocUmeant Publishing, [2024] | Includes bibliographical references and index. | Summary: "Lies and fiction abound about the life of Blackbeard the pirate. Finally, you will learn the truth"-- Provided by publisher.

Identifiers: LCCN 2024019723 (print) | LCCN 2024019724 (ebook) | ISBN 9781957832395 (hardback) | ISBN 9781957832418 (paperback) | ISBN 9781957832401 (epub)

Subjects: LCSH: Blackbeard, -1718. | Pirates--Atlantic Coast (South Atlantic States)--Biography. | Pirates--Atlantic Coast (South Atlantic States)--Biography--Sources. | Pirates--Atlantic Coast (South Atlantic States)--History--18th century. | Pirates--Caribbean Area--Biography. | Pirates--Caribbean Area--History--18th century. | LCGFT: Biographies.

Classification: LCC F257.T422 J33 2024 (print) | LCC F257.T422 (ebook) | DDC 910.4/5092 [B]--dc23/eng/20240513

LC record available at https://lccn.loc.gov/2024019723

LC ebook record available at https://lccn.loc.gov/2024019724

DEDICATION

This book is dedicated to all pirate enthusiasts throughout the world who portray pirates either in costume or from a historically accurate perspective. Each of these reenactors and living historians is devoted to enriching the lives of others through the vitality, culture, history, and imagination of the Golden Age of Piracy. The vast majority of these devoted individuals do this simply for the love of pirate traditions. Each of them often bears a tremendous financial burden, with no reimbursement for transportation, clothing, equipment, and other expenses.

Through their portrayals at pirate festivals, museums, social occasions, charity events, and many other gatherings, they make pirate history enjoyable and entertaining. The educational value of their efforts is immeasurable. By reaching people who otherwise might not be reached, they keep history alive.

ABOUT THE COVER

My special thanks go to Sharon Glaze, an extremely talented Florida artist who specializes in maritime paintings. Six months before publication, I commissioned her to create an original image of the *Queen Anne's Revenge* to be used on the cover of the book. Sharon's inspired rendition of Blackbeard's ship is based on detailed specifications of the original ship. I gathered those specifications during my research at the North Carolina Maritime Museum and from the published report written by Jaques Ducoin on the *Queen Anne's Revenge* for the North Carolina Department of Cultural Resources. From the number of gun ports to the flags atop the masts, Sharon and I collaborated to produce a highly accurate depiction of the most famous pirate ship in the world. Her magnificent work of art helps bring Blackbeard to life.

CONTENTS

Preface XI

Chapter One: The Treasure Hunt Begins 1

Chapter Two: By Any Other Name 4

Chapter Three: Jacobite Pirates 8

Chapter Four: Charles Johnson and the Book that Ignited Legends 12

Chapter Five: Wealth From the Sea: The 1715 Treasure Fleet 21

Chapter Six: Ye Grand Pirate Captain Benjamin Hornigold 39

Chapter Seven: Blackbeard Begins 51

Chapter Eight: The Gentleman Pirate 68

Chapter Nine: The Unlikely Partnership 76

Chapter Ten: Getting the Dates Right 93

Chapter Eleven: The Queen Anne's Revenge 96

Chapter Twelve: Cruising the Lesser Antilles 107

Chapter Thirteen: Death's Head 125

Chapter Fourteen: Cruising the Gulf of Honduras 129

Chapter Fifteen: Nassau and the King's Proclamation 142

Chapter Sixteen: Siege of Charles Town 153

Chapter Seventeen: Topsail Inlet 167

Chapter Eighteen: Life in Bath 178

Chapter Nineteen: Philadelphia 187

Chapter Twenty: I'll Take Two 198
Chapter Twenty-One: William Howard's Arrest 210
Chapter Twenty-Two: Bonnet on His Own 216
Chapter Twenty-Three: Charles Vane's Visit 231
Chapter Twenty-Four: Brand and Maynard's Invasion of a Province 241
Chapter Twenty-Five: Battle of Ocracoke 254
Chapter Twenty-Six: Aftermath in Ocracoke and Bath 266
Chapter Twenty-Seven: Politics and Legalities 272
Chapter Twenty-Eight: The Answers Lie with Cracherode 282
Chapter Twenty-Nine: Chained in Williamsburg 290
Chapter Thirty: Tobias Knight's Ordeal 301
Chapter Thirty-One: Will the Real Israel Hands Please Stand Up? 311
Chapter Thirty-Two: Meet Edward Thache—A Bristol Man Born 315
Chapter Thirty-Three: Meet Black Beard 322
Chapter Thirty-Four: Meet Mrs. Blackbeard 333
Chapter Thirty-Five: Twisted Tales 339
Chapter Thirty-Six: Blackbeard's Treasure 351

Appendix 369
End Notes 379
Glossary 407
Bibliography 419
Index 433

TABLE OF FIGURES

Figure 1: Caribbean 1702 Herman Moll Map	xiv
Figure 2: Portrait of Prince James Edward Stuart by François de Troy	11
Figure 3: Cover of A General History of the Pyrates, 2nd Edition	13
Figure 4: Illustration of Blackbeard from the 1st Edition 1724	15
Figure 5: Illustration of Blackbeard from the 2nd Edition 1724	15
Figure 6: Illustration of Blackbeard from the 1725 Edition	18
Figure 7: Illustration of Stede Bonnet From the 1725 Edition	18
Figure 8: Illustration of Blackbeard from the 1734 Edition	19
Figure 9: 1715 Treasure Fleet Wreck Locations Florida Map	23
Figure 10: Known 1715 Treasure Fleet Wrecks Map	24
Figure 11: Bay Islands to Cuba from Caribbean Map	28
Figure 12: Distance between Cuba and Portobello from Caribbean Map	45
Figure 13: Blackbeard - 1736 Edition of Johnson's A General History of the Pyrates	50
Figure 14: Hispaniola from the 1702 Herman Moll Map of the Caribbean	52
Figure 15: 1793 Close-up from Map of the Bahamas	56
Figure 16: 1720 Map of the North American Coast Line from Carolina to New York	64
Figure 17: Close-up of 1720 Map of Virginia Showing Teches Island	66
Figure 18: Sloop Revenge	70
Figure 19: Close-up from 1709 Map of North Carolina	73
Figure 20: Close-up of Florida and Carolina Coast from 1719 Map of North American Coastline	75
Figure 21: The Boston News-Letter - Monday, October 28 to Monday, November 4, 1717	81
Figure 22: Capes of Virginia and Delaware from a 1685 Map	86
Figure 23: Close-up Showing Blackbeard's Route from Long Island to Martinique	91

Figure 24: Excerpt of a Letter From Capt. Brand dated March 12 1718-1719	95
Figure 25: Bequia to Martinique from Caribbean Map	98
Figure 26: Queen Anne's Revenge	106
Figure 27: Lesser Antilles to Hispaniola from Caribbean Map	109
Figure 28: Illustration of Basse-Terre	111
Figure 29: Painting of Brimstone-Hill	115
Figure 30: Traditional Version of Blackbeard's Flag	125
Figure 31: Black Flag with Death's Head	128
Figure 32: Close-up of Gulf of Honduras - Caribbean Map	131
Figure 33: Blackbeard's Route - Caribbean Map	141
Figure 34: Illustration of Nassau	144
Figure 35: Nassau Harbour New Providence Bahamas US Navy chart town plan 1885 (1924) map	145
Figure 36: Bahamas and South Florida	147
Figure 37: Excerpt from Phenix Log Showing Hornigold and Martin	151
Figure 38: Close-up of Charles Town Harbor - 1690 Map	154
Figure 39: Inset of Charles Town - 1733 Map	157
Figure 40: 1738 James Wimble Map of North Carolina	166
Figure 41: Close-up of Topsail Inlet showing White House used for navigation - 1738 James Wimble Map	168
Figure 42: Close-up of Route to Bath - 1738 James Wimble Map	175
Figure 43: Close-up of Bath Town from an Eighteenth-Century Map	179
Figure 44: Map of Bath Showing Homes of Knight and Eden	182
Figure 45: Close-up of Marcus Hook - 1771 Map of Pennsylvania	191
Figure 46: 1708 Map of Marcus Hook	193
Figure 47: Photograph of the Plank House	196
Figure 48: Benjamin Franklin Map of the Gulf Stream	199
Figure 49: Ocracoke to Bath - 1733 Moseley Map	202
Figure 50: Map of Bath Showing Plum Point	209
Figure 51: Photo of William Howard's Charge Sheet	212
Figure 52: Cover of the Tryals of Stede Bonnet and Other Pirates	216
Figure 53: Close-up of Virginia and Delaware Capes - 1720 Map	220
Figure 54: Close-up of Battle Site at Cape Fear - James Wimble 1738 Map	224
Figure 55: 1709 Map of North Carolina	225
Figure 56: Close-up of Charles Town and Sullivan's Island - 1820 Mills Atlas	226
Figure 57: Close-up of White Point from a 1739 Map of Charles Town	230
Figure 58: Close-up of Edisto to Charles Town - 1820 Mills Atlas of Charles Town	234

Figure 59: 1837 Woodcut Illustration of the Celebration at Ocracoke — 238

Figure 60: HMS Lyme Logbook Entry Showing Ranger and Jane — 249

Figure 61: Close-up of Roanoke Inlet to Ocracoke Inlet - 1733 Moseley Map — 250

Figure 62: Inset of Ocracoke - 1733 Moseley Map — 256

Figure 63: The capture of the Pirate, Blackbeard, 1718. Painting by J. L. G. Ferris — 263

Figure 64: Close-up of Thatches Hole - 1733 Moseley Map — 265

Figure 65: Photo of Knight's Letter - Minutes of NC Governor's Council, May 27, 1719 — 269

Figure 66: Portrait of Alexander Spotswood — 277

Figure 67: Woodcut Illustration of Blackbeard's Head — 280

Figure 68: Close-up of Kecoughton (Kecoughtan) - 1720 Map — 280

Figure 69: List of Killed and Captured at Ocracoke — 283

Figure 70: List of Pirates Killed from the Original Cracherode Report — 284

Figure 71: List of Pirates Captured and Tried - Cracherode Report — 286

Figure 72: Exterior of the Actual Cells in Williamsburg where Blackbeard's men were held — 290

Figure 73: Interior of the Actual Cells in Williamsburg where Blackbeard's men were held — 290

Figure 74: Hezekiah Hands Entry on Cracherode's Report dated June 20, 1719 — 313

Figure 75: 1740 Map of Jamaica — 318

Figure 76: 1690 Map of Charles Town — 323

Figure 77: Close-up Showing Beard Creek and Bath - 1709 Map of North Carolina — 325

Figure 78: West Bath Creek Plantation, Tobias Knight's property, Beard Land, and the Martin Plantation — 328

Figure 79: Family Tree of the Ancestry of Mrs. White — 338

Figure 80: Cover of A Compendious History of the Indian Wars — 341

Figure 81: Close-up of an 1886 map Showing Black Beard Island, Georgia — 342

Figure 82: Map of Bath Showing Governor Eden's Tunnel Property — 348

Figure 83: Close-up of 1709 Map Showing Charles Town and Fripp Island — 354

Figure 84: Blackbeard Buries His Treasure by Howard Pyle — 355

Figure 85: Close-up of Philadelphia to Burlington - Eighteenth Century Map — 364

Figure 86: Illustration of the Vault on Plum Point — 367

PREFACE

With a genuine love of history and a passion for accurately portraying everyday life from past centuries, I have been a reenactor and living historian since 1971. While in high school, I became a member of the 60th Foot Royal American Regiment, reenacting battles of the French and Indian War. A few years later, when the Bicentennial celebrations began, I migrated to Revolutionary War reenacting. Over the decades, I have attended Renaissance Fairs in period-correct clothing, taken part in authentic Mountain Man Rendezvous encampments, and competed in shooting events with Western Gunfighter groups. Being so involved with living history gave me a unique perspective on historical events. It provided me with a first-hand look into the lives of people from the eras I re-created. Eventually, my enthusiasm for historical accuracy led me to pirates.

The first organized pirate group I joined was "Blackbeard's Crew" in Hampton, Virginia. As with most such organizations, the collective historical knowledge within that group was severely limited. This wasn't the result of a lack of research on the part of the members. There just weren't any books or materials available that gave an accurate and complete account of Blackbeard and pirates in general. Over the past three hundred years, the literary market has been flooded with hundreds of books professing to be historically correct—but they aren't! When I began researching Blackbeard for the group, I found that many recent works written about pirates were simply rewrites of those older books.

The Golden Age of Piracy seemed to be the only period in history where most publications dealing with that period were inaccurate, inadequate, and poorly researched. Thirsting for knowledge on Blackbeard, the central figure of our reenactment group, I began to do as much research on the subject of piracy as time would allow. While doing that research, I decided to write my own book on pirates. Ten years later, I published my first book, *A Pirate's Life in the Golden Age of Piracy*. Three years after that, I released my second book, *Pirates of the Florida Coast: Truths, Legends, and Myths*. Both publications included chapters on Blackbeard, but they appeared merely as a small portion of a much larger work. Thus, the details I could put forth on Blackbeard in those works were limited in both scope and space.

Attendees at pirate festivals, book fairs, museums, and other book-signing events have asked me more questions about Blackbeard than any other pirate. Since my research began with him in the first place, I am always delighted to answer them and dispel a few myths. The time seemed right for me to write the book that I always wanted to write, a book that covered every aspect of Blackbeard's fascinating and controversial life. From the onset, I realized that a book like that would have to be far different from all the other books about Blackbeard that were currently on the market. This couldn't be a simple collection of accounts that had been told previously and retold many times by other authors over the centuries, nor could it be a book that twisted the facts to suit the purpose of the author. Furthermore, to the greatest extent possible, this book would have to tell the complete story of Blackbeard's life.

To accomplish this momentous task, I knew that I would have to rely almost entirely on original prime source documents. For those not familiar with the term, an original prime source document is something that was physically written on or shortly after the date that the document mentions. Examples of these are newspaper articles, trial records, letters from officials, and logbook entries. I must point out that even these documents may occasionally be in error. Newspapers get their facts mixed from time to time. Prisoners on trial don't always tell the truth during their testimony. Politicians occasionally write letters that will further their personal agendas. The truth can be determined if everything is considered and logic is applied.

My research took me to the state archives of Virginia, North Carolina, and South Carolina. I also visited libraries in Williamsburg and Charleston, as well as The Marcus Hook Historical Society and the North Carolina Maritime Museum in Beaufort. At each of those locations, I spent hundreds of hours poring over microfilm and verified copies of hundreds of original handwritten prime source documents. Online, I was able to view digital copies of original publications, including *The Boston News-Letter*, and download digital images of other original documents from the French Archives and from sources in Great Britain. Reluctant to rely upon transcriptions of other researchers, all of those documents were read and transcribed by my wife, Anne, who is a master at reading poor penmanship and eighteenth-century cursive writing. In the process, we discovered that many of those documents had been misquoted by various authors throughout the years.

My first two books were well-researched, but not well-referenced. With those books, my intent was to inform and entertain, not to write a paper that would thrive in an academic environment. Often, my sources were explained within the text, especially when quotes from source documents were provided. That being said, I didn't use the traditional footnotes found in many other works. When I first considered writing a book on Blackbeard, I knew that anything I wrote would fall under the most rigorous scrutiny imaginable. Among pirate historians and reenactors, Blackbeard is sacred. Every fact put forth is susceptible to question. Consequently, in *Blackbeard, the Truth Revealed*, everything will be exceptionally well documented and thoroughly notated. I will also provide detailed explanations in the text as to where these documents can be found.

During the course of this book, I mention multiple previously published books and their authors. This is necessary to discuss varying opinions and to identify sources of accounts or documents. I pondered over how I should refer to those authors. A few of them fall under the category of professional historians. Most would be considered amateur historians to some degree, and others are just plain authors. Regardless of their credentials, each one has written a book containing historical facts or legends. For consistency, I settled on *historian authors*. For example, "According to many historian authors . . . " or "Many historian authors believe that . . . "

I've always disliked history books that presented nothing but facts in a dry and boring manner. On the other hand, I've always disliked history books that included more flowery prose than historical facts. I hope to provide a happy balance in *Blackbeard: The Truth Revealed*. Throughout this book, many accounts will first be told dramatically, as if they were fiction or part of a movie script. But those accounts aren't fiction, they are realistic portrayals of the facts conveyed in a manner that brings interest and a sense of realism to the story. Each of those short vignettes will be followed by the prime source documents that support them. My poetic license will never contradict the facts, only enhance them, and make the details easier to comprehend. For example, if the original document mentions a dawn attack, I might describe the sunrise and discuss how the lighting affected the outcome of the events.

Blackbeard is perhaps the most mysterious pirate of the Golden Age, even though historians have studied him more than most other pirates. Their research spawns many questions that can't be answered. Gaps in his life's timeline and theories and speculation about his motivations abound. In truth, historians know almost nothing about his life before he entered the historical record as a pirate captain. This has forced many historian authors to speculate on possible answers and to adopt unverified facts to support their theories. In *Blackbeard, The Truth Revealed*, I shall present all views and perspectives on unanswered questions presented by other historian authors, explain the reasoning, and discuss their conclusions. The ones that are grounded in false facts or illogical conclusions will be exposed as fiction. The others will be discussed and left up to the reader to decide. Of course, I will offer my opinion based on research and logic. Occasionally, I may even put forth an opinion that is based on unsupported facts. To ensure the reader knows when I am presenting historical facts and when I am offering opinion or speculation, I will use a different font. This is an example of the font used for my opinions and speculation.

Blackbeard: The Truth Revealed is the first comprehensive and definitive look at the world's most famous pirate. Every known aspect of his life is explained. In this book, any gaps in his life story result from gaps in the historical record. I hope that everyone, from casual readers to historian authors, will enjoy learning the truth and that they will gain a deeper understanding of the events and motivations surrounding this fascinating pirate.

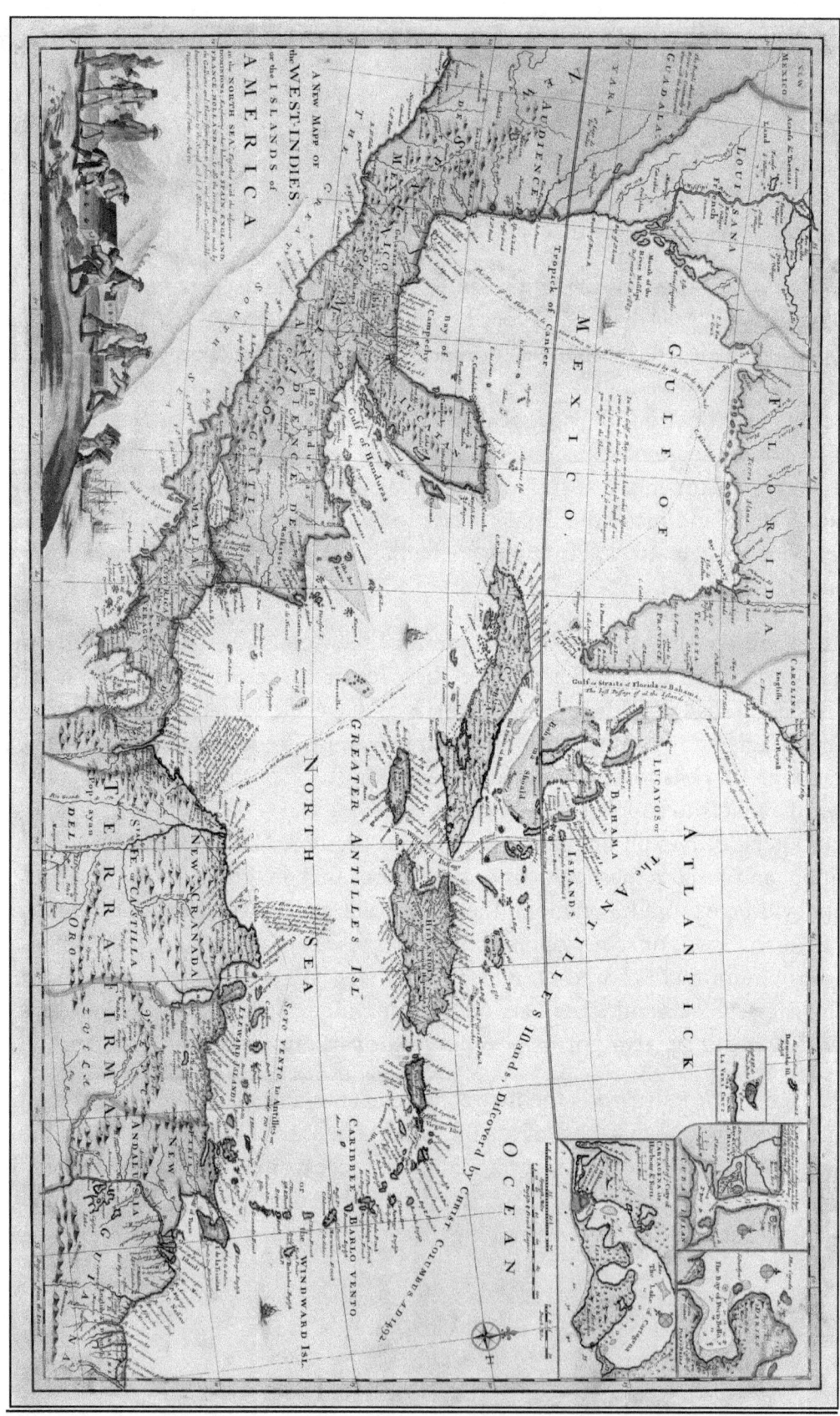

Figure 1: *Caribbean 1702 Herman Moll Map. Courtesy of Barry Lawrence Rudman.*

One
The Treasure Hunt Begins

"The night before he was kill'd, he set up and drank the whole Night, with some of his own Men, and the Master of a Merchant-Man, and having had Intelligence of the two Sloops coming to attack him, as has been before observed; one of his Men ask'd him, in Case any thing should happen to him in the Engagement with the Sloops, whether his Wife knew where he had buried his Money? He answered, That no Body but himself, and the Devil, knew where it was, and the longest Liver should take all."

The paragraph above is a quote from a book written in 1724, titled *A General History Of The Robberies And Murders Of The Most Notorious Pyrates, And Also Their Policies, Discipline and Government, From their first Rise and Settlement in the Island of Providence, in 1717, to the present Year 1724.*[1] In addition to being somewhat dramatic, that paragraph launched a massive hunt for Blackbeard's treasure that persists to this very day. Cities and towns from the Caribbean to New England proclaim their legendary status as the one and only location of Blackbeard's buried treasure.

Over the centuries, it seems that Blackbeard has had more fictional tales invented about his life than any other real historical figure. This is due to a combination of factors. Blackbeard led an action-packed and fascinating life, but his origins are cloaked in mystery. On top of this, there are very few reliable first-hand accounts of his exploits. Many historian authors have embellished the facts when writing about Blackbeard or even fabricated entertaining and imaginative events that have no basis in fact.

Blackbeard's story is perhaps the most complicated pirate tale ever told. There is nothing straightforward about it. Political intrigue abounds. Challenging relationships within his crew and between him and his partners add to the complexity. Blackbeard himself was equally convoluted. Some of the ships he took were burned to the waterline while others were

AUTHOR NOTE:

For consistency, I shall use the term *historian author* to refer to anyone who has written about pirates or pirate history

Chapter One

set free. Throughout his career, there is no evidence that he killed anyone until he found himself engulfed in a desperate struggle for his life, battling Maynard's men at Ocracoke. There are thousands of books, articles, documents, and papers published about Blackbeard. Adding to this is the plethora of television shows and documentaries, movies, podcasts, and blogs. Unfortunately, most of them are severely lacking or outright incorrect. Taking all this into account, it is easily understandable how so many historian authors have missed or confused some of the facts when writing about the world's most famous pirate.

Many historian authors, pirate enthusiasts, and treasure hunters have debated and even argued over the existence of Blackbeard's treasure for over 300 years. So far, to my knowledge, no one has found a single coin. But whether you believe there is a buried treasure somewhere or not, it's undeniable that Blackbeard certainly was a real person. His deeds as a pirate are exceptionally well documented for the brief two-year period between the month when he entered the historical record and the day he was killed. Based on numerous characterizations in books, magazines, movies, and television shows, it is easy to consider him the world's most famous pirate. However, before we can delve into the vague and mysterious world of treasure hunting for any valuables he may have left behind, we must first get to know Blackbeard himself.

Normally, I begin telling every story from the start. In chronicling the lives of pirates, I typically describe where and when they were born, discuss their childhoods, provide details about their early careers, reveal theories about reasons they turned to piracy, and describe in great detail how each of them developed a reputation as one of the world's most famous pirates. However, that won't work in Blackbeard's case.

Everything about his early life still remains unproven, but of the many theories about Blackbeard's origins that have emerged over the centuries, two are verifiably the most credible, and I will describe them in detail. However, these theories are fairly complex and will make better sense to the readers if they are already aware of the known facts regarding Blackbeard's life as a pirate. Therefore, I am compelled to wait until after Blackbeard's complete story has already been told. That way, the theories will make better sense. Hopefully, this approach will empower the readers to put all the pieces together and decide for themselves.

In researching Blackbeard's complete story, as mentioned in the preface, I used extensive documentation taken directly from primary source documents within the historical record or well-grounded research provided by other historian authors. All of my sources will be included in the footnotes.

Additionally, to the greatest extent possible, I shall inform the readers of my primary sources within the actual text.

Occasionally, even source documents can be conflicting and lead to questions about the truth. In some cases, I will provide discussion on both sides and let the readers come to their own conclusions. In cases where I have a strong, well-informed opinion, or when I feel the need to suggest a version of the story that is speculative, I shall use a different font. **As mentioned in the preface, this font will allow the reader to easily differentiate my opinions from well-supported facts.**

At the risk of disappointing the readers who expected to open this book and find a treasure map on the second page, I shall save the thrill and excitement of tracking down Blackbeard's treasure for the final two chapters. Discussion of the various theories on the prospective locations of Blackbeard's treasure won't have any meaning to the readers until the true and complete story of Blackbeard's life as a pirate has been investigated.

Two
By Any Other Name

As we begin our voyage to find Blackbeard's treasure, we must clarify an important point. What is the correct spelling of his legal name? The answer is that no standard spelling existed for the English language between 1716 and 1720. When viewing documents from that time, it is common to find a word or name spelled several different ways, even within the same document.

Blackbeard's legal first name, Edward, isn't in contention. It's his last name that creates the confusion. The most commonly accepted spelling throughout the twentieth century was *T-e-a-c-h*. That's the one most often used in modern books, movies, and television shows. However, nothing is straightforward when dealing with the elusive facts of Blackbeard's life.

Perhaps it was his pronunciation that caused the discrepancy. Based on the various spellings, he most likely introduced himself as Edward /tăch/ (rhyming with patch) or /tāch/ (rhyming with the letter 'h'). Several written variations begin his name with "Th," but this wouldn't have changed the pronunciation. In England, "Th" was often pronounced with a hard "T," such as in *Thomas* or the *Thames River*. Considering the lack of consistency in spelling rules, combined with the unusual accent that Blackbeard may have had, it is not surprising that his contemporaries would have had difficulty figuring out how to spell his name.

Most of the world came to know Blackbeard by reading one of the many editions of Charles Johnson's book, *A General History of the Pyrates*. In his first edition, released in 1724, he spells Blackbeard's name as *Thatch*. Johnson opens his chapter by writing, "Edward Thatch, (commonly called Black-beard),"[1] but just three months later, he released his second edition and changed the spelling to *Teach*. Johnson opens this chapter by writing, "E*dward Teach* was a *Bristol* Man born."[2] That spelling was used in all future editions released throughout the eighteenth and nineteenth centuries, and is the way it is spelled in modern reprints of his book. However, many other spellings exist, as exemplified in the following paragraphs.

Thach. Henry Timberlake was the first person to associate Blackbeard's name with piracy. Timberlake was the master of the *Lamb*, a brigantine taken by two pirate sloops in December 1716. Afterward, he sailed to Jamaica and gave a deposition, in which he provided a detailed description of the incident, including the names of two pirate captains. I will, of course, cover this attack in great detail later in this book, but for now, one of the captains he identified was "Edward Thach Comander of another Sloop."[3] Timberlake probably never saw Blackbeard's name in print, since pirates seldom left written records of their piracies. He likely only heard the name spoken by someone onboard. From that, either Timberlake or the court recorder taking his deposition concluded that the pirate captain's name was spelled *Thach*.

Two years later, Governor Alexander Spotswood of Virginia used this same spelling when he mentioned Blackbeard in an address to his Council. In the minutes of that meeting, his name is written as "Edward Thach."[4] In addition to the Governor of Virginia, some of the naval officers involved with the attack on Blackbeard also used that spelling. Shortly after Blackbeard's death, the logbook of HMS *Pearl*, the ship that provided part of the crew for the sloops that attacked and killed Blackbeard, contains the entry, "This day the Sloop *Adventure* Edward Thach formerly Master . . ."[5] Additionally, Captain Ellis Brand, the naval officer and area commander during that attack on Blackbeard, also spelled his name this way. In an account of the engagement dated February 6, 1719, Brand wrote, "Pyrate Thach alias Blackbeard."[6]

Tache. To illustrate the point of the constant variation of spelling, a year earlier, on December 22, 1718, the same Governor, in a letter to the same Council, had spelled his name differently, writing, "one Capt Tache a noted Pyrate."[7]

Tach. William Howard was Blackbeard's quartermaster. He was arrested in Virginia in September 1718. The order for his arrest contains the statement, "since which one Howard, Tach's Quarter Master."[8] This order was issued by Governor Spotswood, making this the third variation of the spelling of Blackbeard's name used within the same Virginia legislature.

Thatch. Captain Mathew Musson was the second person to use Blackbeard's legal name and identify him as a pirate. The governor of South Carolina sent Musson to the Bahamas in 1717 to make a report on all the pirates who were using Nassau as a base. In that report, Musson wrote, "Thatch, a sloop 6 gunns and about 70 men."[9] His report became part of the minutes of the Council of Trade.

This spelling seems to have been the one most commonly used in official documents within the British government. Two years later, a report made by the Secretary of the Treasury concerning reward payments for Blackbeard's death includes this comment, "Edwd. Thatch a Notorious pirate with his Crew."[10] Two years after that, Captain Gordon, the naval officer who was the overall commander of the personnel who attacked and killed Blackbeard wrote a letter to the Admiralty dated September 14, 1721, referring to the event as "that action with Thatch, alias Blackbeard."[11]

Teach. The most common spelling of his name within the world of publication comes from *The Boston News-Letter Boston Mass,* the only newspaper in the English colonies of North America in the early eighteenth century. There were numerous articles on Blackbeard appearing in this paper from October 1717 until after his death. The issue dated Monday, October 28 to Monday, November 4, 1717, contains, "*From Philadelphia, October 24.* We are informed that a Pirate Sloop of 12 Guns 150 Men, Capt. Teach Commander."[12] Every time Blackbeard's name was mentioned in this publication, it was spelled *Teach*. Leave it to a newspaper to get your name wrong!

This spelling was also used in a letter dated October 24, 1717. It was written by James Logan, an influential Philadelphia merchant, to Robert Hunter, Governor of New York and New Jersey, concerning several pirate attacks on vessels near Philadelphia. Logan wrote, "their Comandr is one Teach."[13]

Tatch alias Blackbeard. Another letter was written as a result of Blackbeard's attacks on Philadelphia shipping. This one, dated October 23, 1717, was from Jonathan Dickinson, the mayor of Philadelphia, to a Jamaican merchant named Joshua Crosby. Dickinson wrote, "Capn Tatch alls Blabeard."[14] This letter is remarkable! It was the first time in recorded history that the nickname *Blackbeard* was used.

Titche. The strangest spelling came from the French. After Blackbeard took *La Concorde*, the French captain, Pierre Dosset, made a statement dated December 10, 1717, in which he said, "hommes commandés par Edouard Titche anglaise."[15] Of course, in addition to this odd spelling, his first name was spelled in the French language too. To quote a humorous line from a famous movie, "That would be the French!"

Theach. Blackbeard's family supposedly lived in Spanish Town, Jamaica, in the early eighteenth century. Most of the church records St. Catherine's Church, Spanish Town, spell the name as *Theach*.[16]

Teech. Captain Pearce of HMS *Phenix* mentioned Blackbeard in a letter writing, "I presume e're this comes to hand their Lordships will hear that one Teech Commander of a pirate."[17]

Thache. This spelling is most likely the one actually used by Blackbeard during his lifetime. A man named Cox Thache lived in Jamaica in the 1720s and claimed to be the half-brother of Blackbeard. In his will dated November 7, 1736, his name is spelled *Thache*.[18] The most conclusive evidence that Blackbeard himself spelled his name as *Thache* comes from Tobias Knight. He was the Customs Collector and Chief Justice in North Carolina in 1718 when Blackbeard was there. During that time, Blackbeard and Knight had several business dealings where his legal name would have been used in the documents. After Blackbeard's death, those business dealings became the subject of a hearing conducted by Governor Eden and his Council. Throughout those minutes of that meeting, dated May 27, 1719, every time Blackbeard's legal name was mentioned, it was spelled *Thache*. An example of this is, "forenamed Tobias Knight—directed to Capt. Edward Thache on board his sloop *Adventure*."[19]

So, it appears that the spelling of Blackbeard's name was *Thach, Tache, Tach, Thatch, Teach, Tatch, Titche, Theach, Teech,* or *Thache*. Throughout this book, I shall try to avoid confusion by using his nickname, *Blackbeard*, as often as possible. On the occasions when I do write his last name, I shall use the same spelling as the source documents I am referencing.

Three
Jacobite Pirates

A term that every Blackbeard enthusiast must become thoroughly familiar with is *Jacobite*. This refers to supporters of James Edward Stuart. Living in exile in France, he laid claim to the throne of England and Scotland after the death of his sister, Queen Anne, in 1714. *Jacob* is Latin for *James*, and all those supporters who were in favor of placing James on the throne were called *Jacobites*. Many pirates sailing out of Nassau and other ports in the Caribbean after 1715 were Jacobites. This is evident by two facts. The first is that they named their ships after James Edward Stuart, his family, and the Jacobite principles. The second is the many reports from victims who were being held prisoner by various pirate groups between 1716 and 1722, who witnessed their captors drinking toasts to James Edward Stuart.

Perfect examples of both of these can be found in the case of Stede Bonnet, Blackbeard's partner. After leaving Blackbeard's company in the summer of 1718, he renamed his sloop the *Royal James*. This is in direct reference to James Edward Stuart.[1] As for toasting, during the trial of Bonnet's crew, James Killing, who was held captive by Bonnet's pirates, testified that the pirates "made Bowls of Punch, and went to Drinking to the Pretender's Health."[2] *Pretender* was a common term referring to James Edward Stuart. Historian author Colin Woodard correctly states that Henry Jennings, Edward England, Palgrave Williams, Charles Vane, and Blackbeard were all Jacobites.[3]

So, what is this all about? Who was James Edward Stuart and what were his grounds for claiming the throne of England and Scotland? To answer that, I must give you a brief and condensed history of the English monarchy and the House of Stuarts in the seventeenth and early eighteenth centuries.

Scotland had been ruled by the Stuarts for hundreds of years, while England had been ruled by the Tudors since 1485. In the last half of the sixteenth century, the queen of England was Elizabeth I, daughter of Henry VIII of the House of Tudor.[4] Her aunt, Henry VIII's sister, had married James V, a member of the House of Stuarts. When Elizabeth died in 1603 without an heir, England looked to her cousin, James VI of Scotland, and he became King of England as James I, while retaining the throne of Scotland, ruling both countries simultaneously.[5]

In 1685, his grandson became King James II of England and King James VII of Scotland and ruled them simultaneously as his grandfather had done.[6] He had two daughters with his first wife, who were members of the Church of England. Their names were Mary and Anne. But his first wife died, he married a Catholic, and eventually converted to that religion. The English people as well as the Parliament weren't very keen on having a Catholic king, and unrest circulated throughout England. However, Scotland wasn't that concerned. The straw that broke the camel's back was the birth of a Catholic son in 1688. This was, of course, James Edward Stuart.[7]

An uprising began in England called the "Glorious Revolution" and James was deposed. His daughter Mary and her husband William became King and Queen of England. Both William and Mary had a strong claim to the throne. Mary was the daughter of King James II and William was his nephew, meaning that William and Mary were first cousins. Therefore, they were referred to as *King and Queen*. When they assumed the throne of England, they also claimed the thrones of Scotland and Ireland as well.[8] But as far as the people of Scotland and Ireland were concerned, James was still their king. They argued that the English had no right to remove James from their throne, so they revolted in what is known as the first of the *Jacobite Rebellions*. The revolt was eventually put down and James fled to France with his infant son, James Edward Stuart, who became known as *The Pretender, The Pretender Across the Sea, Royal James*, and even *James III*.[9]

James II died in exile in 1701, and his 13-year-old son, *James the Pretender*, became a legitimate heir to the throne of England, Scotland, and Ireland in the eyes of his supporters. They hoped that when King William III died, the young James would become king. But young James was a Catholic like his father, so in 1701, the English Parliament passed The Act of Settlement, which forbade a Catholic from becoming king of England. But it went much further than that. It specifically named Princess Anne, Mary's sister, as the heir when William III died. But there's more. It names a successor to Anne if she dies without heirs. That successor was any legitimate heir of James I's daughter, Sophia, who had married the Elector of Hanover.[10]

1603–1625

James I Ruled England

1685–1688

James II, father of James, ruled England

The Act of Settlement, passed in 1701, forbade a Catholic from becoming a monarch in England.

Chapter Three

1702-1714

Queen Anne ruled from 1702 to 1714 and established the union of England and Scotland as Great Britain in 1707.

The following year, King William III died. His wife Mary had died several years earlier, so in 1702, Anne Stuart was chosen to rule England, Scotland, and Ireland as Queen Anne. She was a good choice because she was the second daughter of James II and the older half-sister of James the pretender. Additionally, she was a member of the Church of England and not a Catholic. The English people accepted her, as did the people of Scotland and Ireland. At least she was a Stuart. During her reign, in 1707, the Union Act was passed, officially joining England and Scotland as one nation. That's when they stopped calling the two nations England and Scotland and began calling them Great Britain. When Ireland and all the English colonies were included, it was called The British Empire.[11]

However, when Queen Anne died in August of 1714 with no surviving children, the English Parliament referred back to the Act of Settlement and chose Sophia's son, George Ludwig the Elector of Hanover to be King George I. Because of the Union Act, the English didn't feel it necessary to consult the Scottish government, who also had a significant interest in who their king would be. With that decision, the stage was set for civil war.[12] Scotland's view toward the English Parliament's decision was that they could choose George to be king of England, but they had no authority to choose him as the king of Scotland.[13]

The English government countered with the Union Act and claimed that they did. George had never been to England and didn't even speak English, but now, he was the King of Great Britain. This tore the kingdom apart and divided English politics into two groups: the Whigs who were supporters of George, and the Torys who were opposed. As soon as he became king, George I instituted extensive policy changes that excluded all members of the Tory party from royal favor, and many successful military leaders and statesmen found themselves out of a job. This included the Duke of Ormond, Captain General of the British Army, and Captain George Cammock of the British Royal Navy.[14]

1688-1766

James Edward Stuart was also known as The Pretender, The Pretender Across the Sea, Royal James, and James III.

By 1715, a full civil war was being planned in Scotland. In England, anyone suspected of plotting against George I was arrested. Finally, on September 6, 1715, a Scottish army was ready to march against England and what is called *The 15* began.[15] James Edward Stuart claimed the title of James III and traveled to Scotland to raise support for his army.[16] But he needed a navy too. James III chose Captain George Cammock to command the White Squadron and commissioned him as an admiral on October 17, 1715.[17] During November and December of 1715, thousands of Scottish troops marched on England but they were all defeated in a series of fierce battles. By January 1716, it was clear to James III that his revolution was over, at least for the moment, so he fled back to France.[18]

At the time, the English government painted the supporters of the Jacobite Rebellion as being almost completely Scottish and Irish. They believed that it wouldn't look very good for the government if the general public knew that a great many Englishmen supported the revolution, too. This concept exists to this day. However, many Englishmen did support the Jacobite cause. Because of the tight control the government held over the press and the fact that all who openly voiced support for the Jacobites were arrested, it is impossible to know how many supporters were in England. But in the colonies, where control was less restricted, Jacobite support was far more evident. This is especially true for pirates. Many of them were openly Jacobites. Most pirate historian authors and biographers have commented on the great number of Jacobite pirates operating between 1716 and 1722. In fact, I am convinced that many honest sailors and captains may have turned to piracy simply in order to support the Jacobite cause.

Figure 2: *Portrait of Prince James Edward Stuart by François de Troy. The image is dedicated to the public domain under CC0.*

George Cammock is the key link between the pirates in Nassau and the Jacobite cause. His name is sometimes spelled Camocke. Both spellings appear in contemporary documents, but Cammock is used more often. After the rebellion crumbled in Scotland, Cammock began supplying arms to pirates and pirate-friendly gun runners.[19] Late in 1717, the pirates on Nassau, including Charles Vane,[20] wrote Cammock a letter saying that they "did with one heart and voice proclaim James III for their King," and were "resolved to prosper or perish in their bold undertaking." Additionally, they proposed to raise a navy of pirates with Cammock as their admiral with the title "Captain General of America, by Sea and Land."[21]

1666–1722

George Cammock entered the navy in 1682 and served with great distinction as captain of the *Speedwell* (1702) and *Monck* (1712–14).

The concept of a Jacobite navy based at Nassau appealed to Cammock. As part of his plan, he would grant pardons to all the pirates in the name of King James III. Cammock even planned on purchasing a 50 gun warship from the Spanish and enlisting Jacobites from Scotland and England to be the crew. He wrote, "Employing them against the common enemy will be the only means to make way for a Restoration." However, this plan never materialized.[22] The second paragraph of this chapter clearly shows that Stede Bonnet and most of his crew were Jacobites. As you shall see later on in this book, Blackbeard was a Jacobite, too.

Four
Charles Johnson and the Book that Ignited Legends

For 300 years, most historian authors of books on pirates have relied heavily on one primary source, a book published in 1724 titled, *A GENERAL HISTORY OF THE PYRATES FROM Their Rise and Settlement in the Island of Providence, to the present Time. With the remarkable Actions and Adventures of the two Female Pyrates Mary Read and Anne Bonny*. This book was written by Captain Charles Johnson. Even today, many people assume that since it was written in 1724, it must be accurate and correct. But this is not the case. *Johnson*, which was a pen name, was focused on selling books, not telling the facts. Most of the details described in Johnson's work are highly embellished or simply created by the author. It wasn't until the late twentieth century that most historians realized this. As images and transcriptions of primary-source documents became widely available on the internet, historians began to compare the verifiable facts against those in Johnson's book, and the accuracy of that book came into question. Eventually, it became apparent that his book should be considered more as a word of historical fiction. Although it took 300 years for most to come to that realization, a few people recognized it for what it was right from the beginning. In August 1724, shortly after its release, a contemporary correspondent wrote that "this shim-sham Story of Pyrates" was more of a political allegory rather than a contemporary history.[1]

Even though most of the details described in Johnson's work prove to be incorrect, some of them are correct. This is what makes it so maddening for researchers who use Johnson's book as a source. Which facts can be relied upon and what elements should be completely discounted? The solution is to cross-reference everything with other primary sources. Fortunately, when it comes to Blackbeard, many primary sources are available. I would first like to briefly dispel some of Johnson's most popular stories about Blackbeard.

The battle with HMS Scarborough never happened. Johnson wrote:

> A few Days after, Teach fell in with the Scarborough Man of War of 30 guns, who engaged him for some hours; but she finding the Pyrate well mann'd and having tried her strength, gave over the Engagement, and returned to Barbados, the Place of her Station; and Teach sailed towards the Spanish America.[2]

HMS Scarborough was a 32 gun fifth-rate man-o-war.

The engagement was supposed to take place near Nevis Island. The logbook of the Scarborough makes no mention of such a battle. In fact, the Scarborough wasn't even in the area. HMS Seaford was near Nevis at the time and recorded seeing a large ship they assumed to be a merchant vessel, but there were no shots fired.[3]

Blackbeard's fourteen wives can't be verified either. Johnson wrote this sensationalized and highly provocative account:

> Before he sailed upon his Adventures, he marry'd a young Creature of about sixteen Years of Age, the Governor performing the Ceremony. As it is a Custom to marry here by a Priest, so it is there by a Magistrate; and this, I have been informed, made Teach's fourteenth Wife, whereof, about a dozen might be still living. His Behaviour in this State, was something extraordinary; for, while his Sloop lay in Okerecock Inlet, and he ashore at a Plantation, where his Wife lived, with whom after he had lain all Night, it was his Custom to invite five or six of his brutal Companions to come ashore, and he would force her to prostitute her self to them all, one after another, before his Face.[4]

There are no contemporary records or even comments about any wife of Blackbeard's except a casual comment made by Brand in his February 6, 1719 letter where he wrote "he design'd to be an inhabitant & leave of his Piraticall Life and the sword to put a life so to his designs he married there."[5] As for his Bath Town wedding, an in-depth search of the records

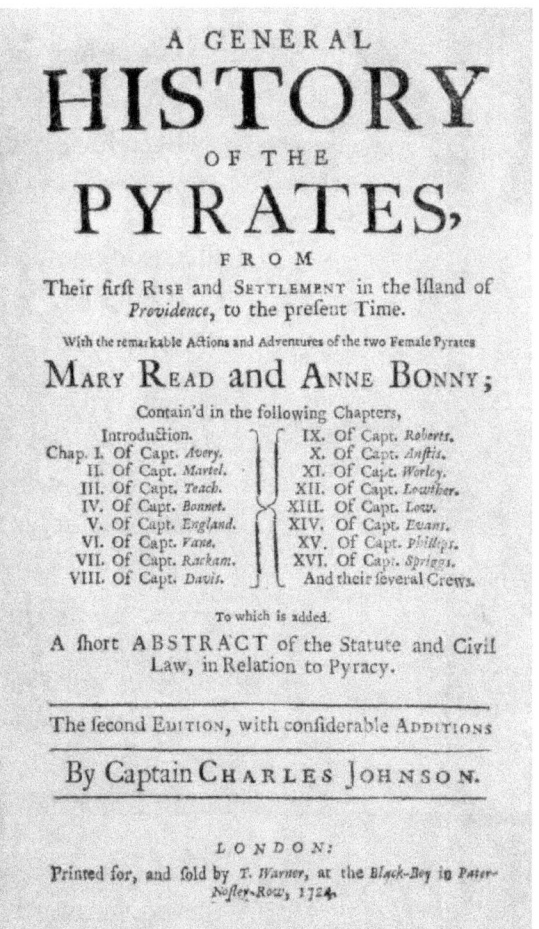

Figure 3: *Cover of A General History of the Pyrates, 2nd Edition. The image is dedicated to the public domain under CC0.*

failed to produce any evidence of any wedding ceremonies performed in Bath by Governor Eden in 1718.

Johnson's details of the battle at Ocracoke are filled with many inaccuracies too. A correct and detailed account of the battle will be discussed in a later chapter of this book. However, there is one event that took place during the battle that I would like to address now. At some point, Blackbeard posted a man below deck with a match and orders to blow up the *Adventure* if it looked as if they were going to lose. This happened, but Johnson chose to embellish it a little, writing:

> Teach had little or no Hopes of escaping, and therefore had posted a resolute Fellow, a Negroe whom he had bred up, with a lighted Match, in the Powder-Room, with Commands to blow up when he should give him Orders.[6]

This actually happened, as it was documented in a letter written by Lt. Governor Spotswood of Virginia, who wrote:

> His orders were to blow up his own vessel if he should happen to be overcome, and a Negro was ready to set fire to the Powder had he not been luckily prevented by a Planter forced on board the night before & who lay in the Hold of the sloop during the actions of the Pyrats.[7]

Johnson's embellishment, the comment that he was a "Negroe whom he had bred up," seems to be a matter of poetic license and implies that this man had been his slave since he was a child. There is no mention from any source that Blackbeard ever owned a slave.

The true story of the man with the match is far more intriguing. Although today he is commonly referred to as *Black Caesar*, Johnson didn't mention his name. In fact, the name *Black Caesar* doesn't appear anywhere in any of Johnson's editions. *Caesar* appears in his books as one of the captured crew, but not *Black Caesar*. However, not even the name of *Caesar* was attributed to this mysterious individual with the match until the mid-twentieth century, in Rankin's 1960 publication titled *The Pirates of North Carolina*. In my research, *Black Caesar* is not mentioned in Blackbeard literature until Angus Konstam's 2006 publication, *Black Beard: America's Most Notorious Pirate*.

Johnson's publication mentioned above was actually the second edition of his book, released in August 1724, just three months after the first edition.[8] This second edition is by far the most popular. It remains in print today under a plethora of titles and is the one most often used as a primary source reference by historian authors. After examining and comparing both editions, I concluded that the differences between the two are negligible, but they are interesting. Both of these editions, as well as several others, are in my personal possession. In the first edition, Blackbeard's name is spelled *Thatch*, and he is reportedly from Jamaica. Also, since the chapter on Stede Bonnet comes before the "Black-beard" chapter, all the escapades they shared as partners are contained within the Bonnet chapter.[9]

In the second edition, and all subsequent editions, Blackbeard's name is spelled *Teach*, and he is reportedly from Bristol. Since the chapter titled "Black-beard" appears before the chapter on Bonnet, the reverse is true. All the escapades they shared as partners are contained within the "Black-beard" chapter.[10] Therefore, many researchers looking at only the "Black-beard" chapter in the first edition will notice that many accounts, such as

Figure 4: *Illustration of Blackbeard from the 1st Edition 1724. The image is dedicated to the public domain under CC0.*

Figure 5: *Illustration of Blackbeard from the 2nd Edition 1724. The image is dedicated to the public domain under CC0.*

Daniel Defoe (1660-1731) was the author of many novels, including *Robinson Crusoe* (1719).

the battle with HMS Scarbrough and the siege of Charles Town, seem to have been omitted. They weren't. They're just in another chapter.

The first two editions only contained three illustrations: one of Blackbeard, one of Anne Bonny and Mary Read, and one of John Roberts. Blackbeard's illustration is the only one of the three to have changed. The two illustrations are very similar, with the most noticeable difference being his hat. What Johnson describes as a "Fur-Cap" in his first edition was replaced with a more traditional hat by the second. The only other change was to Blackbeard's bandoleer.[11]

For almost 300 years, scholars have debated over the true identity of the mysterious Charles Johnson. In 1932, literary scholar John Robert Moore suggested that Daniel Defoe may have been Charles Johnson, based on a similarity in literary style. This was reinforced in the introduction of the 1972 reprint of *A General History of the Pyrates*,[12] when Manuel Schonhorn again declared that Charles Johnson was actually Daniel Defoe. This concept became widely circulated. Many recently republished versions of this book identify Daniel Defoe as the author, and even today, many modern libraries catalog this book under the author Daniel Defoe.

The debates over Johnson's true identity continued until 2004, when researcher Arne Bialuschewski uncovered a treasure trove of fascinating facts concerning the identity of our author in question. After extensive research, Bialuschewski concluded that the actual identity of Charles Johnson is Nathaniel Mist, who began publishing a tabloid paper titled *The Weekly Journal; or, Saturday's Post* on December 15, 1716.[13] This publication was a politically oriented publication that ran articles opposed to the Whig party. Researcher James Markham wrote:

> The Journal's voice was to be more persistent in its attacks and criticisms than that of any other Tory newspaper, and Mist was to be so consistently punished as to become the outstanding printer prosecuted by the Whigs during this decisive period in the history of the fight for press freedom.[14]

Daniel Defoe was heavily connected to Mist right from the beginning of the *Journal* and worked on that publication as part of the staff.[15] As colleagues, Mist and Defoe undoubtedly would have shared many writing tips and developed similar writing styles. As a result, articles written by Mist were often attributed to Defoe. This upset Mist, who was constantly defending his articles and denying that Defoe had written them. Eventually, the two broke off their association.[16]

Mist's *The Weekly Journal; or, Saturday's Post* began featuring stories on pirates in 1717. Due to Mist's personal style and the need to sell papers, the stories in his papers were a mixture of information and entertainment. By December 1719, fictional accounts of pirates began to appear in this publication regularly. This dramatically increased between 1722 and 1723 when even more articles on pirates appeared as Mist drew closer to authoring his first book.[17]

By 1724, all those stories had coalesced into one enormous volume, and Mist was ready for publication. His first edition was titled *A GENERAL HISTORY of the Robberies and Murders Of the Most Notorious PYRATES, and also Their Policies, Discipline and Government, From their first Rise and Settlement in the Island of Providence, in 1717, to the present Year 1724. With The remarkable Actions and Adventures of the two Female Pyrates, Mary Read and Anne Bonny*. It was "printed in London by Ch. Rivington at the Bible and Crown in St. Paul's Church-Yard, F. Lacy at the Ship near the Temple-Gate, and J. Stone next the Crow Coffee-house the back of Greys-Inn."[18]

MAY 14, 1724

The first edition of *A General History of the Pyrates* was released.

Advertising for this upcoming book was spread all over *The Weekly Journal; or, Saturday's Post*. The first copy hit the streets on May 14, 1724.[19] If all those coincidences between Mist's magazine and Charles Johnson's book aren't enough to convince you that Mist and Johnson are one and the same, the ultimate proof can be found among the records of the Stationer's Company. Mist's foreman, John Wolfe, registered Mist's first edition under the name Nathaniel Mist with the Stationer's Company on June 24, 1724. At that time, such a registration was similar to a modern-day copyright.[20]

Most historians today believe that Captain Charles Johnson's true identity was Nathaniel Mist. Some historian authors refer to him in their works as "Johnson/Mist." Since *Charles Johnson* is so ingrained in the minds of the public as the author of this book, I shall set aside the fact that his true identity was *Nathaniel Mist* and continue to use the pen name of *Charles Johnson* throughout the rest of this book.

Charles Johnson was the pen name used by author Nathaniel Mist.

The most famous of all the editions is, of course, the second, *A General History of the Pyrates from Their Rise and Settlement in the Island of Providence, to the present Time. With the remarkable Actions and Adventures of the two Female Pyrates Mary Read and Anne Bonny*. Released in August 1724, it was printed for, and sold by T. Warner, at the Black-Boy in Pater-Noster-Row.[21] The third edition, with the same title as the second, was published in 1725 with the addition of a chapter about "Captain Smith."[22]

Defoe's novel, *Robinson Crusoe*, was also printed by W. Taylor at the Ship in Pater-Noster-Row.

That same year, a rival publisher named Edward Midwinter at the Looking-Glass on London-Bridge, pirated Johnson's work and released a

Figure 6: *Illustration of Blackbeard from the 1725 Edition. The image is dedicated to the public domain under CC0.*

Figure 7: *Illustration of Stede Bonnet From the 1725 Edition. The image is dedicated to the public domain under CC0.*

smaller version with the title, *The General History and Lives of all the most Notorious Pirates, and Their Crews*.[23] Imagine, *A General History of the Pyrates* being pirated! Charles Johnson's name doesn't appear anywhere in the book and no author is credited. The content is basically identical, with a few minor cuts to shorten the chapters. However, each chapter is adorned with a poorer quality woodcut print, one for each pirate mentioned. There are nineteen in all. These prints are the illustrations that one most often sees at museums throughout the United States.

A fourth edition was released in 1726 as a two-volume set with fifteen additional chapters on totally new pirate captains. His printer for this one was Woodward, at the Half-Moon, over against St. Dunstan's Church, Fleet-Street.[24] Due to his continued support for the banished Jacobite court, Johnson finally fled England for France in 1728 and lived in Boulogne from 1730 to 1734. He kept in close contact with his printers back in England and continued to release editions of his most famous book.[25] While Charles was out of the country, his rival printer, Edward Midwinter at the Looking-Glass on London-Bridge, released another pirated edition in 1729.[26]

Figure 8: *Illustration of Blackbeard from the 1734 Edition. The image is dedicated to the public domain under CC0.*

Chapter Four

Johnson's 1734 edition was printed for and Sold by J. Janeway and was titled *A General HISTORY OF THE LIVES and ADVENTURES Of The Most Famous Highwaymen, Murderers, Street-Robbers, &C. To which is added, A Genuine Account of the VOYAGES and PLUNDERS of the most notorious PYRATES. Interspersed with several diverse TALES, and pleasant SONGS. And Adorned with the Heads of the most Remarkable VILLAINS, Curiously Engraven on Copper*. Included some very impressive engravings that don't appear in Johnson's earlier editions. Among them is the most realistic and best-known of all the Blackbeard images.[27]

That edition was followed in 1736 by another edition with the same title as the previous one, only this one but was printed and sold by Olive Payne, at Horace's Head, in Round-Court in the Strand, over-against York Buildings. It has a total of twenty-six engravings, including the same impressive image of Blackbeard, plus a few more.[28]

Changes in the title were made with each new edition. The title of the original began with "*A General History of the Robberies and Murders*," while the second and most reprinted edition began with "*A General History of the Pyrates from Their Rise and Settlement in the Island of Providence*." In the name of simplicity, throughout this book, I shall refer to all of these editions as "*A General History of the Pyrates*."

Even though many editions and reprints would follow throughout the eighteenth and nineteenth centuries, none of them were influenced by Charles Johnson. Nathaniel Mist died in September 1737. He never achieved the financial success he had hoped for.[29] It is pitiful that the author of the world's most recognized and widely referenced book on pirates never knew that his book would become immortal. He had inadvertently ignited the evolution of pirate legends that would continue to permeate the hearts and imaginations of children and adults hundreds of years later.

Five
Wealth From the Sea: The 1715 Treasure Fleet

Piracy in the Caribbean and the Bahamas received an enormous financial boost at the end of 1715 and throughout the following year. This boost came suddenly and unexpectedly when eleven ships of the 1715 Spanish treasure fleet sank in a hurricane off the east coast of Florida. Most of the ships sunk in relatively shallow water and anyone who found the correct spot could "fish the wrecks" by simply diving down to the sandy bottom and picking up millions of dollars of gold and silver coins. News of this "get rich quick" opportunity spread rapidly throughout the English colonies and a sort of "gold fever" spread just as swiftly. However, in reality, recovering anything of value was far more difficult than it sounded. In addition to the depth of the water, riptides, sharks, stinging jellyfish, and shifting sand, Spanish warships would pursue and even fire upon anyone they found in the area. This last obstacle, the thought of Spanish warships, deterred most of the treasure hunters. But it didn't deter pirates!

Before we proceed, I am compelled to explain the connection between the 1715 treasure fleet and the world's most famous pirate, Blackbeard. Although it is conceivable that he may have been among those who came to the Florida coast in search of riches, there are no contemporary documents mentioning him being there. Of course, this applies to thousands of treasure hunters. Only a dozen or so names are mentioned in the historical record. The significant connection is the creation of a pirate port. As you are about to read, the wealth that came from the sea, combined with the piratical circumstances in which that wealth was obtained, resulted in the founding of a new pirate base. A port where pirates could safely gather and resupply. A port that Blackbeard frequented. I'm referring to the pirate port of Nassau in the Bahamas.

Chapter Five

1565–1898

Spain occupied the Philippines.

Establishing a historical perspective of the fleet is important. Spain's enormous wealth predominately came from the New World. The Spanish colonies in Mexico and South America contained dozens of gold and silver mines, which produced thousands of gold and silver bars each year. The gold and silver that wasn't smelted into bars was minted into coins. Most of the Spanish currency used in Spain and Europe was minted in Mexico City and Lima.[1]

The New World also produced vast amounts of jewels, such as emeralds, topaz, amethyst, opals, and many others. Adding to this wealth were extremely valuable pearls, gems, and other goods from the Pacific brought to Mexico by the Manila Galleons from the Spanish colony in the Philippines.[2] As treasure accumulated throughout the year, it would be stored in Havana and then sent to Spain in a yearly treasure fleet.

Concern over the capture of the treasure fleet began in 1701 when the War of the Spanish Succession broke out. This war involved most of the nations in Europe, but at sea, it was mainly between Spain and England. In 1702, very early in the war, a large English naval force won a great victory at the Spanish port of Vigo Bay, and a large number of Spanish warships were sunk or captured. The English sank another Spanish treasure ship in 1708. This caused the Spanish to suspend their treasure fleet until the war was over. However, the gold, silver, and jewels from the mines continued to accumulate. When the long war was finally over in 1714, Spain was just about bankrupt. The king sent a fleet to the Caribbean to bring back all the treasure that was stored in Havana.[3] The 1715 treasure fleet was the richest ever to sail to Spain. It was under the command of Captain General Don Juan Esteban de Ubilla. His flagship was the *Nuestra Señora de la Regala. Second in command was* General Don Antonio de Echevers aboard the *Nuestra Señora del Rosario.*[4]

Twelve ships made up the convoy, eleven were Spanish, and one was French. The eleven Spanish ships, including the two command ships, were the *Nuestra Señora de la Regala*, the *Santo Cristo de San Roman*, the *Nuestra Señora del Carmen*, the *Nuestra Señora de La Popa*, the *Nuestra Señora del Rosario*, the *Urca De Lima* or *Santisima Trinidad*, the *Nuestra Señora de las Nieves*, the *Maria Galante*, the *El Cievro* also called *La Galleria*, the *Nuestra Señora de la Concepción*, and the *El Senor San Miguel*.[5] The French ship was a fourth rate warship called the *Griffon*, commanded by Captain Antoine d'Aire.[6]

To give you some concept of the massive wealth that was aboard these ships, the *Nuestra Señora de la Regala* alone carried thirteen hundred chests of silver coins plus a vast amount of gold and silver bars, jewels, pearls, and valuable Kangxi Chinese porcelain. The overall value of the registered treasure was estimated at fifteen million pieces of eight in 1715, which is equivalent to over $1.5 billion at face value today.[7] That doesn't count the personal property of the many wealthy passengers or the unreported gold and silver that the captains were smuggling into Spain to avoid having to pay taxes. Present-day coin collectors and jewelers appraise the precious treasure at exorbitantly higher rates due to the added historical value. Considering that much of the fortune is still waiting to be found, it is no coincidence that professional treasure hunters and amateur beachcombers continue to flock to Florida's Treasure Coast in search of a piece of the priceless cargo.

Kangxi is a blue and white porcelain produced in China.in 1662–1722.

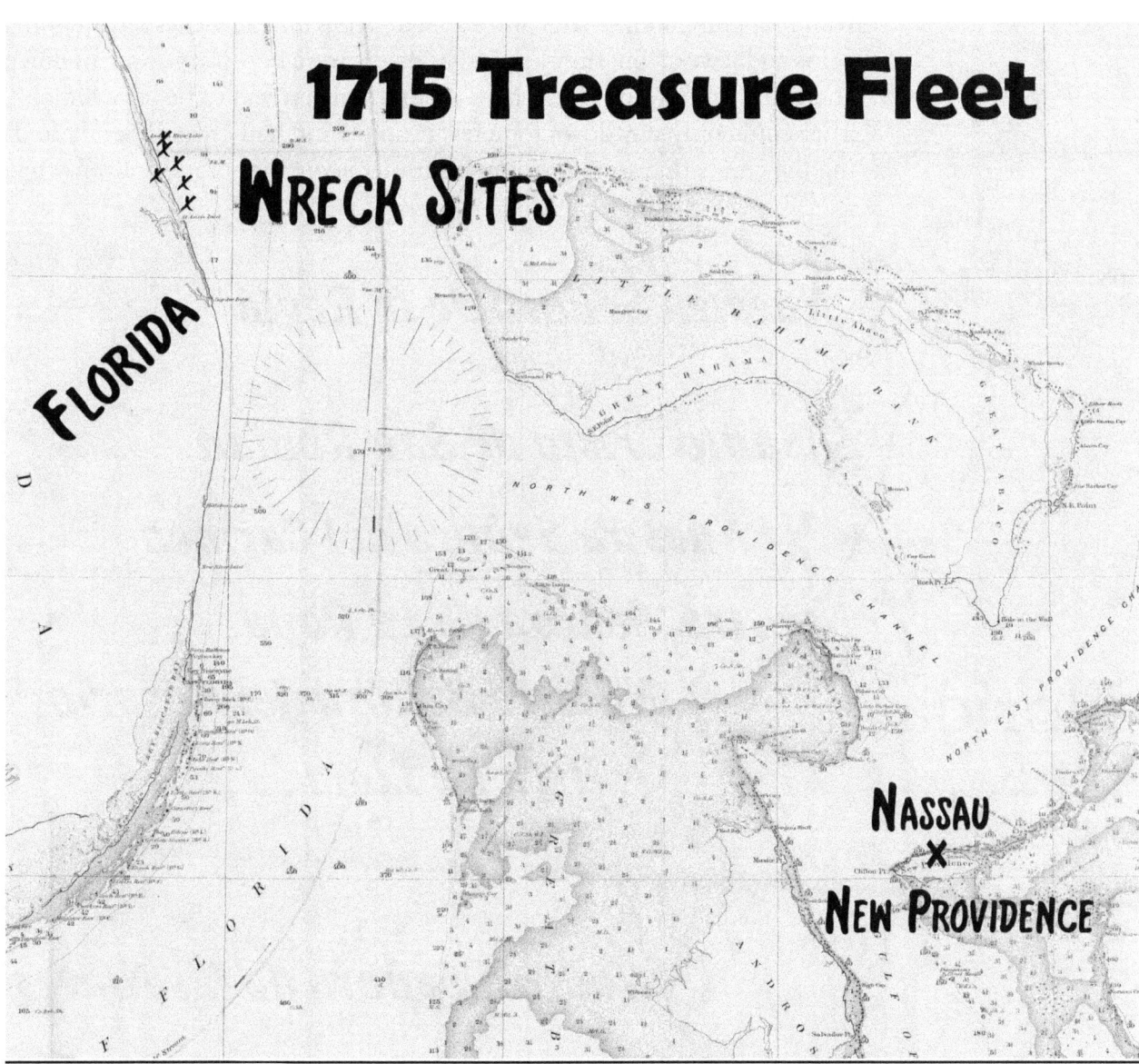

Figure 9: *1715 Treasure Fleet Wreck Locations Florida Map, Courtesy of Barry Lawrence Rudman.*

The fleet left Havana harbor early on the morning of July 24, 1715. It was a beautiful and calm day with a gentle breeze. The ships entered the Gulf Stream and sailed along the Straits of Florida. Everything seemed fine until they noticed strong swells on July 29. Around four o'clock in the morning on July 31, the full force of a hurricane struck. Ubilla's fleet was relentlessly pushed closer and closer to shore until they began to strike reefs and shoals. The only ship to survive the storm was the French warship Griffon, which was faster than the other ships and had sailed on ahead. By the next day, the beaches of la Florida were littered with wreckage and bodies.[8]

By August, the Spanish had rescued the survivors and constructed a salvage camp near one of the most accessible wrecks. They achieved some initial success by simply sailing over the wrecks in sloops, dragging lines with grappling hooks. They were able to bring up a few chests containing coins and jewels,[9] but they knew that divers were necessary to swim down and bring up treasure personally. At first, this proved to be very difficult. They could only stay down for a few minutes and had great difficulty finding the chests that were now buried under six inches of sand. Even when

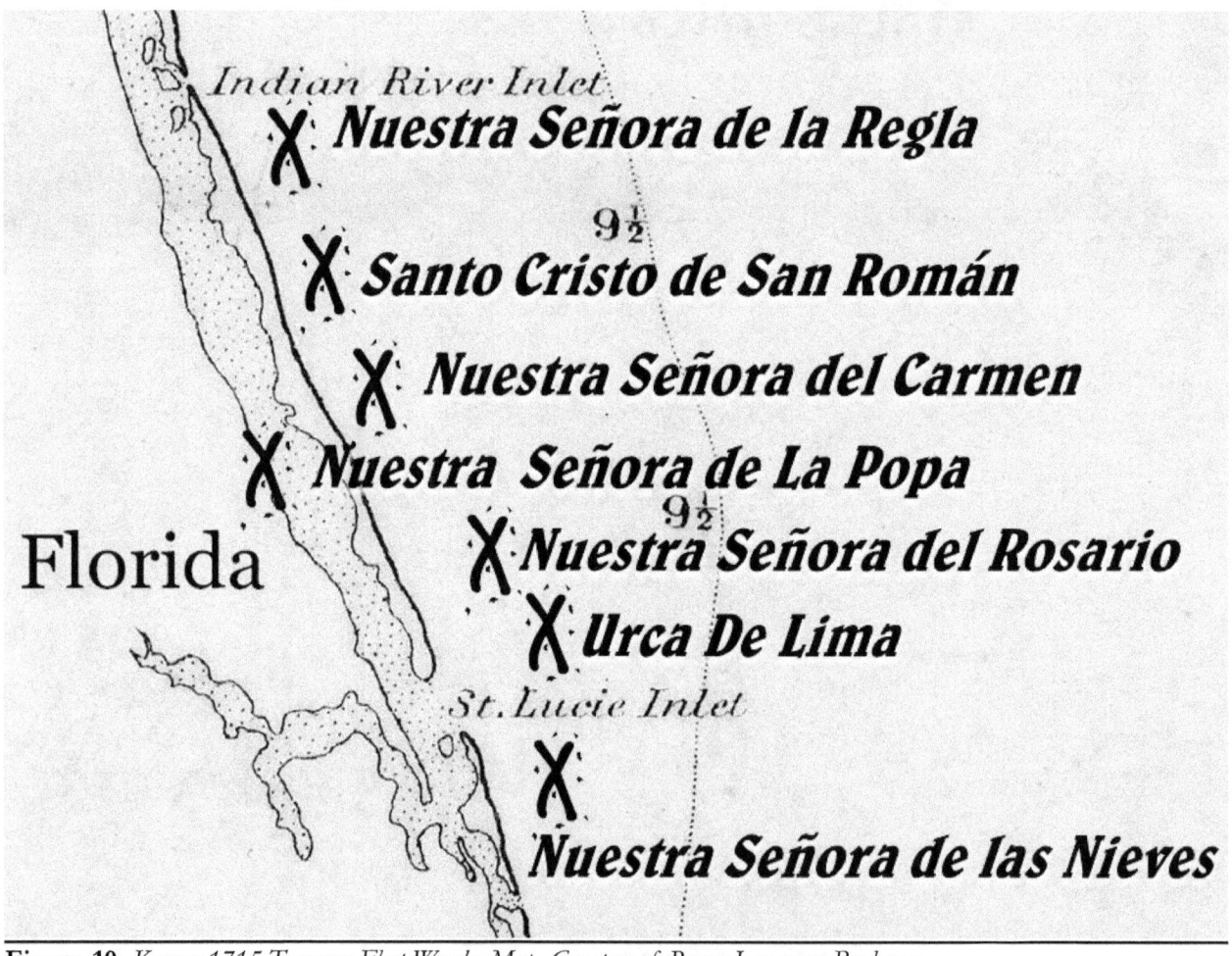

Figure 10: *Known 1715 Treasure Fleet Wrecks Map, Courtesy of Barry Lawrence Rudman.*

they found the chests, they were extremely heavy and had to be hoisted to the surface with ropes. Often, the ropes came loose, and the chests fell back to the ocean floor. In the first week alone, three divers died. Two were killed by falling chests and the third was eaten by sharks.[10]

However, after a few weeks, they finally figured out all the best practices to recover the sunken cargo successfully, and the salvage camp began to fill with chests and wooden boxes loaded with treasure. More ships from both Havana and St. Augustine began arriving on a continual basis. These ships would drop off supplies and take aboard whatever recovered treasure was in the camp. Realizing the need for protection, in September, twenty-five soldiers with a good supply of ammunition were stationed at the salvage camp. Thirty-five additional soldiers arrived shortly after that. By October, the Spanish estimated that they had recovered treasure valued at over 5,000,000 pieces of eight.[11]

Spanish Reales were silver coins that came in denominations of one, two, four, and eight, and were generally called pieces of eight.

Meanwhile, dozens of vessels from just about every English colony rushed to the east coast of Florida to search for the shipwrecks that were scattered along the shore. The Spanish only had resources to salvage a few wrecks at a time. That left the others open for whoever got there first. Some of the wrecks were only twenty or thirty feet deep in the water and were visible under the surface. An article in the December 1715 edition of the *Boston News-Letter* reads, "But last from the Bahamas, says, That the Inhabitants of these Islands have got abundance of Money out of the Spanish Wrecks; and that seven or eight Vessels from Bermuda were gone to those Wrecks."[12]

In Jamaica, the vessels leaving for the Florida coast created a serious problem for captains of ships not destined for the wrecks. These captains were suddenly faced with mass desertions of their crewmen who left their vessels, hoping to hitch a ride with anyone going to fish the wrecks. That included ships from the Royal Navy. Captain Balchen of HMS *Diamond* wrote,

> Since that there has been at least 20 sloops fitted out for the wrecks, and if I had stay'd a week longer, I do believe I shou'd not have had men enough to have brought home, I lost ten in two days before I sail'd being all mad to go a wrecking as they term it.[13]

These English vessels weren't foolish enough to engage any Spanish ships or attempt to attack the salvage camp. There was an exception, of course. The atmosphere of peace and tranquility within the salvage camp was about to come to an abrupt stop. It was December 27, 1715. In the early morning light, just as the sun began to peek over the horizon, one hundred fifty well-armed men calmly walked up to the edge of the camp. Their leader was Henry Jennings.[14]

The Spanish must have been astonished and somewhat confused. Spain was at peace with everyone. The Spanish Commanding Officer approached Jennings and politely asked if their nations were at war again. Jennings answered no and then added that they were there to fish the wrecks.[15] Jennings had arrived in two sloops the night before and quietly landed his men about three miles south of the camp. His sloop was the *Bersheba*, with 8 guns and 80 men. The other sloop, commanded by John Wills, was the *Eagle*, with 12 guns and 100 men. They had sailed from Bluefields, Jamaica in December 1715.[16]

> 120,000 pieces of eight is equivalent to approximately $12 million in today's money.

Objections from the Spanish commander fell on deaf ears. Jennings even ignored an offer of 25,000 pieces of eight if they left immediately. The Spanish watched helplessly as Jennings's men looted the camp of treasure in the value of 120,000 pieces of eight. When he was finished, Jennings peacefully left the camp and returned to his sloops. On the way back to Jamaica, Jennings even took a Spanish ship within sight of Havana.[17] The Spanish believed their camp had finally been attacked by pirates. They were wrong. Jennings and his men weren't pirates at all. They had a commission to raid the camp from the governor of Jamaica, Archibald Hamilton.[18]

These events are well documented in two prime sources. The first is a letter dated March 18, 1716, from Don Juan Francisco del Valle, the Governor of Havana, to Archibald Hamilton, the Governor of Jamaica, which was recorded in the Calendar of State Papers. The gist of the letter was to complain about the piratical acts and to demand reparation. A portion of this letter reads:

> To complain, that he suffered ships to be fitted out in the Island, under pretext of cruising upon pirates, but that instead of that they committed many hostilities on the ships and dominions of the King of Spain ... That one of these vessels ... who in company with another had cast anchor in the Canal of Bahama on the coast of Florida, near the Spanish camp, under Spanish colours, they laid still till night, and then landed their people, who the next morning march'd to the camp with their arms; upon which the Spanish Commanding Officer ask'd them, if it was war, they answer'd no, but that they came to fish for the wrecks, to which the Officer said, that there was nothing of theirs there, that the vessels belonged to his Catholick Majesty and that he and his people were looking for the said treasure; but seeing that his insinuations were of no use, he profer'd them 25,000 pieces of eight, which they wou'd not be satisfy'd with, but took all the silver they had and stript the people taking likewise away four small cannon,

two of them brass, and nail'd two large ones . . . They carried away to the value of 120,000 pieces of eight, besides the wrought silver, this is what the captors own themselves, from whence it is inferr'd, that there was a great deal more . . . [That] in sight of the Havana she met an English ship, who was one of them who had been at the Spanish camp, the English sent on board her, and finding that she was loaden with silver, corn etc. they took her . . . That the silver taken by these two vessels shou'd be returned . . . That the captors shou'd be punish'd.[19]

The second primary source comes from the charges against Governor Archibald Hamilton, who was charged with being financially involved with the pirates and for encouraging "piratical Hostilities upon the Subjects of France and Spain." The details are in the *Articles Exhibited against Lord Archibald Hamilton,* which was published as a book in 1717.[20] When Jennings returned to Jamaica, he discovered that the situation had completely changed.

News of his attack on the Spanish camp had already reached Kingston, and the government officials were very upset, especially Peter Haywood, the Chief Justice. It was their impression that Governor Hamilton had given Jennings and Wills commissions to fight pirates, not to attack the Spanish. Apparently, that wasn't the case. Haywood brought charges against the governor alleging "that the said Lord Archibald Hamilton had been concerned in eighth part in the Voyage of each of the Sloops *Eagle* and *Bersheba,* whom he had commissioned as aforesaid, with intent to go to the said Wrecks,"[21] Another specification was that "several of his Majesty's Subjects were counseled and instructed by the said Archibald Hamilton to go to the said Wrecks, and if stronger than the Spaniards, to beat them off, land take what Mony they could get."[22]

Governor Hamilton, wishing to clear up these allegations, revoked Jennings's commission. In the *Articles Exhibited against Lord Archibald Hamilton,* one statement reads, "Lord Archibald Hamilton's having endeavour'd to get back Capt. Jenning's Commission before he went out on his second Voyage."[23]

Jennings saw the writing on the wall. If he was going to continue his piratical activities, he needed to leave Jamaica and find a new base. On March 9, 1716, Jennings left Jamaica for the last time. He sailed with three other sloops, the *Mary,* Captain Leigh Ashworth, the *Cocoa Nut,* Captain Samuel Liddell, and the *Discovery,* Captain James Carnegie. It is widely believed

1710–1716

Archibald Hamilton, Governor of Jamaica

that among Jennings's crew was a mariner named Charles Vane. He, too, was destined to play a huge role in the life of Blackbeard.[24]

Back in January 1716, as Jennings was taking the Spanish ship and returning to Jamaica, another important figure in Blackbeard's life entered our narrative. His name was Samuel Bellamy. In a very short period of time, he attempted to fish the wrecks, joined Henry Jennings, and then double-crossed him. Bellamy joined forces with Benjamin Hornigold and befriended Blackbeard.

Samuel Bellamy was born on March 18, 1689, in England, and came to Massachusetts to earn his fortune. A professional sailor living in Wellfleet, Massachusetts, he fell in love and needed to get a lot of money quickly in order to marry the woman he loved. The answer to his financial problems came with the news of the sinking of the treasure fleet. He received assistance from Palsgrave Williams, a local goldsmith who financed the purchase of a small vessel and the hiring of a crew. Both Bellamy and Williams formed a partnership and set sail for the Florida Straits.[25]

Bellamy and Williams arrived at the Florida coast sometime in January 1716 and found the wreck site of one of the ships. Unfortunately, several other English ships had gotten there ahead of them and lay at anchor over the wreck site. By now, there was almost no treasure left at the sites that

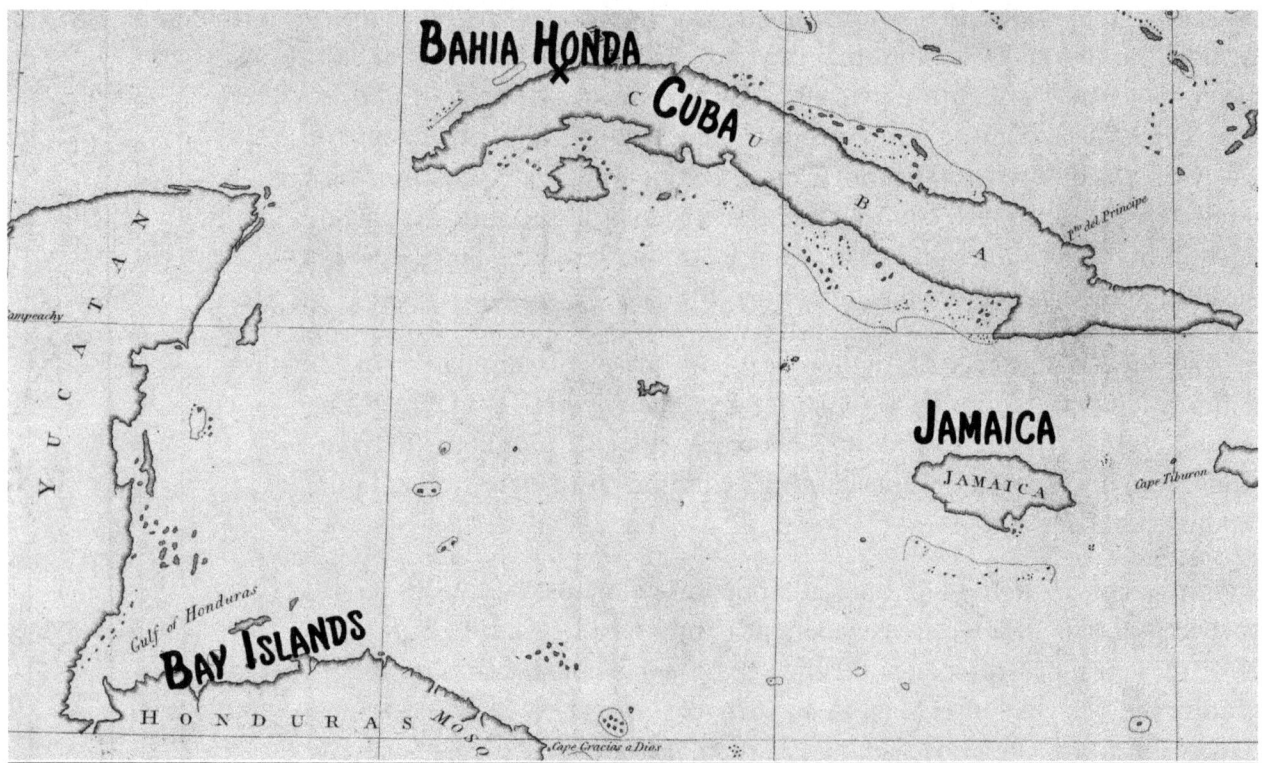

Figure 11: *Bay Islands to Cuba from Caribbean Map, Courtesy of Barry Lawrence Rudman.*

were easy to pinpoint. To make matters worse, the Spanish had learned their lesson from Jennings. They sent warships to patrol the area and drive off any English ships. With nothing in the offing, Bellamy and Williams had no other choice than to become pirates. First, they needed a larger crew. The few crewmen they had on their vessel were insufficient for piracy. In February 1716, the best place to recruit a pirate crew was the Gulf of Honduras, so Bellamy and Williams headed south.[26]

At that time, there were several small English settlements in the Gulf. Some were along the coast of modern-day Belize and others were on the Bay Islands, about thirty-five miles north of the coast of Honduras. These tiny pockets of English settlements had been around since the days of Henry Morgan in the 1660s, when English buccaneers used them to raid Spanish shipping. By 1716, the only business they had was the exportation of logwood. However, the recent war and the news of the treasure fleet caused business to plummet. In desperation, many of them were preparing to make the trip to the wrecks themselves until the Spanish attacked and destroyed their vessels. Life there seemed hopeless until Sam Bellamy arrived.[27]

It was easy for Bellamy to gather a pirate crew, and he was now ready to begin his career as a pirate. However, he only had two periaugers with which to attack vessels. A periauger is a basic term used for a two-masted vessel generally used along the coast or in rivers. There is a reconstruction of one at the North Carolina Maritime Museum in Beaufort. In my research, I've seen the vessel and have spoken with the craftsmen who built it. Periaugers were fairly sturdy and always had a course of oars, so they could either be sailed or rowed, allowing the vessel to travel effectively in light wind or even against the wind. It could hold a crew of about twenty men at most. Obviously, Bellamy didn't travel from Boston to Honduras in a periauger. He must have had a sloop of some sort. However, the historical record doesn't reveal its fate.

Bellamy's first prize was a Dutch vessel. One of the crewmen, Peter Cornelius Hoof, was forced to join Bellamy's crew. He remained with Bellamy to the end and was one of the pirates who was caught and tried in Boston in 1717. In that trial, he testified that "he was taken by Bellamy in a Peri'aga, he belong'd to a Ship whereof one Cornelison was Master."[28] The second vessel they captured was an English sloop under the command of Captain Young. After taking control of the vessel, Bellamy had his periaugers tied alongside and ordered the captain to take them north toward the wrecks. Near the west end of Cuba, Bellamy's captured sloop with the two periaugers in tow came within sight of Henry Jennings's fleet. Eventually, Captain Ashworth of the *Mary* boarded, and Captain Young told him he

1663–1671

With Port Royal, Jamaica as a base, Henry Morgan was the most successful privateer in the Caribbean.

Chapter Five

had been taken by "two maroon periaguas, and had obliged him to tow them over from the bay of Honduras."[29]

Details of the next series of events come from three letters and three depositions. The letters were all written by French officials to Governor Hamilton to complain about the loss of their vessels. They include Captain D'Escoubet's letter dated 4 April 1716,[30] and two letters dated June 18, 1716, one from Michon, the Intendant of the French Settlement in Hispaniola, and one from Le Comte de Blenac, Governor of the French Settlement in Hispaniola.[31]

One of the depositions was given by Samuel Liddell, captain of the *Cocoa Nut*.[32] When he realized that Jennings was committing acts of piracy, Liddell left the pirate group and sailed to Jamaica to tell the authorities about Jennings and avoid arrest. Another deposition was made by Joseph Eels, who was Ashworth's quartermaster on the *Mary*. Upon his return to Jamaica, he was arrested and turned "evidence for the king."[33]

The final deposition was made by Jennings's quartermaster, Allen Bernard. He had just joined Jennings's crew before they sailed and was made quartermaster, which, according to Jennings, "would be an easy Post for him."[34] After leaving Jamaica, Bernard became sick and was confined to the great cabin for most of the voyage. His account comes mostly from what he heard through the bulkhead or what someone told him shortly afterward. Not really being a pirate, Bernard also wanted to clear his name upon returning to Jamaica.

In the following narrative, I have woven details and quotes from the six accounts together to create a sequential report of their actions.

Jennings's fleet of four sloops had been scattered by strong winds after they left Jamaica. They had planned to rendezvous along the northwest coast of Cuba, and by April 2, 1716, three of the sloops, Jennings's *Bersheba*, Ashworth's *Mary*, and Liddell's *Cocoa Nut*, had already joined. About six leagues (thirty-five miles) from Bahia Honda, the lookout on *Bersheba* cried out, "A Sayl." There was a flurry of activity as the men on deck rushed to the rail to see what vessel had been spotted. In the distance, they saw a sloop with two periaugers alongside. As Jennings's sloops grew nearer and nearer, their crews were surprised to see the two periaugers preparing to sail away. The sloop they sighted was, of course, Captain Young's captured sloop with Bellamy's pirates onboard. Apparently, Bellamy believed the fast-approaching sloops were pirate hunters and that they would be arrested if they were caught. The pirates frantically threw all the valuables onto the periaugers, and when they were finished, they cut loose and began to row away.

Uncertain of what was really happening, Jennings ordered his sloop to pursue the two periaugers. The *Bersheba* came about and trimmed the sails to gain speed. Unexpectedly, the two mysterious periaugers headed directly into the wind. This was a brilliant tactic. Sailing vessels can't sail straight into the wind. If they try, they quickly stop. They have to tack back and forth, taking a long time to travel a short distance. However, the periaugers were using oars and could head directly into the wind without issue. Within a short amount of time, the periaugers were safely out of sight.

Jennings pulled the *Bersheba* alongside of the sloop and had the two vessels tied together. Captain Young came aboard the *Bersheba* to thank his rescuer. They were soon joined by Captain Ashworth of the *Mary*. Captain Young explained that the men who rowed away in the periaugers were pirates who had taken his sloop in the Bay of Honduras and had forced him to sail to Cuba. In Bernard's words, "the People that went away from them in ye two Periaugas were a parcell of Villains that had obliged him from ye Bay to bring them over there." Young's sloop stayed with Jennings's group for a while. At that time, Jennings wasn't taking English vessels, so Young may have accompanied Jennings willingly. However, considering that Jennings would later burn his sloop, perhaps he wasn't so willing.

The four sloops sailed the thirty-five miles to Bahia Honda on the north coast of Cuba. Captain Carnegie, on the *Discovery*, still hadn't arrived. Inside the bay, Jennings saw a frigate rigged French ship laying at anchor. It was the *Mary of Rochell*, under command of Captain D'Escoubet. Jennings launched the ship's boat and sent George Dossitt, the companies quartermaster, and two other men over to the imposing ship to do a little reconnaissance. They returned with a startling report. The ship had about 16 guns and a crew of forty-five.

Occasionally, a vessel would have two quartermasters, one responsible for the operation of the vessel and the other responsible for the discipline of the crew, which were called the ship's company. The "Companies Quartermaster" was responsible for the ship's crew.

Upon hearing of the strength of the ship, Jennings was reluctant to attack, but some of the crew weren't. An argument began. George Dossitt said, "What are you come out for? To look upon one another and return with your fingers in your mouth?" To which Jennings replied, "The Sloop Bersheba must not go to be taken. And that if they went alongside, they probably might be sunk." Eventually, the crew persuaded Jennings that a successful attack was possible and he called a meeting with his two other captains, Liddell and Ashworth, to discuss the matter. It was two against one. Jennings and Ashworth were for the attack while Liddell argued against it. His primary concern was that this ship didn't appear to be a legitimate prize and that by taking it, they would be committing an act of piracy. Apparently, this didn't concern the others. The captains returned to their sloops and Liddell watched in abhorrence as the *Bersheba* and the *Mary* positioned themselves for the attack.

Frigate Rigged was an eighteenth-century term that was replaced with the term *full-rigged* by the nineteenth century.

Chapter Five

By now, it was about ten o'clock at night. However, it wasn't dark. The bright moon shined through the pure air, which allowed the crews of the vessels to clearly see the ship. Suddenly, the silence of the night air was disrupted by the shouts and cheers of all the pirates on the sloops. Simultaneously, they began shouting, "Hurra! Hurra!" and "One and All," over and over again. Something very exciting was happening. The two periaugers who had escaped into the wind the day before had unexpectedly reemerged and were engaging the French ship. What motivated the pirates to cheer was Bellamy's men in the periaugers. They were totally naked except for the cartridge boxes they wore about their waists. These attackers were shouting and waving cutlasses in the air while firing pistols at the French. This sight was awe-inspiring. None of Jennings's men had ever seen such a thing. Neither had the French.

In the excitement, one of Jennings's men fired a pistol at the French. "Somebody asked, 'Who fired?' And some others answered he ought to be cutt down, others said it was done by accident." As the periaugers came close to the ship, the French captain shouted, "Where are you coming?" They answered, "Aboard, where do you think?" With that response, all of Jennings's men resumed their cheering. It must have sounded like a football game as one of the home team's players was making a 90-yard run for a touchdown.

The French ship launched a boat, perhaps to escape or to ask for quarter. The naked men from one of the periaugers swarmed aboard the ship while the other periauger pursued the boat. Suddenly, one of the great guns on the *Bersheba* fired. Bernard heard someone shout, "Who fired it?" The answer was that it was accidentally fired; however, it frightened the French into believing this was a combined assault. Captain D'Escoubet hailed the *Bersheba* and begged them not to fire again. He added that the attacking pirates threatened all of them with death if anyone fired upon them. A few minutes later, it was all over. Samuel Bellamy had taken *Mary of Rochell*.

There is nothing in the record that points to a meeting between Bellamy and Jennings prior to the attack. The apparent lack of effective coordinated fire from either *Bersheba* or *Mary*, coupled with the reaction of their crews, indicates that they hadn't met. I like to think that Bellamy's appearance was a total surprise. At any rate, shortly after Bellamy had secured the *Mary of Rochell, the Bersheba, and the Mary* came alongside and put some of their men aboard. Bellamy must have been relieved when he learned that Jennings was a fellow pirate, not a pirate hunter. They agreed to join forces, but this partnership wouldn't last long.

The next morning, Captain Carnegie on the *Discovery* finally pulled into the bay. He was delighted to see the French ship at anchor with the *Bersheba* and the *Mary* tied alongside and immediately positioned his sloop to join them. Meanwhile, Captain Liddell had had enough. He was no pirate and wasn't going to continue in the company of pirates. However, twenty-three of his men, including his quartermaster, had disagreed and jumped overboard to join one of the other crews. Liddell quietly prepared to get underway.

Later that morning, a Spanish periauger entered the bay and came alongside the *Mary of Rochell*, unaware that these other sloops were pirates that had taken the ship. The Spaniard was delivering a message to the French ship from another French vessel anchored at Porto Mariel, about twenty-five miles to the east. This was the sloop *Marianne*, under the command of Ensign le Gardew, an officer of the garrison who was there to deliver letters to Spanish officials in Havana. Upon hearing this news, Carnegie decided to set sail immediately and take it. The *Discovery*, accompanied by one periauger, sailed out of Bahia Honda shortly afterward. Following Carnegie out of port was Captain Liddell on the *Cocoa Nut*. Whereas Carnegie headed east to take the French sloop, Liddell sailed west for Jamaica. His short career as a pirate was over.

Onboard the *Mary of Rochell*, Samuel Bellamy was greatly dismayed at the treatment of the French prisoners at the brutal hands of Jenings's men. Before the engagement, as a routine precaution, the French had hidden a chest containing 30,000 pieces of eight ashore. Believing that the French had hidden some treasure, Jennings "tormented the crew to inhumane degree" in the "vilest manner." Eventually, the terrified captives told the pirates where the chest was buried. After witnessing the brutal treatment of the French captives, Bellamy began looking for an opportunity to put an end to this hasty partnership. Throughout the rest of his career, once a crew had surrendered, Bellamy's captives were humanely treated. This trait was shared by Bellamy's future friend, Blackbeard.

The next morning, Jennings and his men were surprised to see the periauger returning alone, without the *Discovery* or the French sloop. Coming alongside the cluster of vessels that were still tied together at anchor, the men on the periauger gave Jennings a startling report. The French sloop *Marianne* had already been captured by another pirate before they got there. The news was disappointing to Jennings until he learned who had taken the sloop. His disappointment turned into rage as he heard that the sloop was taken by his old adversary, Benjamin Hornigold.

Chapter Five

The orders quickly came to weigh anchor and raise sail. Drifting free of the ship, the sails of the *Bersheba* and the *Mary* slowly began to fill with the wind, and a few moments later, they were speeding north through the bay toward the open sea. Just as they neared the entrance of the bay, then they saw two sets of sails in the distance heading west. Jennings immediately recognized one of the sloops. It was Hornigold's *Benjamin*, and he surmised that the other must be the French sloop, *Marianne*. Jennings furiously ordered his sloops to catch them, but Hornigold had too much of a lead. After a while, Jennings gave up and came about to return to the bay.

While Jennings and Ashworth were chasing Hornigold, their men aboard the *Mary of Rochell* worked to get the sails up, too. However, this was challenging. There weren't enough of them to work a big ship. The French prisoners were all held below deck, and Bellamy's pirates seemed reluctant to assist. The reason for their reluctance soon became apparent. Bellamy gave some sort of prearranged signal and his pirates quickly overpowered Jennings's men. With the ship secured, Bellamy's men hurriedly threw sack upon sack of silver coins into one of their two periaugers. Joseph Eels, Ashworth's quartermaster, who was among the men and was easily overpowered by Bellamy's pirates, estimated the value of those coins to be 28,500 pieces of eight. With Jennings and Ashworth still far in the distance chasing Hornigold, Bellamy sailed out of sight with all the cash from the French ship.

As Jennings and Ashworth closed the distance, they saw Captain Young's sloop and the *Mary of Rochell* lying motionless in the middle of the bay. They must have wondered what was taking so long for the ship to get underway. It wasn't long before they were "acquainted" with the truth. As they came within shouting distance, the two captains heard frantic calls from the men that they had put onboard. One can only imagine Jennings' reaction when he learned that Bellamy had departed with all the loot. In a fit of rage, he ordered Bellamy's other periauger to be cut to pieces and Young's sloop to be burnt.

The next morning, Carnegie, onboard the *Discovery*, returned. He took possession of the French ship *Mary of Rochell* as his new pirate vessel and gave his old sloop to the French crew. Jennings, Ashworth, and Carnegie weighed anchor and set sail. This time, their destination wouldn't be Jamaica. These captains realized that they had crossed over the line from privateer to pirate. They knew that they would be arrested as soon as they set foot on Jamaican soil. They needed a new port, one that was friendly to pirates. They chose Nassau on the island of Providence.

Nassau had no real government. Years earlier, the Spanish had destroyed the fort, driven off many of the settlers, and eliminated any semblance of an organized administration. Located near all the main shipping lanes, it made an ideal pirate base.[35] Throughout the first five months of 1716, dozens of pirates were gathering at Nassau with the thousands of pieces of eight they had taken by fishing the wrecks. Jennings was certainly among the most influential of these pirates. Eventually, they used this money to build Nassau into a pirate haven, with taverns, brothels, and anything else that a pirate would need. On July 3, 1716, Lt. Governor Spotswood of Virginia wrote a letter to the Council of Trade and Plantations referring to Nassau as, "A nest of pirates are endeavouring to establish themselves at Providence and by the addition they expect of loose dis-orderly people from the Bay of Campeachy, Jamaica and other parts, may prove dangerous to British commerce."[36]

And what of Samuel Bellamy? He joined Ben Hornigold, who gave him command of the French sloop, *Marianne*.[37]

In case you wish to read these accounts in the original words, here are excerpts from the three depositions.

Excerpts from the Deposition of Joseph Eels:[38]

> Joseph Eels of Port Royall . . . last March sett sail on board the Mary sloop, Capt. Leigh Ashworth commander, and soon after arrived at blewfields, where they found Capt. Jennings, Capt. Carnigee, and Capt. Liddal, and from thence sail'd in company with them designing for the wrecks. About six leagues from Baya Honda they spyed a sloop with two periaguas putting from her, and found her to be Capt. Young's, who told Capt. Ashworth they were two maroon periaguas, and had obliged him to tow them over from the bay of Honduras . . . A periagua commanded by a Spaniard informed them that there was in Porto Mariel a French ship a trading, whereupon Carnigee went to seek her, but next morning the periagua which had followed him reported that Hornigold had taken the French ship, whereupon Jennings and Ashworth weighed anchor to go after them, but not being able to overtake them stood in again to the Bay, and came to an anchor, the ship being in the offing, one of the periaguas being on board ship and several of her men halled her alongside and threw the money being about 28,500 odd peices of eight into the periagua and immediately went away with it. Soon afterwards the ship came in again and acquainted Jennings and Ashworth the money was gone,

and then by order of Jennings one of the periaguas was cut to peices and Young's sloop burnt. Next morning Carnigie halled aboard the ship and hoisted out of his sloope into the ship all his guns ammunition provisions and stores, and going on board with his men took the command of her without controul. Jennings, Ashworth and Carnigie weighing anchor in order to go to Providence.

Excerpts from the Deposition of Samuel Liddell:[39]

With Capt. Jennings in the Barsheba and Capt. Ashworth in the Mary . . . Came to the Bay of houndo where they found a french Ship and there Came to anchor, and afterwards Capt. Jennings sent a small Canoa with the Companie's Quarter Master and Two Men More on board the Said French Ship who returning again gave an Account that she had about fourteen or Sixteen Guns and forty five Men or thereabouts on board her, after which they Endeavoured to prevaile with the said depont. to Joyne with them to Attack her . . . and beleived she was there a trader on a Lawfull occation. Capt. Jennings thereupon answered he designed to be a Long side of her that Night, and this Depont. desired him Not to Attack the said french Ship . . . afterwards about ten a Clock that Night Capt. Jennings of the Barsheba, and Capt Ashworth of the Mary were Towed in by Two piraguas that came from the Bay . . . that about Thirty Small Arms were fired by the assailants upon the French Ship and one Great Gun which was said to be fired by Capt. Jennings himself, and the French fired not So Much as a Pistoll . . . Captain Carnegie remained without the Barr, but Came in the next Morning and Joyned with the Barsheba and the Mary and lay along Side the French Ship, and that twenty-three of the said Deponent's Men left him, and Joyned with the rest of the Assailants . . . after Capt. Carnegie was in the harbour he was towed out, and it was reported he went to take a french Sloop or vessel at Marian but what further happened this Deponent Knowed Not, he having left them, and afterwards returned to Jamaica.

Excerpts from the Deposition of Allen Bernard:[40]

Captn. Henry Jennings who told him he should be his Quarter Master which would be an easy Post for him. He thereupon accepted of it and engaged himself to go on Board the Sloop Bersheba . . . they entred in Consortship with Captn. Lyddall Captn. Ashworth and Captn. Carnegie to fish upon ye Wracks . . . heard some persons Cry out, A Sayl . . . after which he heard them say that there was two Periauga's putt off ye said Sloop and Rowed directly in ye Wind's Eye . . . they were making their escape with ye money . . . And that ye Commander of her came on board of ye Bersheba who was one Captn. Young from ye Bay who informed them that the People that went away from them in ye two Periaugas were a parcell of Villains that had obliged him from ye Bay to bring them over there. After which they made ye best way towards the Bay of honda and came to an Anchor without the Barr . . . And that after coming to an Anchor in ye Bay of honda he heard ye Voice of George Dossitt and Francis Charnock (the first being ye Companies Quartermaster) and heard these words spoken among ye People What are you come out for? To look upon one another and return with your fingers in your mouth? At which Capt. Jennings said the Sloop Bersheba must not go to be taken. And that if they went alongside they probably might be sunk . . . Sometime after he heard a General hurra from the Severall Vessells after that One and All. That some short time after a Pistol or Small arm went off upon which somebody asked who fired? And some others answered he ought to be cutt down, others said it was done by accident . . . that the two Periauga's were coming all in their Skins or Buff with their Cartouch Boxes and naked Cutlaces & Pistols And that he had never seen such a sight before Afterwards he heard somebody hayl from ye Ship and asked Where they were coming? and some voice made answer, Aboard, where do you think? and then a hurra again . . . And [a] boat was putt off from the ship and that one of the Periauga's was on board [the French Ship] and one [periauga] rowed after that Boat which putt off from ye Ship. At ye same time a great Gun went off from on board ye Bersheba, and somebody asked who fired it? Some made answer, it was an accident . . . heard a voyce hayl from ye Ship not to fire for all was well . . . say afterwards among ye People of ye Periaugas That they told them that in case a Gun was fired they would give no Quarters.

And that Captn. Carnegie being the worst Sayler did not come in that night but came in ye next day And to the best of his Knowledge joyned them again. Afterwards Captn. Carnegie and one of the Periaugas sayled out of ye Bay of Honda in Quest of a French Sloop

Chapter Five

... at which time we saw two Sayl off ye Barr who was said to be Captn. Hornegole and a French sloop.

At ye same time they weighed in generall with ye Ship and went out ... after Hornigole ... Some little time after she came in and those people on board her hailed and said they had done finely to leave them, for ye Marooners had risen upon them and the Frenchmen and had carried away all ye money, at which there was a murmuring among ye People. At ye same time Captn. Carnegie's Sloop was halled alongside ye Ship and everything was taken out of her and putt on board ye Ship which he afterwards comanded, his sloop being given to ye French Captain.

Six
Ye Grand Pirate Captain Benjamin Hornigold

Benjamin Hornigold is the man most often associated with Blackbeard. Many historians, as well as pirate enthusiasts, think of Hornigold as the highly experienced pirate master who took the young and upcoming Blackbeard under his wing and molded him into the world's most infamous pirate captain. This is precisely the way he has been portrayed in fictional literature and television productions involving Blackbeard since the late twentieth century.

Here's how the story usually goes. Ben Hornigold had been one of the most successful privateer captains during the war against the Spanish and the French in the early eighteenth century. Privateers had letters of marque, which were documents issued by colonial governors legalizing the taking of enemy shipping. However, when the war ended in 1714, the British government stopped issuing any letters of marque, and all the privateers suddenly found themselves out of a job. Hornigold couldn't conceive of any profession other than the one he knew so well, and crossed the line into piracy. He established his headquarters at Nassau in the Bahamas where he became the leader of the pirate community and was looked upon as the pirate master. Remaining loyal to England, Hornigold continued as if the war was still on, only attacking Spanish and French shipping, and leaving Dutch and English vessels alone. There is historical evidence that this assessment may be accurate. An article appeared in the May 21 to May 28, 1716 issue of *The Boston News-Letter,* which reads:

> Capt. Evertson from Providence, who says there are 13 Vessels from Jamaica, and 12 Bermudians upon the Wreck, some of whom are said to have taken a French Ship of 24 Guns, with 60 or 70 Thousand Pieces of Eight on Board, and that they will not permit either French

Chapter Six

1704-1776

The Boston News-Letter was published from 1704 to 1776 and was heavily subsidized by the British government, with all copies being approved by the Royal Governor.

1716

Blackbeard is believed to have joined Hornigold in 1716.

or Spaniards to come there. That there is an English Pyrate Sloop about the Bahema Islands, with 150 Men on board, who say they meddle not with English or Dutch, but that they never consented to the Articles of Peace with the French and Spaniards.[1]

This article is most interesting because it references the French ship that Jennings captured at Bahia Honda and sailed to Nassau in April 1716, although the amount of money on board was greatly exaggerated. As stated in the previous chapter, if this is indeed the French ship taken by Jennings, Sam Bellamy got away with all the cash. The second part of the article that mentions an "English Pyrate Sloop" who "meddle not with English or Dutch" obviously refers to Hornigold. Contemporary evidence that Hornigold was considered a pirate master exists as well. Thomas Walker was a leading citizen in Nassau before the pirates began arriving in force in the spring of 1716. He was forced to flee with his family and wrote a letter of complaint to the Council of Trade and Plantations in August 1716. In this letter, he refers to Hornigold as "ye grand Pirate Capt. Benja. Hornigold."[2]

Blackbeard enters the narrative as Hornigold's protégé. In the traditional narrative, Blackbeard became a member of Hornigold's crew before 1716 and looked upon Hornigold, the pirate master, as a mentor. Seeing that Blackbeard possessed unique and special abilities, Hornigold eventually gave him a command of his own. Together, Blackbeard and Hornigold sailed as partners. Eventually, they went their separate ways, but Blackbeard always retained a special admiration for his old boss. There is no doubt that Blackbeard and Hornigold were sailing together as partners in their own sloops in November and December 1716, but everything else may be a combination of speculation and poetic license. On the other hand, it may be completely true. Historian author Baylus Brooks disagrees with the concept that Blackbeard was some sort of apprentice to Hornigold and suggests that they sailed together as equals. He adds that perhaps Blackbeard may even have been the far more experienced mariner.[3] If this were true, one would expect Blackbeard to be the captain of his own vessel, but there is no evidence that he commanded a vessel before November 1717. Let's first examine the facts.

Little is known about Hornigold's origins for certain, but according to Brooks, records from Ancestry.com may give us a window into his earlier life. There is a marriage record between Ben Hornigold and Sara Mosse in Saint Matthew's Parish, Ipswich, Suffolk County, England dated January 8, 1679. This record indicates that Hornigold was born between 1655 and 1661. This would put Hornigold in his late 50s when he became the pirate master. Additionally, there is a will of a man named Nathaniel Morse,

"Mariner of Ipswich" who left twenty shillings to his "loving Brother in Law Benjamin Hornigold of Ipswich aforesaid Mariner." Another record from Massachusetts lists Mary, Robert, and Sara Morse Hornigold as siblings and names Sara's husband as Benjamin Hornigold.[4]

Where this is most compelling, it might not be the same Ben Hornigold. However, thus far, no other evidence for Hornigold's early life has surfaced.

The earliest mention of Hornigold as a pirate can be found in a letter written by the same Thomas Walker mentioned above and dated March 12, 1715. At the time, Walker was living in the town of Nassau on Providence Island. His letter was a warning to the Lords Proprietors informing them of the rise of pirates on Eleuthera Island, the very large and oddly shaped island just to the east of Providence. He wrote:

> List of men that sailed from Ileatheria and committed piraceys upon the Spaniards, on the coast of Cuba, since the Proclamation of Peace . . . Strangers that sailed from lleatheria a piratting:—Benja. Hornigold, Thomas Terrill, Ralph Blankershire, Benja. Linn. An account of what they took from the Spaniards in two voyages in the sloop Happy Return, etc. The inhabitants pray the Lords Proprietors to order the inhabitants of Ileatheria, through the next Governor, to settle and strengthen Providence, etc.
>
> Signed, 'Tho. Walker, Inhab. and Setler of Providence, March 12, 1715.[5]

Many pirates who fished the wrecks of the 1715 treasure fleet used Nassau as their home port. After all, it was the closest English port to the east coast of Florida, and the lack of any formalized government made it an ideal spot for pirates to come ashore. Hornigold was among them, arriving in November 1715 aboard his sloop *Mary*. Shortly afterward, he took a Spanish sloop off the Florida coast and brought it back to Nassau, but Hornigold's captured Spanish sloop was taken from him by Henry Jennings, captain of the sloop *Bersheba*. There was likely bad blood between these two captains for many years; however, if there weren't any issues between them before, there certainly were after this incident. In January 1716, Hornigold returned to the Florida coast to capture and keep another Spanish sloop. He was successful and brought this new sloop back to Nassau to fit it out as a pirate sloop. Once ready for action, he named his sloop *Benjamin* and sent his sloop *Mary* back to its original owner in Jamaica. In March 1716, Hornigold sailed out of Nassau. This is documented in a deposition of

Chapter Six

John Vickers, a resident of Nassau, whose statement is contained in the *Calendar of State Papers*.

> In Nov. last Benjamin Hornigold arrived at Providence in the sloop Mary of Jamaica, belonging to Augustine Golding, which Hornigold took upon the Spanish coast, and soon after the taking of the said sloop, he took a Spanish sloop loaded with dry goods and sugar, which cargo he disposed of at Providence, but the Spanish sloop was taken from him by Capt. Jennings of the sloop Bathsheba of Jamaica. In January Hornigold sailed from Providence in the said sloop Mary, having on board 140 men, 6 guns and 8 pattararas, and soon after returned with another Spanish sloop, which he took on the coast of Florida. After he had fitted the said sloop at Providence, he sent Golding's sloop back to Jamaica to be returned to the owners: and in March last sailed from Providence in the said Spanish sloop, having on board near 200 men, but whither bound deponent knoweth not ... but upon the coming of Hornigold to Providence ... Hornigold saying that all pirates were under his protection. It is common for the sailors now at Providence (who call themselves the flying gang.) [6]

Pattararas are swivel guns, mounted on the rail of a vessel.

As detailed in the previous chapter, Bellamy and his men had taken a French ship while Jennings's men cheered them on. The two crews appeared to be working together to secure and loot the French ship when they heard of a nearby French sloop, the *Marianne*. One of Jennings's captains set out to take this coveted prize.

Meanwhile, after leaving Nassau, Hornigold had sailed to Porto Mariel, where he arrived and took the *Marianne* before Jennings's men could get there. In an act of boldness, Hornigold brazenly flaunted his success by sailing his sloop *Benjamin* and the captured French sloop *Marianne*, right past Jennings's fleet.

Jennings set out in pursuit of Hornigold, but Bellamy and his men jumped at the opportunity to overpower the other pirates onboard the French ship and sail away with all the loot. To add insult to injury, a few days later, Hornigold formed a partnership with Bellamy and gave him command of the prize sloop *Marianne*.

This brings us up to the point in time where the last chapter left off. The best source documentation describing these events is the deposition of Jeremiah Higgins, who had been a member of Hornigold's crew and was later being tried for piracy.

> That before they came to the wrecks, one Capt. Hornigold Commander of a Pyrate Sloop called the *Benjamin* came on board their sloop and after some time Desired the Examinate and some other of the Men belonging to the Sloop Blackett to row him on board the said Sloop *Benjamine* . . . That the said Sloop *Benjamine* afterwards sailed to the coast of the Havana Having upwards of Eighty Men on Board, and off the Coast of the Havana at a place called Porta Maria, they took the Sloop *Mary Anne* then belonging to the French & Spanish . . . put some of the company on Board the *Mary Anne* and chose on Samuel Bellamy to be Commander.[7]

No source documentation has yet surfaced to indicate that Blackbeard was present as a member of Hornigold's crew at that time. However, the circumstantial evidence that he was indeed part of Hornigold's crew is quite compelling. As mentioned in the previous chapter, once Blackbeard heard the news of Bellamy's death and the subsequent execution of seven of his men in Boston, Blackbeard began a personal vendetta against all vessels from Boston. He destroyed each one that he captured, explaining to their captains that it was because their vessel was from Boston, the town that hanged his friends. Specific examples of this will be highlighted in several chapters to come. The only logical reason that explains Blackbeard's actions is that he had formed a close personal relationship with Bellamy and some of his crew.

The movements and actions of both Bellamy and Blackbeard are well documented from November 1716 to the end of each of their lives. During those months, there was no point in time when they could have encountered each other. If Blackbeard and Bellamy were friends, it must have happened in the spring and summer of 1716, which means that Blackbeard would have been among either Hornigold's or Bellamy's crew. Since Blackbeard was Hornigold's partner in November 1716, it seems highly likely that Blackbeard was with Hornigold's crew when he took the *Marianne* and gave it to Bellamy to command. Most modern historian authors accept this last statement as true.

Blackbeard, Hornigold, and Bellamy are believed to have sailed together

Much has been written about the reasons Hornigold chose Bellamy, a man he had just met, to be the captain of his prize sloop, *Marianne*. Some historian authors rationalize this appointment by suggesting that Hornigold must have been impressed with this young man's confidence, or that he saw a special quality of character or level of leadership within him and decided he would make the best captain. Some have even pointed out that Hornigold chose Bellamy above his protégé, Blackbeard. However, one doesn't have to look that deeply to understand why Bellamy was chosen.

Chapter Six

Bellamy had successfully led a relatively small force of pirates aboard two small periaugers in an attack on a much larger and well-armed French ship. Then, he double-crossed Henry Jennings and made off with 28,500 pieces of eight. Those actions speak for themselves. And if they weren't enough, we must remember that Bellamy had 28,500 pieces of eight. None other of Hornigold's men could offer that much capital.

Once we accept that Blackbeard was with Hornigold on the *Benjamin* in March 1716, it becomes easy to believe that he was with Hornigold long before that. Colin Woodard suggests that Blackbeard was with Hornigold right from the start. He wrote, "Benjamine Hornigold was one of the very first to turn to this other course of life and he took Edward Thatch with him."[8] This might be true; however, I am unaware of any source documents that would prove the assumption.

Some historians and enthusiasts suggest that the main source for the details on the famous association between Hornigold and Blackbeard comes from Charles Johnson's *A General History of the Pirates*. However, Johnson actually provides very little information on Hornigold. In his first edition, Hornigold isn't mentioned at all. In the second and subsequent editions, Hornigold is mentioned exactly eight times, three of those being in the introduction among a list of pirates at Nassau. Hornigold is only mentioned five times in the text within his chapter on Blackbeard, as shown in the following excerpts:

> [Blackbeard] was never raised to any Command, till he went a-pyrating, which I think was at the latter End of the Year 1716, when Captain *Benjamin Hornigold* put him into a Sloop that he had made Prize of, and with whom he continued in Consortship till a little while before *Hornigold* surrendered.

> In the Spring of the Year 1717, *Teach* and *Hornigold* sailed from *Providence*, for the Main of *America*. . . .

> [Blackbeard] made Prize of a large *French Guiney* Man, bound to *Martinico*, which by *Hornigold*'s Consent, *Teach* went aboard of as Captain, and took a Cruize in her; *Hornigold* returned with his Sloop to *Providence*, where, at the Arrival of Captain *Rogers*, the Governor, he surrendered to Mercy, pursuant to the King's Proclamation.[9]

When tracking down the sources on the mentor/protégé relationship between Hornigold and Blackbeard, the details aren't in Johnson's book. It is true that Hornigold is portrayed as the senior officer, as he gave Blackbeard command of, or "put him into," a sloop, and later gave his

The value of 28,500 pieces of eight is equivalent to approximately $2.9 million in today's money.

consent for Blackbeard to become captain of the "French Guiney Man." But Johnson stops there. It appears that elaboration on their relationship is a creation of historian authors beginning in the late twentieth century. Prior to that, the books about Blackbeard either simply restate the information in Johnson's book or don't mention Hornigold at all.

Back to the action! After Jennings failed in his pursuit of Hornigold and was double-crossed by Bellamy, he sailed to Nassau with the French ship he gave to Captain Carnegie. Afterward, he returned to fish the wrecks with his fleet. This was confirmed in a letter written by John Vickers.

> About 22nd April last, Capt. Jenings arrived at Providence and brought in as prize a French ship mounted with 32 guns which he had taken at the Bay of Hounds . . . and then went in the said ship to the wrecks where he served as Comodore and guardship.[10]

While Jennings was cooling his heels in Nassau, Hornigold on his sloop, *Benjamin,* and Bellamy on his sloop *Marianne*, continued to cruise the waters around Cuba. Some accounts state that they began their partnership by sailing to Portobelo on the north coast of the Isthmus of Panama and then back to Cuba. The source of this idea comes from the deposition of Peter Hoof, who was taken off the Dutch ship that Bellamy captured at the start of his career. He said:

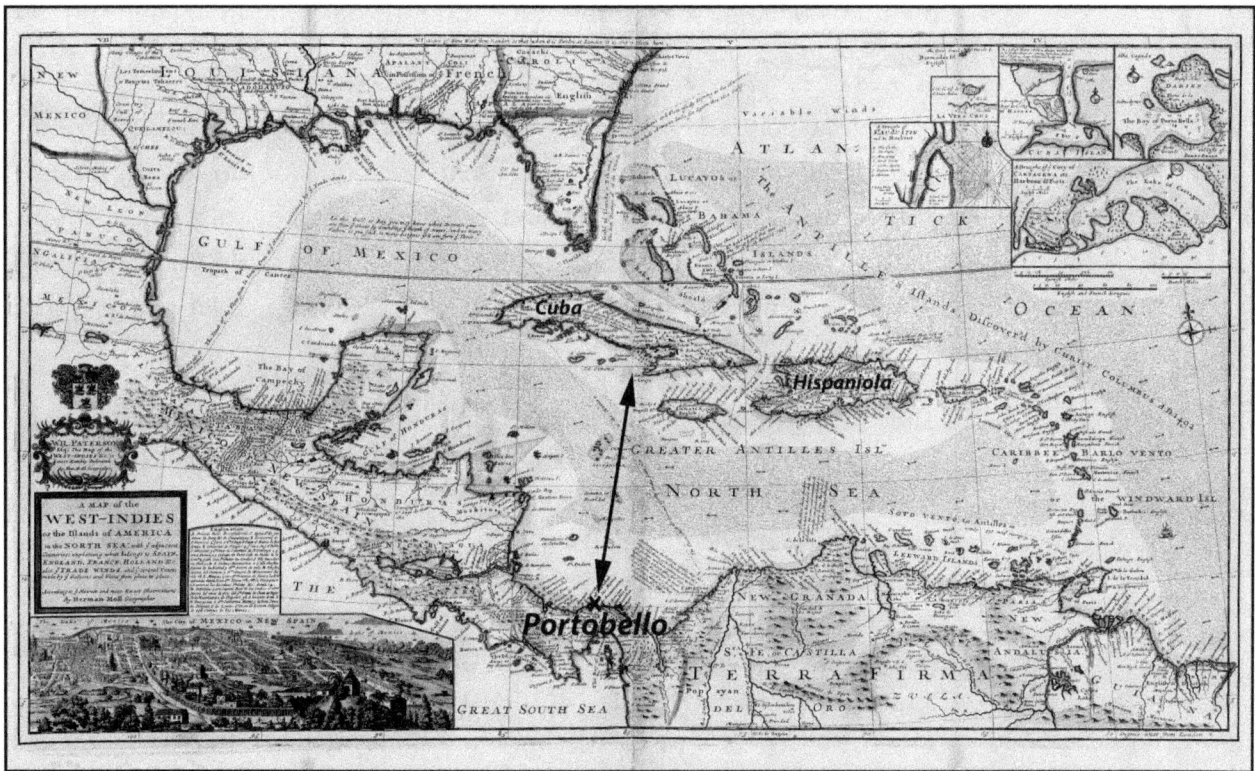

Figure 12: *Distance between Cuba and Portobello from Caribbean Map by Herman Moll. The image is dedicated to the public domain under CC0.*

Chapter Six

Olivier Levasseur (1668–1730), known as La Buse, was executed in 1730 on Réunion Island in the Indian Ocean for piracy.

> Three Weeks after he was taken they went to Portobello in a French Sloop with 60 Men on board; then stood for the Havana, and from thence to Cuba, where they met with a Pink, an English-man Master, and took out some Powder and Shot, and some Men.[11]

The trip to Portobelo seems problematic. The round trip from Cuba to Portobelo is over 2,000 miles. When reading his entire deposition, he earlier stated that "He Sail'd for the most part with the Dutch on the Coast of Portobelo." However, he was taken near the Bay of Honduras. Perhaps to him, Portobelo represents the entire coast of Central America, which would make their journey far more reasonable. Either way, upon their return to Cuba, they stopped an English pink, out of which they took some gunpowder and shot, along with some of the crew.

Somewhere along the coast of Cuba, they met another pirate sloop flying a distinctive flag. This was the *Postillion*, commanded by the French pirate Olivier Levasseur, known as La Buse. Unlike most pirate flags, La Buse hoisted a white flag with a black skeleton on it. This unique flag was described in a letter to the Council of Trade and Plantations as "a white Ensign with the figure of a dead man spread in it."[12]

According to the deposition of Abijah Savage, one of La Buse's victims, the *Postillion* was a sloop with 8 guns and about 80 or 90 crewmen. He also mentions Bellamy's *Marianne*. Of course, the spelling in his deposition is not exactly accurate.

> One, called the *Mary Anne*, was commanded by Samuel Bellamy who declared himself to be an Englishman born in London, and the other, the Postillion, by Louis de Boure a Frenchman, who had his sloop chiefly navigated with men of that Nation. Each sloop was mounted with 8 guns, and had betwixt 80 or 90 men. The *Mary Anne* was chiefly navigated with Englishmen.[13]

The three pirate captains formed a partnership and sailed together for the next four months. Shortly after their partnership began, they took a vessel near the southeast tip of Cuba, where sailor John Brown joined La Buse's crew. They continued westward along the south coast of Cuba until they reached Cabo Corrientes on the southwestern tip of Cuba. Spotting a Spanish brigantine sailing off the coast, they attacked and easily took the prize without a fight. After looting the brigantine, the pirates demanded a ransom for the safe return of their vessel. However, the Spanish refused to pay, so the crew was put ashore and the brigantine was burned. From there, they sailed east for about 100 miles to the Isle of Pines, where they careened their sloops with the assistance of several English sloops.

Afterward, they sailed to Hispaniola, arriving near the end of May 1716, where they remained throughout the long, hot summer. This information comes from the deposition of John Brown. After serving onboard the *Postillion* for a while, he joined Bellamy and was among the six pirates who were hanged in Boston in 1717. An excerpt from his deposition given at his trial reads:

> [He] was taken to the Leeward of the Havana by two Piratical Sloops, one Commanded by Hornygold and the other by a Frenchman called Labous, each having 70 Men on Board. The Pirates kept the Ship about 8 or 10 Days and then having taken out of her what they thought proper delivered her back to some of the Men, who belonged to her Labous kept the Examinate on board his Sloop about 4 Months, the English Sloop under Hornygolds command keeping company with them all that time. Off Cape Corante they took two Spanish Briganteens without any resistance laden with Cocoa from Maraca. The Spaniards not coming up to the Pirates demand about the ransom were put a-shore and their Briganteens burn'd. They Sailed next to the Isle of Pines, where meeting with 3 or 4 English Sloops empty, they made use of them in cleaning their own, and gave them back. From thence they Sailed to Hispaniola in the latter end of May, where they tarryed about 3 Months.[14]

At some point, probably in early June, Hornigold disposed of his sloop *Benjamin* by selling it to a captain named Perrin. Hornigold had originally taken that sloop from the Spanish. It is unclear what Hornigold chose as a replacement for the sloop, but he certainly had a wide selection to choose from. Perrin sailed the sloop north to the Florida straits, but he never profited from his timely purchase. He was intercepted by Captain Mathew Musson, who had been commissioned by the governor of South Carolina to hunt down pirates. The Spanish sloop was confiscated and Perrin was arrested. In July 1716, Musson's report was included in a deposition made by Robert Daniell, Deputy Governor of South Carolina, which reads:

> I renewed the commission of Captn. Mathew Musson to take pirates etc., the commission he had from the Lord Hamilton being nearly expired, and he intending to cruise about Cape Florida a station now Perrin from Virginia on board a sloop in which Hornigold the pirate sailed and which he took last winter from the Spaniards, having on board sundry goods which Perrin pretended to have bought of Hornigold. Musson seized the sd. sloop and sent Perrin in a sloop

Chapter Six

properly owned by him and the sd. goods under command of Joseph Carpenter to this Governmt. in order to be prosecuted.[15]

Discord among such a partnership was inevitable. By the end of August, a mutiny occurred and Benjamin Hornigold was forced to leave. The argument was over Hornigold's reluctance to plunder English vessels. As you may recall, he made this position clear just six months earlier while he was in Nassau, "who say they meddle not with English or Dutch, but that they never consented to the Articles of Peace with the French and Spaniards."[16] Bellamy and La Buse had no such convictions. Neither did most of their crew. In the pirate tradition, when the crew opposed the will of their captain, a vote was held. The result was that Captain Hornigold lost his command and was forced to leave the company of pirates. Twenty-six men remained loyal to Hornigold and left with him. Hornigold and his supporters were given one of the prize sloops and Sam Bellamy took command of Hornigold's vessel. Bellamy's original partner, Palsgrave Williams, was given command of the *Marianne*.[17]

AUGUST 1716

Hornigold parts ways from Bellamy and La Buse.

The primary source for this once again comes from the deposition of John Brown. His is the only source for Hornigold's refusal "to take and plunder English Vessels" and that he "departed with 26 hands in a prize." In his own words:

> The Examinate then left Labous and went on board the Sloop Commanded formerly by Hornygold, at that time by one Bellamy, who upon a difference arising amongst the English Pirates because Hornygold refused to take and plunder English Vessels, was chosen by a great Majority their Captain & Hornygold departed with 26 hands in a prize.[18]

Although not as detailed, there are two other collaborative statements made by the only two members of Bellamy's crew who were there: Peter Hoof and Jeremiah Higgins. Hoof's deposition reads:

> A difference arising amongst them about taking Prisoners; Some being for one Nation and some for another; and having at that time Two Sloops and about 100 Men, Hornygold parted from them in One of the Sloops, and Bellamy and Labous kept company together.[19]

Jeremiah Higgins, who joined Hornigold before they took the *Marianne*, didn't stay with Hornigold when he departed. Instead, he stayed with Williams aboard the *Marianne*. In his deposition, he stated that the Hornigold and Bellamy "consorted together with the said Sloops until a quarrel happened among the Company and then they gave the said

Sloop *Benjamin* to the said Hornigold and some Company and parted from him detaining the Examinant on board the said Sloop *Mary Anne*." He seems a little confused as to which sloop Hornigold retained.[20] Higgins stayed with Williams on the *Marianne* until he was arrested in New York on May 15, 1717. An article appeared in *The Boston News-Letter*, Monday, June 17 to Monday, June 24, 1717 edition that reads, "On the 15th in the Morning Richard Caverly, late Master of the Pirate Sloop commanded by Paul Williams, and Jeremiah Higgins, late Boatswain of the same Sloop, were apprehended here, and are in Irons in the Fort."[21]

What about Blackbeard? If he had been with Hornigold and Bellamy since they first formed their partnership, he must have been among the twenty-six men who left with Hornigold. Not everyone agrees. Of course, since no prime source documentation has yet been found, just about anything is possible. Blackbeard may have remained with Bellamy for a month or two and then rejoined his old friend Hornigold at Nassau. Other scenarios are plausible, too. However, the cleanest and most likely scenario is that he left with Hornigold as one of the twenty-six men.

Chapter Six

Figure 13: *Blackbeard from the 1736 Edition of Johnson's* A General History of the Pyrates. *The image is dedicated to the public domain under CC0.*

Seven
Blackbeard Begins

It was a dark night at sea as a brigantine peacefully cut through the calm waters of the Caribbean. The only light came from the stars that shone brightly in the cool December sky. They were passing Cap Dame Marie on the far western tip of Hispaniola, which was less than two hundred miles away from their final destination, Jamaica. The brigantine was the *Lamb*, under the command of Henry Timberlake. They had left Boston twenty-eight days earlier. Now the crew knew that their long voyage was almost over. However, their journey was about to be interrupted by an event that every sailor feared the most. The *Lamb* was about to be taken by pirates.

Another sail appeared in the distance, but instead of maintaining course, as a merchant vessel would do, it unexpectedly changed course and made its way directly toward the *Lamb*. The crew watched anxiously as the unidentified vessel gradually closed the distance. As it grew nearer in the dark, the details became clear. It was a sloop with 8 guns and its deck was crowded with men holding weapons. By now, the *Lamb's* crew realized they were about to be taken by pirates. The silence of the night sea was suddenly shattered by the blast of several cannon shots. The rounds splashed near the bow of the *Lamb*, a signal to heave to. It was about eight o'clock at night on Thursday, December 13, 1716.

"Heave to" is a nautical term meaning to stop your vessel.

Captain Timberlake ordered his brigantine to head into the wind and come to a stop. The pirate sloop skillfully pulled alongside, about twenty-five yards away. A call came from the sloop demanding that the captain come onboard. The *Lamb's* boat was put into the water, and Captain Timberlake climbed over the rail into the boat. Although terrified and extremely apprehensive, Captain Timberlake maintained his composure as he rowed over to the sloop and bravely climbed aboard. Once on deck, Timberlake was greeted by an imposing figure. It was obvious that this

Chapter Seven

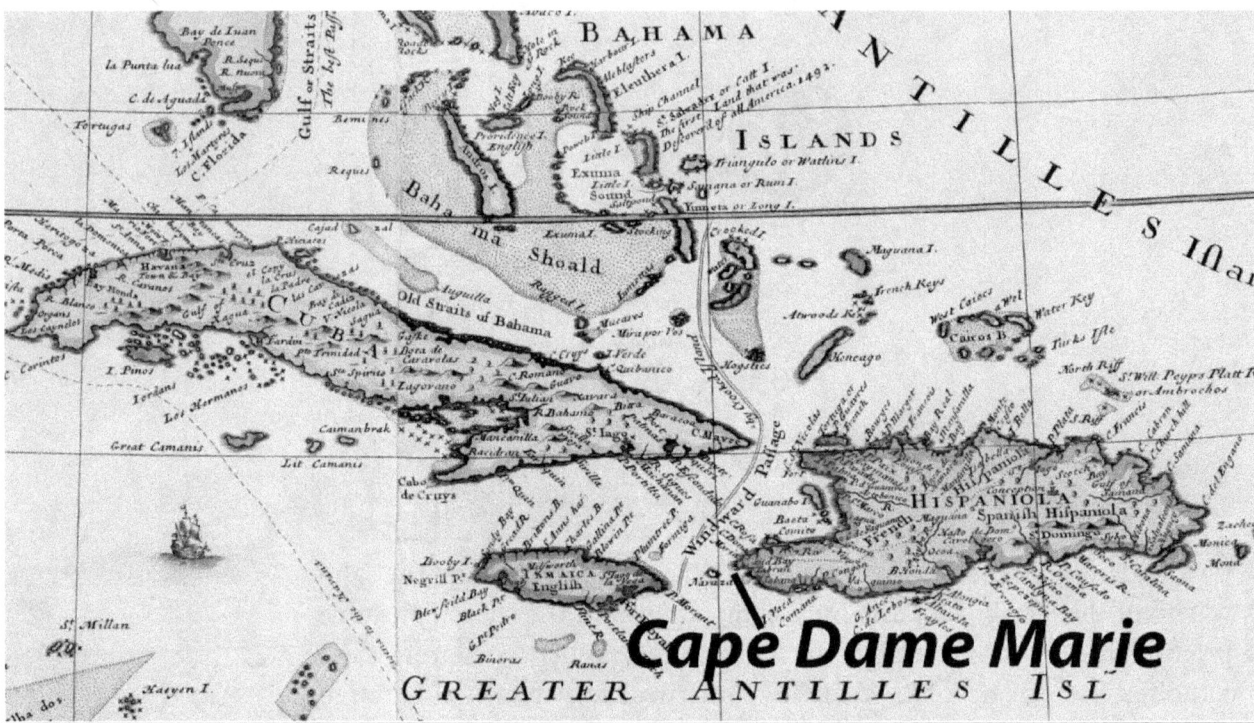

Figure 14: *Hispaniola from the 1702 Herman Moll Map of the Caribbean, Courtesy of Barry Lawrence Rudman.*

pirate had taken many vessels before and knew exactly how to handle these situations. After all, the prize crew had already surrendered, and it was time for the pirates to foster their cooperation. At that point, reassuring his victims would be far more effective than intimidating them. The pirate politely introduced himself as Captain Benjamin Hornigold, of the sloop *Delight*.

About a dozen of the pirates launched a boat from the *Delight* and rowed over to the *Lamb* to loot it. During that time, Hornigold and Timberlake engaged in polite conversation. Hornigold told him he had taken a Spanish ship of 40 guns one week earlier. Timberlake noticed the infuriation building in Hornigold's voice as he explained that afterward, fifteen of his men had secretly loaded 40,000 pieces of eight into a small boat and sailed away. Hornigold took a deep breath, calming himself, and continued to tell Timberlake of another ship from Bristol, captained by a man named Quarry, that the pirates had taken a week before as they sailed from Jamaica. Wistfully, he added that, when Quarry returned to Jamaica with the news that his cargo had been taken by pirates, the authorities suspected him of willingly giving his cargo to an accomplice and did not believe his story. The unfortunate captain had been arrested. Hornigold urged Timberlake to tell the authorities in Jamaica that Quarry had been wrongfully accused and was indeed forced to hand over his cargo.

An hour after Timberlake boarded the *Delight*, their conversation was abruptly interrupted by the sight of another sloop pulling alongside, armed with 8 guns, and carrying about ninety men. The unexpected appearance of this other sloop in the darkness surprised Timberlake, but not Hornigold, who glanced at the approaching sloop and smiled. Timberlake immediately understood Hornigold's reaction—that he expected the arrival of this other sloop and that its captain was a friend and a partner. As this other sloop came to rest near the *Delight*, its captain climbed aboard. This man's appearance was both terrifying and impressive. Well-armed with several pistols hanging about his bandolier, he was tall with a swarthy completion and had an exuberant black beard of extraordinary length covering his whole face and twisted into numberless tails; the ends tied with bows in brightly covered ribbons. The captain walked up to Timberlake and calmly introduced himself. He said, "Good evening, my name is Edward Thach."

This dramatic and somewhat poignant description of Blackbeard may be completely accurate or might be exaggerated. It was taken from Henry Graham Ashmead's 1884 book, *The History of Delaware County*.[1] There were only two descriptions of Blackbeard made during his lifetime. One came from a captive named Henry Bostock, who described him as "a tall Spare Man With a very black beard which he wore very long."[2] The other came from Lt. Maynard, who mentioned that "he let his Beard grow, and tied it up in black Ribbons."[3]

This account was derived from Henry Timberlake's deposition, the earliest primary source that unarguably identifies Edward Thach as a pirate captain. Although my narrative is intentionally colorful to increase reader interest, it is the result of an in-depth interpretation of the facts taken from the actual deposition. Based upon the descriptive accounts that Hornigold told Timberlake, their conversation must have been cordial and relatively relaxed. Hornigold's polite treatment of Timberlake and his crew once they surrendered is echoed in later reports about Blackbeard, who politely greeted his captives and treated them the same way. This will be described in detail in later chapters.

Blackbeard's sloop also sent men in a boat over to the *Lamb* to loot it. The fact that Hornigold's men had no objection proves that they were working together. Hornigold explained to Timberlake that they only wanted provisions, but his men took more than that. They took all their clothes and threw part of his cargo overboard. The provisions they took consisted of three barrels of pork, one barrel of beef, one barrel of peas, three barrels of salted mackerel, five barrels of onions, and several caggs of oysters. They left Timberlake with about forty biscuits and a small quantity of meat, just enough to get to Jamaica. After the pirates had made several trips back

A cagg was a small cask, usually ten gallons or less. "several dozen caggs of oysters."

Chapter Seven

and forth between their sloops and the prize brigantine, they signaled to Hornigold that they were finished. It was about three o'clock in the morning. Hornigold sent Timberlake back to his brigantine and tranquilly sailed away. When Timberlake arrived in Jamaica, he made the following deposition:

> That about the Sixteenth day of november last he Set Sail on board the said Brigantine from the Port of Boston in New England bound for this Island. That about the thirteenth of December instant about Eight Leagues off of Cape Donna Maria on the west end of Hispaniola about Eight a Clock at night a Sloop mounted with Eight Guns and manned with about ninety men as they told him called the Delight Benjamine Hornigole Comander came up with this Depont. fired Several Shot at him, obliged him to bring too and then Comanded him with his Boat on board and this Deponent and two of his men coming on board him, hornigole told this Deponent he had taken a Spaniard with forty Guns the Thursday before and a Bristol man that Sailed from this Island the week before, but gave this Deponent no further account of either of those Vessels and acquainted this Deponent the week before fifteen of his men had run away with their own Canoa and carried forty Thousand peices of eight with them. That hornigole Said to this Deponent, give my Service to the Captn of the man of warr and tell him I design to have his Ship from him if I meet him. That hornigole hoisted out his Boat with about a dozen hands and Boarded this Deponents Said Brigantine, this Deponent remaining on board the Sloop till about two or three of the Clock in the morning. That hornigole's Boat returning twice or thrice loaden with provisions from the Brigantine this Deponent asked them why they used him so they answered they wanted provisions and this Deponent further Saith That in about an hour after hornigole Boarded him Edward Thach Comander of another Sloop, the name whereof this Deponent knows not mounted with Eight Guns & manned with about ninety men came alongside the said Brigantine and Sent their Canoa with Several hands on Board her and plundered her That the said hornigole and the said other Sloop took from this Deponent Three Barrills of Porke, one of Beef, two of pease, three of Markrill five Barrills of onions Several Dozen Caggs of oysters most of his Cloaths and all his Ships Stores Except about forty Biskets and a very Small quantity of meat just to bring them in and threw Some of

December 17, 1716

Edward Thach (Blackbeard) mentioned in the historical record as a pirate for the first time their Staves over board. That this Deponent was cheifly loaden with Staves and Shingles and that he beleived the loss he and his comdrs sustained by the Said Pirates might be about Sixty pounds Jamaica mony That hornigole about three in the morning Sent this Deponent in his own Boat on board That this Depont. Soon after arrived in Port Royal harbour and further this Deponent Saith that the said hornigole told him that he understood by the Bristol Ship afore mentioned that Captn Quarry was in Goal for being concerned in a Pyracy with him but Said he was wrongfully accused therein for that Quarry did not act or concerned himself and was by him forced to be in their Company & declared that it was him and his Crew alone that had robbed that Spaniard. Hornigole further declared that if he thought that him this Deponent would not So Soon as he got into Jamaica declare and make known that Quarry was not concerned in that Pyracy he would not Suffer him to go from them and further Saith not.

Henry Timberlake. Sworn this 17th of December 1716 before his Excellency Peter Heywood. Jamaica[4]

It is interesting to note Hornigold's apparent change of heart. The previous chapter ended with an argument between Hornigold and Bellamy. According to John Brown's statement, they argued over Hornigold's refusal to take English vessels, which subsequently led to his dismissal from their pirate group, along with twenty-six others who shared his views.[5] Based upon his taking of the Bristol merchant ship and the *Lamb*, Hornigold's convictions had changed. At any rate, Blackbeard and Hornigold were now sailing together, each with his own sloop. Charles Johnson wrote that Hornigold gave Blackbeard this command. The accuracy of his account may come into question; however, Johnson was correct about the month and the year. In his second edition, Johnson wrote:

[Blackbeard] was never raised to any Command, till he went a-pyrating, which I think was at the latter End of the Year 1716, when Captain Benjamin Hornigold put him into a Sloop that he had made Prize of, and with whom he continued in Consortship till a little while before Hornigold surrendered.[6]

There is no reliable prime documentation that Hornigold gave Blackbeard this command. However, if the two of them had been sailing together for the past year with Hornigold always in the position of captain, he likely did. The only other logical alternative is that Blackbeard wasn't with Hornigold

Chapter Seven

1696

The Commissioners for Trade and Plantations was formed in 1696 to promote trade and to inspect and improve the plantations of the British colonies. Historian John William Fortescue introduced the term "Council of Trade and Plantations" in the early nineteenth century while serving as Royal Librarian and Archivist at Windsor Castle.

before, and they had just met as two captains who decided to form a partnership. Of course, that flies in the face of Blackbeard's friendship with Bellamy as being the catalyst for his vengeance on vessels from Boston.

In the meantime, Captain Thomas Walker, who had been chased out of Nassau by the pirates, wrote a letter to the Council of Trade and Plantations about the tremendous pirate growth in Nassau. Since there was no government at Nassau, the island was "open to be a recepticall and shelter of pirates" and that the locals were all now heavily engaged in buying the good the pirates brought to the island as well as selling them everything they needed and even "entertaining and releiveing" with the pirates. He adds:

> The pirates daly increse to Providence and haveing began to mount ye guns in ye Fort for there defence and seeking ye oppertunity to kill mee because I was against their illegall and unwarrantable practices and by no means would consent to their mounting of guns in ye Fort upon such accots.[7]

Figure 15: *1793 Close-up from Map of the Bahamas. The image is dedicated to the public domain under CC0.*

As mentioned in the previous chapter, Captain Matthew Musson was instructed by Robert Daniell, the Deputy Governor of South Carolina, to sail to the Bahamas to investigate the pirate activity and to make a report. However, he only reached Harbour Island when his ship went aground. While on shore, Musson realized that Harbour Island served as the main source of provisions for the pirates on Nassau. He also met Captain Thomas Walker, who had been chased out of Nassau a year earlier with his family. Walker gave him the vital information he needed, which he put into his report to the Council of Trade and Plantations. The most fascinating part of his report mentions Blackbeard, or more precisely, Thatch, who has a sloop of 6 guns and a crew of 70 men. Musson's report reads:

> They advis'd him that five pirates made ye harbour of Providence their place of rendevous vizt. Horngold, a sloop with 10 guns and about 80 men; Jennings, a sloop with 10 guns and 100 men; Burgiss, a sloop with 8 guns and about 80 men; White, in a small vessell with 30 men and small armes; Thatch, a sloop 6 gunns and about 70 men.[8]

This report evokes an image of Hornigold, Jennings, Burgiss, White, and Thatch all sitting together in a Nassau tavern having a pleasant drink of ale while sharing tall tales of their adventures. Some historian authors have even made the mistake of physically placing Blackbeard in Nassau in March 1717 when Musson was on Harbour Island, gathering his information. However, Musson didn't actually see any of the pirates he mentioned in his report. He was told about them by Walker, who hadn't been in Nassau for months. Walker's intent was to relay to the governor via Musson that these pirates were known to frequent Nassau, not necessarily to report that they were actually there at the time. Blackbeard and Hornigold may have been far away from Nassau in March 1717, taking a snow from Jamaica.

March 1717 Dr. John Howell was forced to join Hornigold's crew.

The next piece in the jigsaw puzzle of Blackbeard's complicated life comes from a totally unexpected source, the testimony of the trial of Dr. John Howell. The good doctor was aboard the sloop *Bennet* when it was taken by Hornigold in March 1717. The pirates needed a surgeon, so they forced Dr. Howell to join their crew. Eventually, he was released and took up residence in Nassau, where he lived peaceably as a law-abiding citizen. However, when Woodes Rogers was replaced by George Phenny as governor of the Bahamas in 1721, the door opened for our old friend, Thomas Walker, to bring charges against anyone suspected of having ever been a pirate. The trial record states that "Mr. Walker in express and plain Words did charge sd Mr. Howell with Pyracy."[9] In his defense, six former members of Hornigold's crew testified on his behalf. The trial was held on December 22, 1721, and each of the witnesses stated that Dr. Howell was taken by force and that he never shared in the profits. As a result, Dr. Howell was found not guilty and sent back to England.[10] The testimony of six of Hornigold's former crewmen provides a wealth of information. The following account comes directly from the trial record.[11]

Aligning the testimony with the facts can be a bit confusing. Hornigold's sloop was named the *Adventure* when he captured Dr. Howell. Later on, he captured the sloop *Bennet* and swapped sloops, keeping the *Bennet* and giving the *Adventure* to the captured crew. In the testimony, one pirate stated, "Hornigold exchanging his Sloop *Adventure* for the *Bennet*." However, reading the testimony, some of the former pirates testified that they belonged to the sloop *Bennet*, giving the impression that this was their vessel when they forced Dr. Howell aboard. For example, Richard Nolan said, "He has known sd John Howell upwards of four Years when belonging to the Sloop Bennet, Beja. Hornigold Comander." This wasn't the case. They eventually were on the *Bennet* with Dr. Howell, just not at first. Below, I summarize the events.

William Howard was Hornigold's quartermaster aboard their pirate sloop *Adventure* in March 1717. Cruising off the Cape of Florida, near modern-day Miami, they took a snow from Jamaica sailing under the command of a man named Blake. Dr. Howell was aboard that snow. Needing a surgeon, Hornigold forced Dr. Howell to join his crew. Near the western end of Cuba, Hornigold joined the pirate captain Napping and together they sailed their two sloops south toward the Friends Islands, near Portobelo. They arrived by April 1, 1717, when they took the sloop *Bennet*, commanded by Captain Hickinbottom. On board was a sailor named Robert Brown, who by coincidence, already knew Dr. Howell from Cork, Ireland. Brown willingly joined Hornigold's pirate crew. As usual, the pirates searched the *Bennet* for valuables. To their great surprise, they found a chest of gold belonging to the Assiento Company that contained 400,000 pieces of eight. This amount, as well as the date of the attack, are in a letter from Lt. Governor Bennett to the Council of Trade which we shall soon see.[12]

Hornigold liked the prize sloop better than his. Perhaps it was larger and faster. He kept the *Bennet*, and he gave the *Adventure* to Hickinbottom and his crew. Dr. Howell pleaded with Hornigold to be allowed to leave with Hickinbottom on the *Adventure*, but his request was denied. Following are selected excerpts from their testimony:[13]

> Richard Noland. . . He has known sd John Howell upwards of four Years when belonging to the Sloop Bennet, Beja. Hornigold Comander
> . . . that He sd Wright together with one William Howard Quartr.

> Msr. and others of the same Crew forcibly took sd John Howell from on board a certain Snow belonging to Jamaica, Blake Comander, to serve on board sd Sloop Bennet as Surgeon ... Hornigold afterward took two Dutch Ships at what Time Mr. Howell was on board Sloop Bennet.,
>
> William Howard, late Qr. Mastr. ... Hornigold met a Snow come from Jamaica, Benja. Blake Comander; off Cape Florida; and wanting a Surgon ... Howard with nine others arm'd went on board sd Blake, and thence forced sd Howell with his Medicines to serve on board sd Hornigold.
>
> Robert Brown ... Brown since belonging to the Sloop Bennet one Hickinbottom Comander was met with by Hornigold & Knapping two Pirates off of Friends Islands near Peurto Bello, the latter of which took them, and a Chest of Gold belonging to the Assiento Company ... Hornigold exchanging his Sloop Adventure for the Bennet gave Leave to Captn. Hickinbottom to go for Jamaica, at what Time Mr. Howell being Surgeon to Hornigold press'd him very much to have Leave to go with him.
>
> Pearce Wright ... He belongd to Captain Hornigold, and was one of the Boat's Crew that forct sd Howell from Captn. Blake, And well remembers that Mr. Howell prayd at that time Capt. Blake to certifie to the World the manner of his being forct, being much troubld at his Misfortune.

Hornigold's partnership with the pirate captain Napping is also documented in a letter dated July 19, 1717, and written by Captain Bartholomew Candler of HMS *Winchelsea*. In March 1717, his ship was cruising near Cuba looking for pirates. He was told that Hornigold and Napping were near the western (leeward) tip of Cuba, but the pirates found out that he was in the area and left Cuban waters before April. Captain Candler wrote:

> And to Leeward they were Two Sloops, having about 100 men in each Commanded by one Hornigold, & one Napping, those latter I followed over to the So Caios of Cuba ... they soon heard that we were in quest of them, for they are all gone from hence, for since Aprill there has not been one seen about this Island.[14]

Aboard his new sloop *Bennet,* Hornigold sailed toward Jamaica, with Napping sailing in consort. On the way, they took the sloop *Revenge* on

Hornigold and Napping sailed together as partners.

Chapter Seven

April 7, 1717. The *Revenge's* capture, as well as the 400,000 pieces of eight profit, are chronicled in a letter dated July 30, 1717, from Lt. Governor Bennett to the Council of Trade and Plantations. It reads:

> From Jamaica, I am inform'd that a sloop called the Bennett was taken on the 1st of April last by two pirates soon after she came out of Porto Bell where she had been tradeing, and on the 7th following a sloop called the Revenge from the same place was also taken by the same pirates near Jamaica, to where they were both bound: It is said that in those vessells were 400,000 pieces of eight great part belonging to the Asiento Company.[15]

What about Blackbeard? After all, this book is about him, not Hornigold. In the letter above, Lt. Governor Bennett mentions, "two pirates." There was no other vessel. So far, no reliable source documentation mentioning Blackbeard's location from December 1716 to September 1717 has surfaced. This includes the name Thatch by any of the spellings. Logically, there are only two options for the whereabouts of our favorite pirate during the spring and summer of 1717. Either Blackbeard was once again sailing with Hornigold onboard his sloop without a vessel of his own, or Blackbeard was somewhere else. Before exploring these possibilities, it is important to establish Hornigold's whereabouts.

In the early summer of 1717, Hornigold returned to Nassau, where he finally released Dr. Howell. However, a few months later, a French pirate named Jean Bondavias needed a surgeon and demanded that Dr. Howell go with him. This quickly turned into quite a disturbance, with a townsman named William Pindar intervening. Dr. Howell much preferred Hornigold to Bondavias, and Pindar helped him sail with Hornigold for a second time. Once again, this is evident from the testimony of the trial of Dr. Howell:[16]

> Richard Noland . . . Capt. Bonadvis Crew being very strenuous to take John Howell for their Surgeon; who had left Hornigold and livd then upon the Island . . . Howell applyed himself to sd. Noland complaining that He would rather serve the English than French, if He was compelld to make choice of Either, Accordingly went with the Crew of Hornigold a Second Time.

> William Pindar . . . Howell wth Tears in his Eyes told Pindar that He would rather choose to go wth Hornigold than those French Men who dealt so hardly with Him.

Robert Brown ... The sd Brown also declard, that he has heard Hornigolds Quarter Mr. John Martin swear that Howell should go when Mr. Howell went the Second Time.

The date of this incident is not specified in the testimony; however, Dr. Howell's second voyage with Hornigold must have occurred around mid-September 1717. That would place Hornigold's return to Nassau in the early summer of 1717. The key to establishing these dates lies with William Howard and John Martin, who both served as Hornigold's quartermasters. William Howard held the position until early September 1717, when he left Nassau to sail with Blackbeard, and John Martin took his place. In Robert Brown's testimony, John Martin is identified as Hornigold's quartermaster at the time of the incident, verifying that it took place after Howard's departure.

So, what was Blackbeard doing while all this was going on? As mentioned above, he might have been with Hornigold the entire time between March and early June 1717. If he wasn't with Hornigold, Nassau was the most likely choice for Blackbeard's location. It isn't difficult to imagine him simply staying in town for three months while his partner, Ben Hornigold, was cruising the Caribbean. Either way, they could have joined forces for their cruise in the late spring of 1717. Charles Johnson provides information on Blackbeard and Hornigold's next activity.

> In the Spring of the Year 1717, *Teach* and *Hornigold* sailed from *Providence*, for the Main of *America*, and took in their Way a Billop from the *Havana*, with 120 Barrels of Flower, as also a Sloop from *Bermuda*, *Thurbar* Master, from whom they took only some Gallons of Wine, and then let him go; and a Ship from *Madera* to *South-Carolina*, out of which they got Plunder to a considerable Value.[17]

You must have noticed by now that once again I am using a book that was referred to as a shim-sham story of pirates and that I called mostly a work of historical fiction. However, I also noted that even though most of the details described in Johnson's work prove to be incorrect, occasionally his facts are precise. The solution to determining which is which lies in cross-referencing with other primary sources. In this case, there is another collaborating primary source, the Monday, July 29 to Monday, August 5, 1717 edition of *The Boston News-Letter*, which reads:

> *New York July* 29. Last week arrived here Billop in a Sloop from the Havana, who soon after he left that Port, was taken by Hornygold a Pirate, who took from him 120 Barrels of Flower, and several

other things This Day arriv'd Thurbar from Jamaica, who was also taken by the said Hornygold, who only took from him some few Gallons of Rhum:

That a Snow bound from Plymouth to South Carolina, and a Ship from Madera were taken by Pirates, and grievously plundered; the first by a Ship of 26 Guns, 200 Men, commanded by a French Man, call'd Labous; the other by a Sloop, thought to be Paul Williams; the Indian War at South Carolina was pretty well over.[18]

Johnson obviously saw a copy of this publication or some sort of extract. He ran the same story in the March 15, 1718 issue of *The Weekly Journal or Saturday's Post*, which, as you may recall, was Johnson's earlier tabloid magazine published under his real name, Nathaniel Mist.[19] This story was resurrected when he wrote his book, *A General History of the Pyrates*, six years later. When comparing the two, it is clear that Johnson made a few errors. Captain Thurbar's sloop was from Jamaica, not Bermuda, and the ship from Madera was taken by La Buse, not Hornigold. But more importantly, *The Boston News-Letter* does not mention Blackbeard.

Once again, historian authors and Blackbeard enthusiasts are confronted with a choice. For those who require firm and indisputable prime source documentation, we must conclude that, thus far, there is no evidence that Blackbeard sailed with Hornigold in the summer of 1717. For those who are comfortable dealing with circumstantial evidence and informed speculation, the summer of 1717 offers many intriguing possibilities for the Hornigold/Blackbeard partnership.

Charles Johnson must have had a few reliable sources in London who provided him with information about pirates. There are too many accurate details sprinkled amongst his embellishments for the stories to be entirely concocted in his imagination. Since Johnson, who was really Nathaniel Mist, was a former sailor living in a major seaport, he likely spent several evenings sitting on the docks listening to sailors tell of contemporary encounters with pirates. If Blackbeard was once again sailing in consort with Hornigold, the possibility of one or more London sailors passing the information to Johnson is fairly considerable. However, there is far more substantial evidence that they sailed together. This comes from *The Boston News-Letter*.

Two articles appeared in two separate issues of *The Boston News-Letter* that clearly pair Blackbeard with Hornigold. The first article is from the Monday, November 18 to Monday, November 25, 1717 edition, and states that Teach and Hornigold took a vessel at latitude 36 and 45, which in current navigational language, means latitude 36.45. This is the approximate latitude of the Virginia-North Carolina border. It also provides the date of October 18, 1717. The article reads:

> Pritchard from St. Lucie, who on the 18th of October in Lat. 36 and 45 was taken by Capt. Teach, in Compa, with whom was Capt. Hornygold, they took from him about 8 Cask Sugar and most of their cloaths at the same time, they took a Ship from London for Virginia, out of which they took something and let them go.[20]

The second article is from the Monday, December 30 to Monday, January 6, 1718 edition, and states that Teach and Hornigold took a ship from Maryland near the Virginia Capes. This event took place fourteen days before the arrival of the ship in Philadelphia, December 10, 1717, which would make November 26, 1717, the date the ship was taken. It reads:

> Philadelphia, December 10th . . . We are told from Maryland that a Ship from London was arrived there, who about fourteen days ago was taken off the Capes of Virginia by Teach & Hornigold, that took out of him a New Suit of Sailes and Rigging.[21]

The dates of both of these attacks mentioned in the articles must be incorrect. Numerous reports chronicled in *The Boston News-Letter* and letters mentioning Blackbeard that were written by officials in Philadelphia clearly place Blackbeard in the Delaware River and along the shore of Long Island in October and early November 1717. He couldn't have been off the Virginia shoreline on the 18th of October. The second article is far more problematic for those who believe that Blackbeard and Hornigold were off Virginia in November. Indisputable French documents undoubtedly place Blackbeard on Bequia Island in the Grenadines on November 26, 1717.[22] That's about 1,900 miles away.

Since Blackbeard couldn't have been in two places at once, there must be an explanation for this mistake. There are only two rational explanations for this conflict. Colin Woodard suggests that people were used to the idea that Blackbeard and Hornigold were sailing together, and when Hornigold boarded them, they just assumed the other sloop sailing with him belonged to Blackbeard.[23]

If this was the case, it would certainly give credence to the concept that Blackbeard was indeed with Hornigold when he attacked Captain Billop's vessel, containing 20 barrels of flower, and Captain Thurbar's sloop. Perhaps he was and *The Boston News-Letter* failed to mention him because he wasn't as well known. Of course, a simpler explanation would be that

Chapter Seven

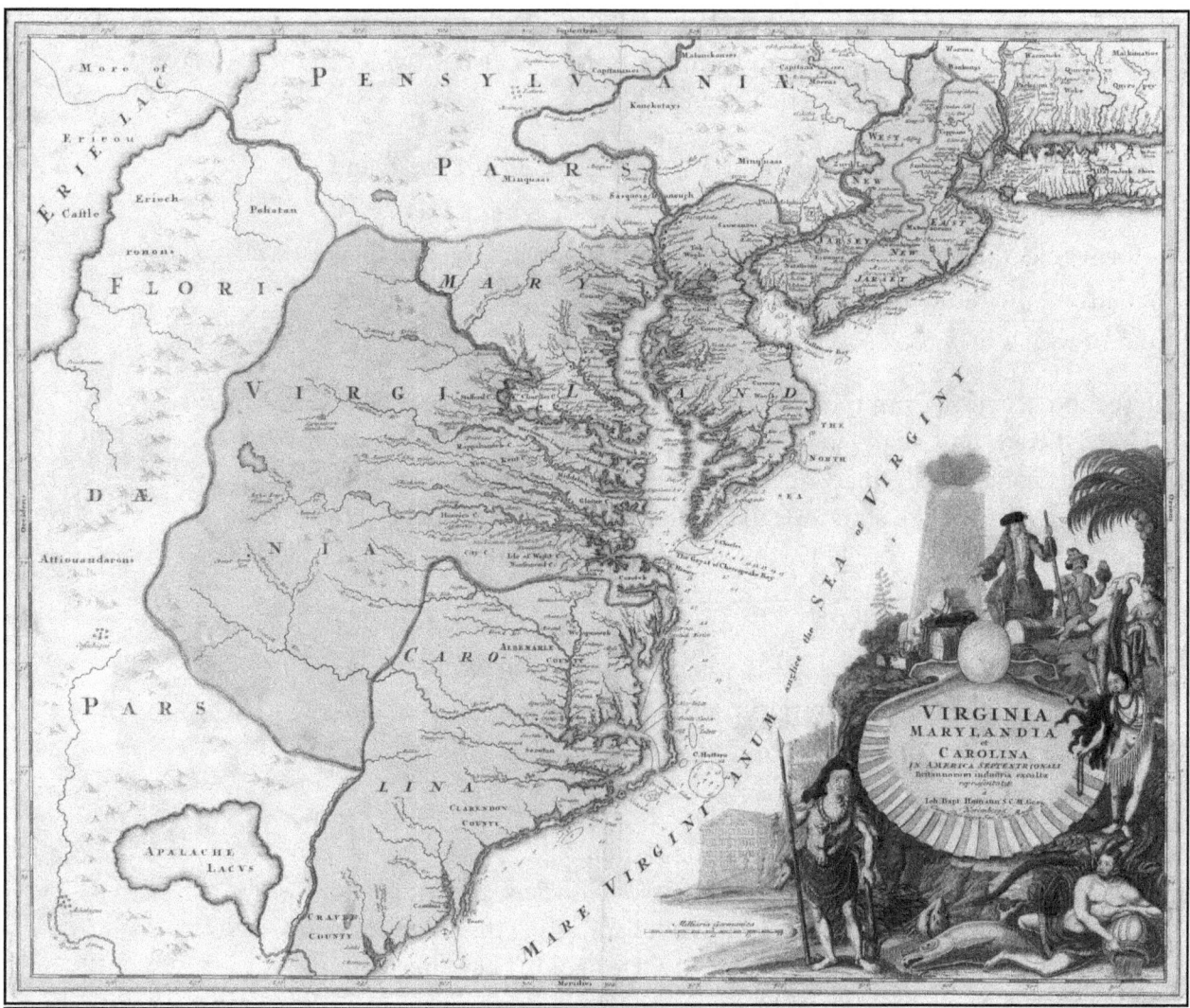

Figure 16: *1720 Map by Homann of the North American Coast Line, from Carolina to New York, Courtesy of Barry Lawrence Rudman.*

The Boston News-Letter simply got the dates wrong. A likely explanation is this:

> I suggest that these reports may have come in to 'The Boston News-Letter' months earlier, but Teach wasn't well known and Hornigold was yesterday's news. After Teach's spectacular assault on the Delaware Capes, his actions became a much bigger story. After a moment or two of reflection, the editors may have intentionally changed the dates and added these stories to boost circulation and capitalize on the success of their earlier editions. How shocking! Imagine, a newspaper journalist changing the facts simply to sell more newspapers. No modern journalist would do such a thing, would they?

If this explanation is correct, Blackbeard and Hornigold were raiding the Virginia coast in July or August of 1717. And if this were the case, Charles Johnson's next assertion becomes much more feasible. Sometime in the summer of 1717, Blackbeard and Hornigold sailed to Accomack, Virginia, to careen their vessels. Johnson wrote, "After cleaning on the Coast of *Virginia*, they returned to the *West-Indies*."[24]

All vessels must be cleaned from time to time. Every vessel accumulates barnacles and seaweed below the waterline. This can greatly reduce the speed of the vessel, a bad thing for pirates. More importantly, to the health of the vessel, in warm water like the Caribbean, teredo worms bore into the wooden hull and can eventually destroy it. To clean a vessel properly, it must be careened. In order to do this, the vessel must first be physically dragged up onto a beach. While laying on one side, the hull is scraped and then treated with a mixture of sulfur and tar, which kills the worms and waterproofs the cracks. When one side is finished, the vessel must be pulled back into the sea, turned around, and dragged back onshore to careen the other side. This is a painstaking process that can be very dangerous for pirates. If they are discovered by any unfriendly vessel, they are helpless to defend themselves. Consequently, pirates must find isolated locations to careen. They need a gradually sloping beach so they can pull their vessel ashore. Accomack, Virginia was ideal. It had been used for careening by pirates for decades. Pirate Captain Cook careened there in 1682, according to one of his crew members, William Dampier, who later wrote a book on his piratical activities called *New Voyage Round the World—1697*.[25] There are two pieces of evidence beyond Johnson and Dampier's historical perspective that suggest Blackbeard careened there. The first is local tradition, and the second is a 1720 map.

Thomas T. Upshur's "Eastern-Shore History" was an article that appeared in The Virginia Magazine of History and Biography, Volume 9, in 1901. It was never published as a book.

Although often ignored by some historians, local tradition can sometimes provide valuable and accurate accounts of the events of the past. In modern times, people are less likely to rely on oral tradition, but in the eighteenth and nineteenth centuries, this wasn't the case. People passed on stories from one generation to another. In the nineteenth century, oral tradition led to the discoveries of many of ancient Egypt's archaeological treasures as well as dozens of other long-lost sites throughout the ancient world.

Historian author Thomas T. Upshur wrote and published *Eastern-Shore History* in 1901, which was primarily based upon oral tradition from the locals who lived there. Prior to writing his article, he had traveled throughout the eastern shore of Virginia, where he interviewed locals and collected a wide variety of their stories. This is what he had to say about Blackbeard:

Chapter Seven

You have all heard of the famous character Preeson Richards, who figured in your court records early in the present century, and have heard from your infancy of "Black Beard," the pirate; but you may not have heard that "Black Beard" was a native of Accomack County, and that his name was Edward Teach. The rendezvou of his men was on Parramore's Beach, Revell's Island, Hog Island, and Rogues' Island. The latter island received its name from being the hiding-place of the band. His depredations became so frequent and his raids so daring that finally the Virginia authorities equipped vessels to put a stop to them. His Eastern-Shore haunts soon became too hot for his safety, and he removed his headquarters to North Carolina, up in Albemarle Sound, whence he continued his excursions; and, as North Carolina did not molest him or, at any rate, did not break up his piracy—Virginia sent her vessels into Albemarle Sound, sank his schooner, and killed him and all his men, except a few who were sick, or who were on shore and escaped. This intrusion of the armed vessels of Virginia into North Carolina's waters was resented by that State, who said she was abundantly able to attend to her own affairs. The matter was amicably settled between the States, but was unpleasant for a time. This item concerning Black Beard is history, but it has given rise to many stories and traditions which, if true, would be worth recording. Unfortunately, however, traditions, when pierced by the ethereal spear, become flaccid, and are usually worthless.[26]

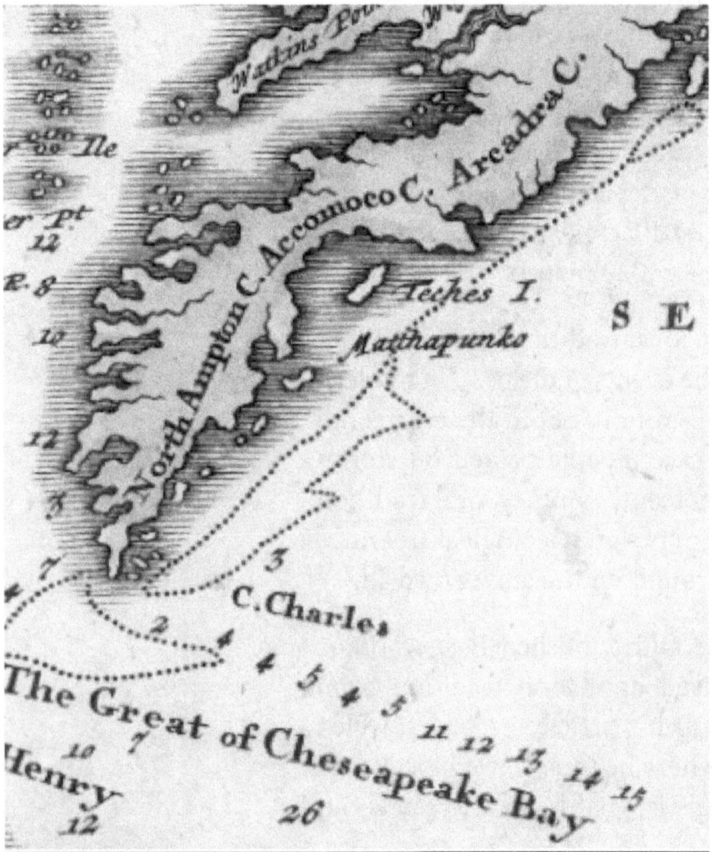

Some historian authors have put forth the idea that Upshur had made a serious assertion that Blackbeard was actually from Accomack. This doesn't seem to be the case. The paragraph above represents the full content written about Blackbeard within Upshur's book. What is disappointing is that Upshur completely discredits the account in his last sentence. Whereas Upshur offers no evidence to

Figure 17: *Close-up of 1720 Map of Virginia Showing Teches Island, Courtesy of Barry Lawrence Rudman.*

support the notion that Blackbeard was at Accomack, he does offer proof that the locals believed he was.

Stronger evidence comes in the form of a 1720 map. In Accomack is a long, narrow island with a great beach for both swimming and careening. *Cedar Island is* its modern name, but in the eighteenth century, it was called Teches Island. JB Homann published a map in 1720 of the North American coastline, from Carolina to New York. Taking a close look, one can see Teches Island clearly illustrated on his map. This wasn't a fluke. Every map printed between 1720 and 1794 that I have examined in which the names of the Accomack islands are included, shows Teches Island precisely where Cedar Island is on modern maps.

Putting all these clues together, I believe that Charles Johnson had learned far more about Blackbeard on the coast of Virginia than just the single line he wrote: "After cleaning on the Coast of *Virginia,* they returned to the *West-Indies.*" I believe that this is precisely what Blackbeard and Hornigold did.

From the Virginia Capes, Blackbeard and Hornigold returned to Nassau. If they hadn't heard the news of Sam Bellamy's death before, they assuredly learned about him in Nassau. The news was everywhere. It had been run in *The Boston News-Letter* back in May.[27] As mentioned in a previous chapter, in April 1717, Bellamy's pirate ship *Whydah* capsized and sank in a storm off Cape Cod. Nine men survived and were arrested when they reached the shore.[28] One was sold into slavery and the other eight were tried for piracy, resulting in two being acquitted and six found guilty and hanged. The transcript of that trial was published in Boston in 1718.[29]

1717 marks the beginning of a course that will lead to Blackbeard's immortality as the world's most famous pirate. His old friend and former partner, Ben Hornigold, will return to the sea as a pirate captain accompanied by his new quartermaster, John Martin.[30] William Howard will leave Hornigold's company and become Blackbeard's quartermaster. And Blackbeard will form an association with an incredibly unlikely accomplice. Together, they will gain tremendous recognition and notoriety. Of course, this is none other than the infamous "gentleman pirate," Stede Bonnet.

Eight
The Gentleman Pirate

A gentleman, a man of honor, and a man of fortune, Stede Bonnet was the most unlikely and unique pirate of the eighteenth century. Before he turned to piracy, he was a wealthy plantation owner, a husband, a father, a major in the militia, and of course, a gentleman. At the start of his pirate career, he purchased his sloop rather than stealing one. Bonnet's one-and-a-half-year career can be divided into three parts. Each of those periods is marked with initial success but ends in disaster. The last of these resulted in his arrest and subsequent trial for piracy in Charles Town, South Carolina. Bonnet's deeds attracted so much public attention that the transcript of his trial was published in London under the title, *The Tryals of Major Stede Bonnet, and Other Pirates*. Shortly after its release, it became a best-selling book. At the start of the trial, the prosecuting attorney, Richard Allein, admonished the jury saying:

> I am sorry to hear some Expressions drop from private Persons, (I hope there is none of them upon the Jury) in favour of the Pirates, and particularly of Bonnet; that he is a Gentleman, a Man of Honour, a Man of Fortune, and one that has had a liberal Education. Alas, Gentlemen, all these Qualifications are but several Aggravations of his Crimes. How can a Man be said to be a Man of Honour, that has lost all Sense of Honour and Humanity, that is become an Enemy of Mankind, and given himself up to plunder and destroy his Fellow-Creatures, a common Robber, and a Pirate?[1]

The Bonnet family had been well-established in Barbados for three generations. His great-grandfather was one of the island's original settlers. They owned a four-hundred-acre plantation southwest of Bridgetown that grew tobacco and sugar.[2]

Stede's parents were Captain Edward Bonnet and Sara Whetston. His mother was the daughter of Barbados's Deputy Colonial Secretary, John Whetstone. Stede was born in 1688, along with his twin sister, Frances. Church records show that they were both baptized on July 29, 1688. His father died in 1694 when Stede was just six years old. He became the ward of his grandfather, John Whetstone, who was the son of Sir Thomas Whetstone, who was a lieutenant onboard the *Swiftsure*, William Penn's flagship during his invasion of Jamaica in 1655.[3] Stede married Mary Allamby on November 21, 1709 at St. Michael's Church in Bridgetown. The couple had three sons, Allamby, Edward, and Stede Junior, between 1712 and 1714. Allamby died as a child, but Stede's other two sons were alive when he left Barbados to sail the seas as a pirate.[4]

In the summer of 1717, Stede Bonnet purchased a sloop, which he named *Revenge*. Initially, it was reported to have 6 guns, but that number varies in subsequent reports. Bonnet's start is documented in a letter dated July 19, 1717, written by Captain Bartholomew Candler of HMS *Winchelsea*. Within this same letter, Candler also mentions, "Two Sloops, having about 100 men in each Commanded by one Hornigold, & one Napping." Candler wrote:

> Over on the Coast a Pirate Sloop from Barbados Comandd by one Major Bennett, who has an Estate in that Island and the Sloop is his Own, this Advice I had by Letter from thence, that in Aprill last He ran away out of Carlisle Bay in the night he had aboard 126 men 6 Guns & Armes & Ammunition Enough.[5]

Bonnet's beginning is also mentioned in the preface of his trial record, although it mentions ten guns instead of six. It reads:

> Major Stede Bonnet, alias Capt. Edwards, alias Thomas, late of Barbadoes: who, it seems, at his own Cost and Charges fitted from thence a large Sloop called the *Revenge* with ten Guns, and about eighty Men, And after his leaving Barbadoes committed several Piracies.[6]

The reason Bonnet turned to piracy has been debated over the years. The only contemporary source that addresses this issue is Charles Johnson's book, *A General History of the Pyrates*. Johnson suggests that he had some sort of disorder of the mind, which was coupled with marital problems. Johnson wrote:

> The Major was a Gentleman of good Reputation in the Island of Barbadoes, was Master of a plentiful Fortune, and had the Advantage

1625

British colony first established on Barbados

Chapter Eight

Figure 18: *Sloop Revenge, Artist's Concept Illustrated by Ginger Marks.*

of a liberal Education. He had the least Temptation of any Man to follow such a Course of Life, from the Condition of his Circumstances. It was very surprizing to every one, to hear of the Major's Enterprize, in the Island were he liv'd; and as he was generally esteem'd and honoured, before he broke out into open Acts of Pyracy, so he was afterwards rather pitty'd than condemned, by those that were acquainted with him, believing that this Humour of going a pyrating, proceeded from a Disorder in his Mind, which had been but too visible in him, some Time before this wicked Undertaking; and which is said to have been occasioned by some Discomforts he found in a married State; be that as it will, the Major was but ill qualify'd for the Business, as not understanding maritime Affairs.[7]

With no other plausible reason found in contemporary records, this sentiment has been echoed by many historian authors throughout the years. However, there is a far more believable reason that is supported by well-documented circumstantial evidence. Stede Bonnet was a Jacobite! Many honest seamen became pirates after George I was crowned King of England. Barbados was among the first English colonies established in the Caribbean. It was founded under the reign of King Charles I. He, of course, was a Stuart, and many of the island's early settlers must have been loyal to his family. As you may recall from an earlier chapter, the entire Jacobite rebellion was over James Edward Stuart, the grandson of King Charles I. Evidence that Bonnet was a Jacobite comes from the testimony of his crew at his trial. Bonnet renamed his sloop the *Royal James*, in direct reference to James Edward Stuart. As mentioned throughout his trial record, "This Pirate Sloop was commanded by that noted Pirate Major Stede Bonnet, and formerly called the Revenge, now the Royal James."[8] Additionally, one of his victims, Mr. Hepworth, testified that "they made bowls of punch, and went to drinking of the Pretender's health."[9] James Edward Stuart was most commonly referred to as the *pretender*.

Regardless of his true motivation, Bonnet was sailing under the black flag in April 1717. However, Charles Johnson's inaccurate account must first be addressed before continuing to Bonnet's actual first acts of piracy. Charles Johnson's account reads:

Stede Bonnet sails as a pirate

> His first Cruize was off the Capes of Virginia, where he took several Ships, and plundered them of their Provisions, Clothes, Money, Ammunitions, &c. in particular the *Anne*, Captain *Montgomery*, from *Glascow*; the *Turbes* from *Barbadoes*, which for Country sake, after they had taken out the principal Part of the Lading, the Pyrate Crew set her on Fire; the *Endeavour*, Captain *Scot*, from *Bristol*, and the *Young* from *Leith*. From hence they went to *New-York*, and off the *East End* of *Long-Island*, took a Sloop bound for the *West-Indies*, after which they stood in and landed some Men at *Gardner's Island*, but in a peaceable Manner, and bought Provisions for the Company's Use, which they paid for, and so went off again without Molestation.[10]

This account places him near the Virginia Capes and off New York where he took the *Anne*, the *Turbes*, the *Endeavour*, and the *Young*. The problem with this account is that these vessels were all mentioned by Lt. Governor Spotswood of Virginia, in a letter to the Council of Trade and Plantations, as being taken by Sam Bellamy. Johnson's vessel he calls the *Turbes* is, in fact, the Agnis, whose captain was Andrew Turbett. It was common in those days to refer to vessels by the name of their captain. Spotswood's letter reads:

> Information of Andrew Turbett, Master, and Robert Gilmor, supercargo of the Agnis of Glasgow, 17th April, 1717. The Agnis was taken and sunk by a pirate, Saml. Bellamy, five leagues off Cape Charles, 7th April. On the same day they took the Ann galley of Glasgow and the

Chapter Eight

Endeavor pink of Brighthelmstone, and on the 12th a ship belonging to Lieth, all bound for Virginia. The greatest part of the pirates crew natives of Great Britain and Ireland ... They declared they intended to cruise for 10 days off Delaware Bay, and 10 days more off Long Island.[11]

In reality, Bonnet only managed to take two vessels during his first venture into piracy. They were both taken on the same day, August 26, 1717, within sight of the entrance to the harbor at Charles Town, South Carolina. The following enhanced account of how Bonnet came to take those vessels is based on facts from Bonnet's trial as well as an article that appeared in *The Boston News-Letter*.

Captain Thomas Porter's brigantine had a relatively easy journey from New England to its final destination, Charles Town. The captain was relieved to finally see the harbor in the distance. Porter noticed another sloop was cruising nearby. There was no reason for him to be concerned. This was a busy harbor, and it was usual to see lots of vessels in these waters. However, this sloop was coming too close for comfort. Before he knew what was happening, the sloop was upon him and Porter's brigantine was taken by pirates. Wishing to conceal his true identity, Bonnet introduced himself as Captain Edwards and told Porter that he was originally from Rhode Island. As the pirates rummaged through Porter's brigantine, they sighted another sloop heading for the harbor entrance. Leaving some of his men on board the brigantine, Bonnet's *Revenge* pushed off and began pursuit.

This other vessel was a sloop carrying slaves, rum, and sugar. The captain's name was Joseph Palmer, and he also believed that he would be docking at Charles Town within the hour. His sloop had just completed its long journey from Barbados. Palmer must have seen the *Revenge* alongside the brigantine as he approached. He may have dismissed it as simply one vessel assisting another. However, as the unidentified sloop pushed away from the brigantine and made sail directly toward his sloop, Palmer must have become suspicious. If Palmer attempted to escape, it was to no avail. In a short amount of time, he too fell prey to Bonnet's sloop *Revenge*. As Bonnet climbed aboard Palmer's sloop, he again introduced himself as Captain Edwards. This time, it didn't work. Captain Palmer had just left Barbados, where Major Stede Bonnet was well known. Bonnet was recognized, and his name was entered into the long roll of North American pirates.

Bonnet's men were in complete control of both prizes. With the *Revenge* trailing close behind, the two prizes were sailed out of the busy waters of Charles Town toward the more remote waters of the North Carolina coast and came to anchor in an isolated inlet near Cape Fear. All the

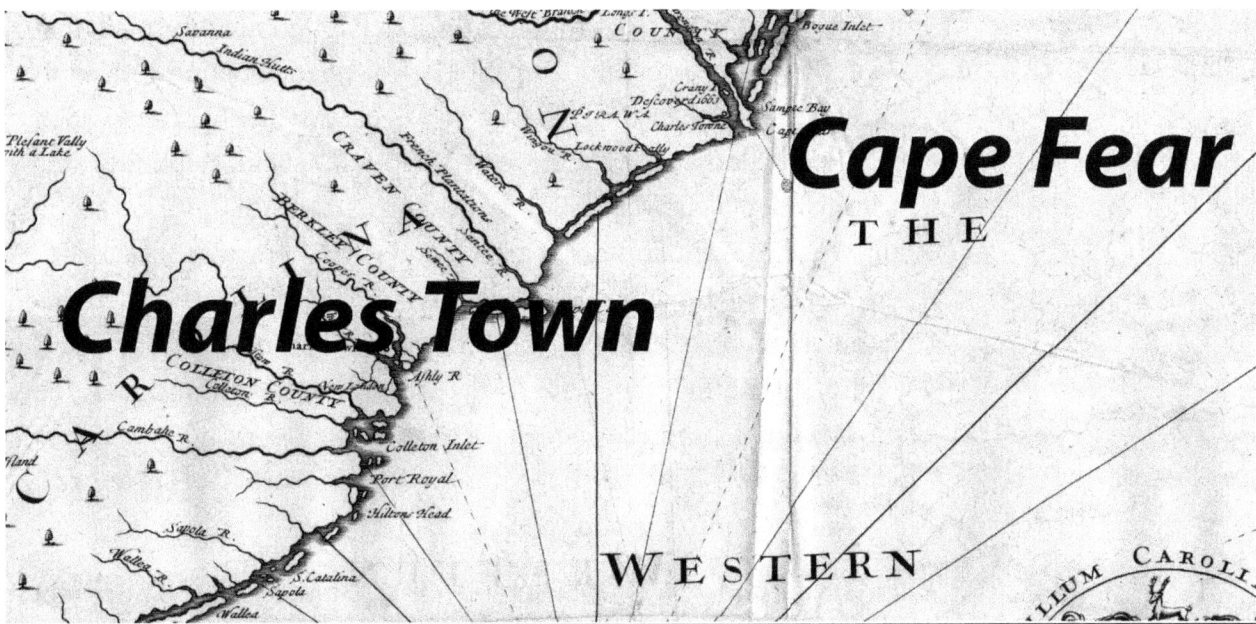

Figure 19: *Close-up from 1709 Map of North Carolina Courtesy of the University of North Carolina at Chapel Hill.*

frightened prisoners were put aboard the larger brigantine, where they could be securely watched. Then the pirates looted both prizes, putting everything they wanted on the *Revenge*, including all the cargo and food. Finally, Porter's brigantine was stripped of all its rigging, its cables, and most of its sails. They even took the anchor. The brigantine was permitted to leave; however, with only one small sail, they had a difficult time sailing anywhere. With no provisions left onboard, Porter was forced to put most of his unfortunate passengers into his rowboat and send them ashore. Eventually, Porter managed to sail to safety and sent help to those passengers who had been stranded on the Carolina coast. Bonnet's pirates used the sloop to careen the *Revenge* and then burnt it.

The events described in the record of Bonnet's trial also appear in Johnson's *A General History of the Pyrates*. Apparently, Johnson bought a copy of *The Tryals of Major Stede Bonnet, and Other Pirates* because the two accounts are almost identical. The trial transcript reads:

> And came in August, 1717. off of the Bar of South Carolina, and there took two Vessels bound in; one a Sloop with Negroes, Rum, and Sugar, Capt. Joseph Palmer from Barbadoes; the other a Brigantine, Capt. Thomas Porter from New England, whom, after he was plunder'd, they dismiss'd: but the Sloop they went away with, and at an Inlet in North Carolina careen'd by her, and then burnt her.[12]

There is a more detailed account in the Monday, October 21 to Monday, October 28, 1717 edition of *The Boston News-Letter*, which reads:

Chapter Eight

> By Letters from South Carolina of the 22d past, we are informed, that Capt. Thomas Porter in a Briganteen from hence bound for South Carolina, was taken the 26th Day of August last in sight of the Bar, by a Pirate Sloop call'd the Revenge of Rhode-Island, Capt. Edwards Commander alias Major Bennet, Capt. Palmer was also taken by him the same day, and carried his Sloop with them; and Porter they stript of all his Rigging, Sails, Anchors, Cables and Cargo, put all his Prisoners on board Porter, carry'd him as far as North Carolina, and left him hardly so many Sails as to carry him in, and no Anchor; he was forc'd to let the Prisoners have his Boat to go on Shoar, else they would all have been Starv'd for want of Provisions; at last Porter with difficulty run into Carolina so Save their Lives.[13]

After leaving the waters of North Carolina, Bonnet sailed south, where his first voyage as a pirate ended in catastrophe. The *Revenge* encountered a Man of War and a battle ensued. The pirate sloop was no match for the bigger and better-armed vessel, and the *Revenge* was blasted full of holes. Bonnet himself was severely wounded and many of his crew became casualties. However, there is very little information to go on. The only contemporary documentation of this battle comes from two short items in *The Boston News-Letter*.

Bonnet fights a Spanish man-o-war off Florida

These "one-line items" don't give a location for the battle. To make matters worse, they differ in their description of the nationality of the victorious Man of War. In the first article, it's from Jamaica, and in the second, it's Spanish. The first article comes from the Monday, October 14 to Monday, October 21, 1717 edition and the second comes from the Monday, November 4 to Monday, November 11, 1717 edition. They read:

> *Monday October 14 to Monday October 21, 1717, New York, October 14.* Arrived . . . yesterday . . . a vessel from Bermuda by whom we are informed . . . That Major Bennet who belonged to Barbadoes and 240 Men on board his Sloop, was taken by a Man of War from Jamaica and carried in thither, which said Bennet Dy'd of his wounds in three Days after.[14]

> *Monday November 4 to Monday November 11, 1717.* Major Bennet . . . was not well of his wounds that he received by attacking of a Spanish Man of War, who kill'd and wounded him 30 or 40 Men. After which putting into Providence, the place of Rendevouze for the Pirates.[15]

Bonnet's attacker was most likely a Spanish Man of War. The first article identifying the ship as Jamaican seems to have been based upon hearsay from another captain, and rushed to print the next day with incomplete information. Afterward, as more details became known, the story was corrected and the nationality of the Man of War was changed to Spanish. The only Man of War coming from Jamaica would have been a ship of the British Royal Navy. If this had been a naval ship, there would have been some sort of logbook entry of the battle. So far, no such entry has been found. However, in 1717, Spanish warships were still patrolling the Florida coastline in search of pirates who were attempting to dive on the wrecks of the 1715 treasure fleet. Bonnet's course, south from the Carolinas, would have taken him right past that location. If he had decided to try his hand at fishing the wrecks, he may have attracted the attention of a patrolling Spanish Man of War and been fired upon. This scenario also makes sense, as he put into Nassau shortly afterward, which is less than two days' sail from the wreck site.

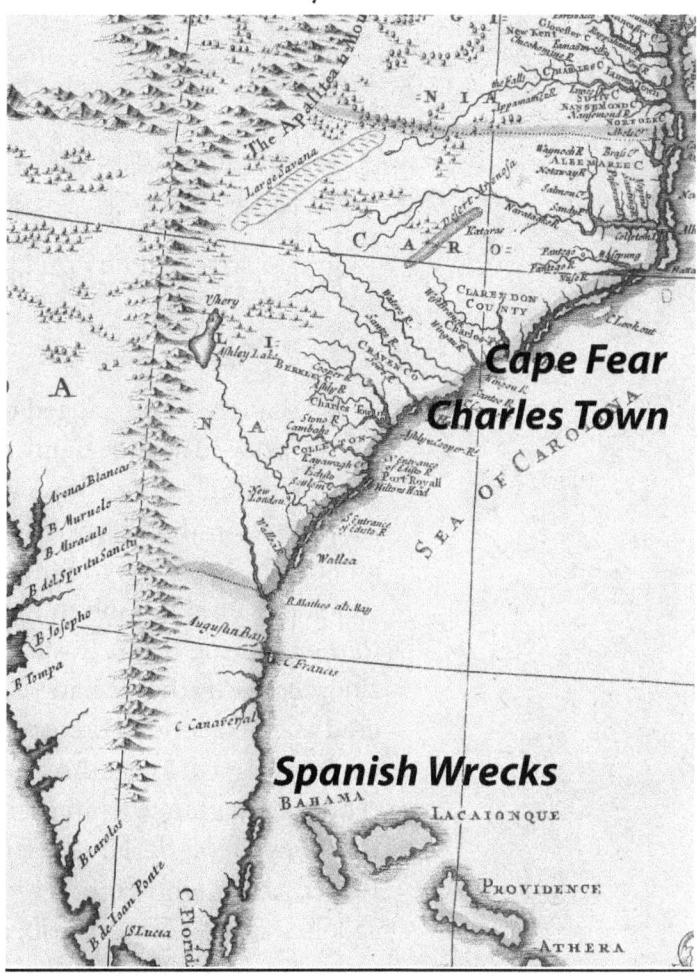

Figure 20: *Close-up of Florida and Carolina Coast from 1719 Map of North American Coastline, Courtesy of Barry Lawrence Rudman.*

The common conception put forth by other historian authors is that Stede Bonnet's ill-advised and foolish assault resulted from poor leadership and that Bonnet proved to be both inexperienced and overconfident when he rashly engaged a much larger vessel. However, this might not be true at all. If the Spanish caught him fishing the wrecks, the battle might have resulted from bad luck rather than incompetence. Either way, the *Revenge* had suffered substantial damage and put into Nassau for repairs in early September 1717.

While in Nassau, Stede Bonnet's life took a major turn. His aspiration to become a successful pirate would become a reality. However, Bonnet wouldn't be the captain, at least not for a while. Bonnet would form a partnership with the man destined to become the world's most famous pirate. Stede Bonnet would become partners with Blackbeard.

Nine
The Unlikely Partnership

A battered and ragged sloop slowly drifted to a stop alongside the dock. The sides and rails were splintered and pockmarked from multiple gunshots. Sails peppered with powder burns from grapeshot luffed in the breeze. The blood-stained deck smelled of gunpowder and death. There were many casualties aboard. A few exhausted crewmen stepped from the sloop to the dock and securely tied their crippled vessel to the mooring posts. It was obvious to anyone watching this beleaguered sloop come in that it had recently been in a fierce battle. A crowd gathered as the wounded crewmen hobbled off the sloop in search of medical attention or a drink at the nearest tavern. Those with more serious wounds required assistance from their crew mates. A man with the look of defeat in his eyes wandered out from the cabin. He watched as his crewmen helped those who couldn't walk come ashore. Steadying himself on the rail, he reached out for help, as he was among those with the most severe wounds. He was the captain. He was the man who had just lost a battle. He was Stede Bonnet.

There are no contemporary accounts of Bonnet's arrival at Nassau in September 1717, but this imaginative description seems a likely one. The previous chapter ended with Bonnet's sloop, *Revenge*, being blasted in battle by a much larger Spanish Man of War that killed or wounded dozens of the crew. Two days later, the *Revenge* limped into Nassau. There are no reliable contemporary accounts of what happened next. The only thing that can be said with certainty is that Stede Bonnet and Blackbeard formed a partnership and sailed from Nassau into fame and immortality.

Unfortunately, the only contemporary account of how Stede Bonnet came to form a partnership with Blackbeard comes from Charles Johnson's *A General History of the* Pyrates. The first edition differs from all subsequent editions. In Johnson's original version, there wasn't actually a meeting.

Blackbeard was already a member of Bonnet's crew as a "foremast man." Once the rest of the crew realized that Bonnet was unfit for command, he was voted out and Blackbeard was voted in. This mutiny is described in the chapter on Blackbeard as well as the chapter on Bonnet, and they are essentially the same thing. The following is an excerpt from Johnson's first edition chapter on Blackbeard:

> Edward Thatch, (commonly called Black-beard) was born in Jamaica . . . he was never raised to any Command. The first of his going upon the pyrating Account, was with Major Bonnet . . . With him Thatch ship'd himself as a Foremast Man; The Pyrates soon found Major Bonnet, to be a Person unfit for his Condition of Life; and tho' the Sloop was his own, yet they deposed him from the Command, and by common Consent, placed this Thatch in his Room.[1]

Johnson's second, and all subsequent editions, relays this account somewhat differently. This version is the one that is most often quoted and used as a prime source reference in other publications. As before, Johnson describes their meeting in two separate chapters, the one on Blackbeard and the one on Bonnet. However, in this version, Blackbeard and Bonnet meet at sea. Johnson has Hornigold as Blackbeard's only partner until they take a large French ship. Afterward, Hornigold leaves to accept the king's pardon and Blackbeard happens to spot Bonnet's sloop sailing by. As pirates who meet at sea often do, they form a partnership. The following is Johnson's description from his second edition chapter on Blackbeard:

> In his Way he met with a Pyrate Sloop of ten Guns, commanded by one Major *Bonnet*, lately a Gentleman of good Reputation and Estate in the Island of *Barbadoes*, whom he joyned; but in a few Days after, *Teach*, finding that *Bonnet* knew nothing of a maritime Life, with the Consent of his own Men, put in another Captain, one *Richards*, to Command *Bonnet*'s Sloop, and took the Major on aboard his own Ship.[2]

The following is Johnson's description from his second edition chapter on Bonnet:

> The Major was no Sailor as was said before, and therefore had been obliged to yield to many Things that were imposed on him, during their Undertaking, for want of a competent Knowledge in maritime Affairs; at length happening to fall in Company with another Pyrate, one Edward Teach, (who for his remarkable black ugly Beard,

was more commonly called Black-Beard:) This Fellow was a good Sailor, but most cruel hardened Villian, bold and daring to the last Degree; and would not stick at the perpetrating the most abominable Wickedness imaginable; for which he was made Chief of that execrable Gang, that it might be said that his Post was not unduly filled, Black-beard being truly the Superior in Roguery, of all the Company, as has been already related.[3]

In all of Johnson's accounts, their partnership was formed in the Caribbean, not Nassau. Additionally, Johnson completely omits all their well-documented piracies committed at the Virginia Capes, Cape May, and Long Island from September to November 1717. In recent times, as more research surfaced describing these events, historian authors writing on Blackbeard could no longer ignore the evidence that pointed to their partnership forming at Nassau. Nor could they ignore the fact that they sailed out together on the *Revenge* and stayed together through the capture of a large French ship off Martinique in November 1717. With this revelation, the standard Johnson narrative was brushed aside, at least that part of it, and a new problem arose for many historian authors. With no source documentation, they needed to invent a story describing their meeting and explaining why Bonnet would give command of his sloop to a perfect stranger.

Modern literature contains a multitude of speculative details regarding how, where, and why Blackbeard and Bonnet met. A few historian authors simply omit it and offer no explanation as to how this unlikely partnership formed. Others suggested that they simply became friends and Bonnet gave command of his sloop to a man he had only known for a few days. A common thread among those who offer a more detailed description of the meeting echoes one aspect of Johnson's account—that Bonnet was incapable of command. These historian authors contend that when Bonnet arrived in Nassau, Hornigold and Blackbeard offered to give him some help. As Colin Woodard wrote in his 2007 book, *The Republic of Pirates*, "Bonnet, now suffering from both mental and physical pain, was hardly in a position to refuse."[4]

None of those scenarios ever made any sense to me. I believe there has to be a more complex reason. One that may explain a far deeper connection between these two men. But the readers will have to wait for me to reveal my opinion in a future chapter toward the end of the book. For now, the most obvious common link was that both Blackbeard and Bonnet were Jacobites.

About five weeks after they sailed from Nassau, Blackbeard and Bonnet took a sloop near Philadelphia, which they looted and then released. When the victims reached port, some of them provided statements describing Bonnet, which were then reported in the Monday, November 4 to Monday, November 11, 1717 edition of *The Boston News-Letter*. Their description gives us a glimpse into the association between these two captains, at least during the first several months. The article reads:

> On board the Pirate Sloop is Major Bennet, but has no command, he walks about in his Morning Gown, and then to his books, of which he has a good Library on Board, he was not well of his wounds that he received by attacking of a Spanish Man of War, who kill'd and wounded him 30 or 40 Men. After which putting into Providence, the place of Rendevouze for the Pirates, they put the aforesaid Capt. Teach on board for this Cruise.[5]

Bonnet is mentioned in *The Boston News-Letter* as wearing a "Morning Gown."

The piratical rampage of Blackbeard and Bonnet during the months of late September, October, and November of 1717 is summed up rather neatly by Ellis Brand, captain of HMS *Lyme*. His ship of 20 guns had just recently arrived at its new station, the small port of Hampton, Virginia, near the entrance of the Chesapeake Bay. On December 4, 1717, Brand wrote a letter to the admiralty saying:

> Since my Arrival in Virginia I have heard but of one pyrot sloop, that was run away with, from Barbadoes commanded by Majr Bonnett, but now is commanded by one Teach, Bonnet being suspended from his command, but is still on board, they have most infested the Capes of delaware and sometimes of Bermudas, never continuing forty eight hours in one place, he is now gone to the Soward.[6]

Sloop *Betty* is taken off Cape Charles.

The *Revenge* sailed north through the Florida straits and up along the coast of the Carolinas in late September 1717. It seems that Blackbeard and Bonnet were looking for prizes, but they didn't sight a single vessel. At least there are no documented reports of them taking any vessels until they reached Virginia. Off of Cape Charles, the new pirate partnership finally had its first success. The date was September 29, 1717, and their first victim was the sloop *Betty*, loaded with pipes of Madeira wine. After taking what he wanted, Blackbeard sank the *Betty*. This was the first vessel that he sank as the captain of his pirate band. It wouldn't be the last.

Details on the capture of the *Betty* come to us through the charges against Blackbeard's Quartermaster, William Howard. He was arrested and put on trial in Williamsburg ten months after he cruised the waters of the

In the eighteenth century, a pipe was a large barrel-shaped container holding 126 gallons of liquid. A hogshead is half the size of a pipe, holding 63 gallons, and a barrel is half the size of a hogshead, holding 31.5 gallons.[7]

Virginia Capes and Cape May aboard the *Revenge*. A copy of the charges against Howard is preserved in the Virginia archives. For the reader's convenience, I removed the repetitive legal language in this quote.

> Articles exhibited for Pyracy against William Howard . . . That the Said Wm Howard . . . Join and Associate him self with one Edward Tach . . . the said Willm Howard did together with his Associates and Confederates on or about the 29th day of Septr . . . near Cape Charles in this Colony . . . attack & force a Sloop Call'd Betty of Virginia . . . did then and there Rob and plunder of Certain Pipes of Medera Wine and other Goods . . . did Sink and destroy the said Sloop with the remaining Part of the Cargo.[8]

This is supported by a letter dated November 14, 1717, from James Logan of Philadelphia to a Jamaican merchant named John Ayscough, in which he mentions the pirates at the Capes. This letter still exists and is in the Pennsylvania Historical Society and reads:

> I hope Annis is with you by this time the Pirates left Capes just before he came on them, & between Virgia & our Capes took a Sloop the Same day he went out wch was first of their return we are told from New York he was chased but we know not the truth of it.[9]

The sloop mentioned in this letter might be the *Betty*, or it might be another sloop we will discuss later in this chapter. However, there is also the possibility that there is another sloop that we know nothing about. The important part of this letter is that it chronicles where Blackbeard went next. From Cape Charles, the *Revenge* sailed north to the waters of Cape May, which was one of the busiest areas for merchant vessels in North America. This cape is in New Jersey and borders the east side of the Delaware Bay, which becomes the Delaware River, the entrance to Philadelphia. Sailing north along the east side of this cape, one quickly arrives at another major seaport, New York. At this vital confluence of maritime trade, Blackbeard took at least sixteen vessels. This number varies from book to book and may be much higher.

Most of the details of the vessels Blackbeard and Bonnet took at the Delaware Capes come from *The Boston News-Letter*. Every edition of *The Boston News-Letter* covered a period of one week, from Monday to Monday. The first edition that mentions these pirates was the issue dated Monday, October 28 to Monday, November 4, 1717. This article lacked details, misspelled some names, and even got some of the facts completely

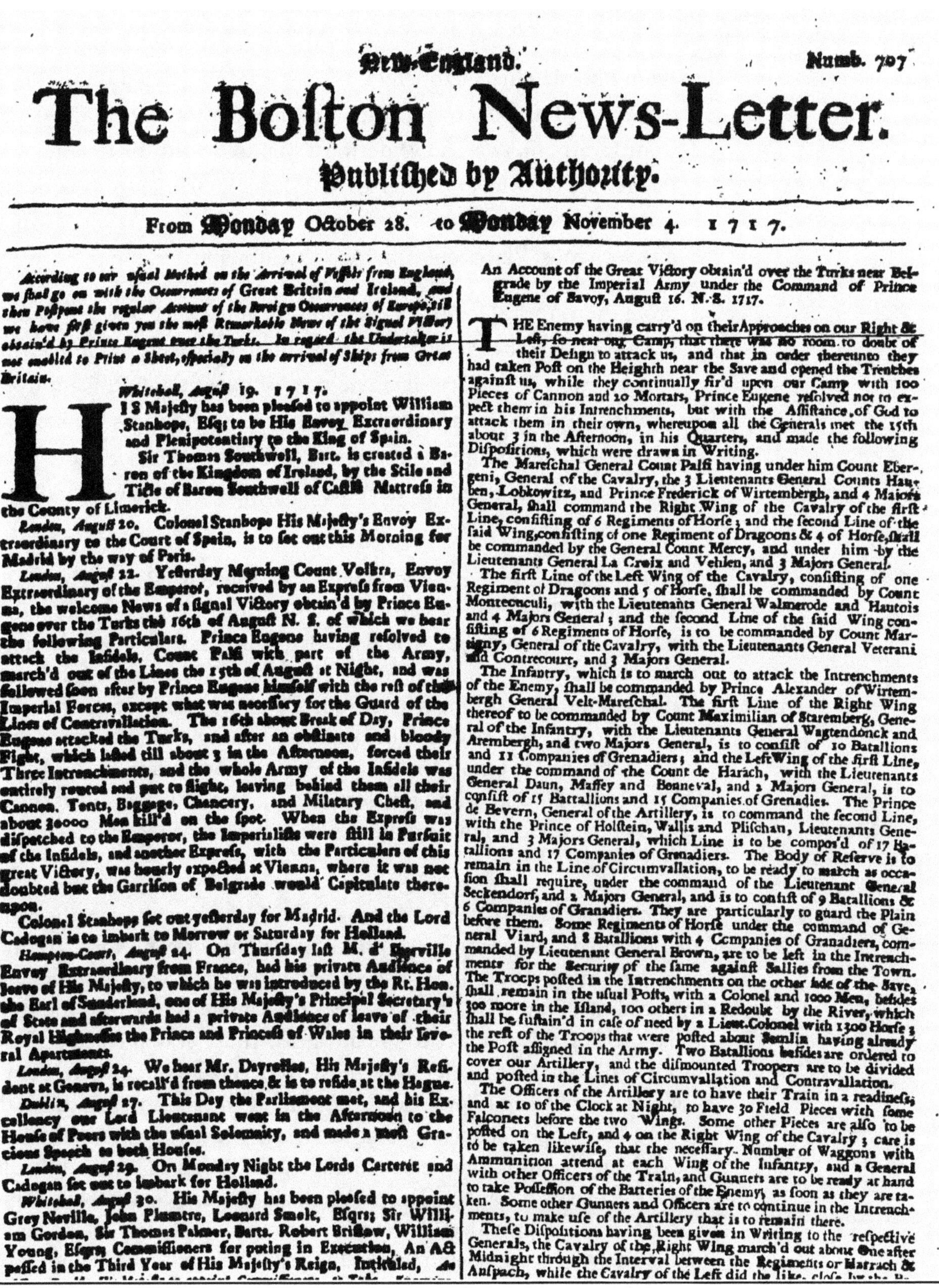

New-England. Numb. 707

The Boston News-Letter.

Published by Authority.

From Monday October 28. to Monday November 4. 1717.

According to our usual Method on the Arrival of Vessels from England, we shall go on with the Occurrences of Great Britain and Ireland, and then Postpone the regular Account of the Foreign Occurrences of Europe, till we have first given you the most Remarkable News of the Signal Victory obtain'd by Prince Eugene over the Turks. In regard the Undertaker is not enabled to Print a Sheet, especially on the arrival of Ships from Great Britain.

Whitehall, August 19. 1717.

HIS Majesty has been pleased to appoint William Stanhope, Esq; to be His Envoy Extraordinary and Plenipotentiary to the King of Spain.

Sir Thomas Southwell, Bart. is created a Baron of the Kingdom of Ireland, by the Stile and Title of Baron Southwell of Castle Mattress in the County of Limerick.

London, August 20. Colonel Stanhope His Majesty's Envoy Extraordinary to the Court of Spain, is to set out this Morning for Madrid by the way of Paris.

London, August 22. Yesterday Morning Count Volkra, Envoy Extraordinary of the Emperor, received by an Express from Vienna, the welcome News of a signal Victory obtain'd by Prince Eugene over the Turks the 16th of August N. S. of which we hear the following Particulars. Prince Eugene having resolved to attack the Infidels, Count Palfi with part of the Army, march'd out of the Lines the 15th of August at Night, and was followed soon after by Prince Eugene himself with the rest of the Imperial Forces, except what was necessary for the Guard of the Lines of Contravallation. The 16th about Break of Day, Prince Eugene attacked the Turks, and after an obstinate and bloody Fight, which lasted till about 3 in the Afternoon, forced their Three Intrenchments, and the whole Army of the Infidels was entirely routed and put to flight, leaving behind them all their Cannon, Tents, Baggage, Chancery, and Military Chest, and about 30000 Men kill'd on the spot. When the Express was dispatched to the Emperor, the Imperialists were still in Pursuit of the Infidels, and another Express, with the Particulars of this great Victory, was hourly expected at Vienna, where it was not doubted but the Garrison of Belgrade would Capitulate thereupon.

Colonel Stanhope set out yesterday for Madrid. And the Lord Cadogan is to imbark to Morrow or Saturday for Holland.

Hampton-Court, August 24. On Thursday last M. d' Iberville Envoy Extraordinary from France, had his private Audience of leave of His Majesty, to which he was introduced by the Rt. Hon. the Earl of Sunderland, one of His Majesty's Principal Secretary's of State and afterwards had a private Audience of leave of their Royal Highnesses the Prince and Princess of Wales in their several Apartments.

London, August 24. We hear Mr. Dayrolles, His Majesty's Resident at Geneva, is recall'd from thence & is to reside at the Hague.

Dublin, August 27. This Day the Parliament met, and his Excellency our Lord Lieutenant went in the Afternoon to the House of Peers with the usual Solemnity, and made a most Gracious Speech to both Houses.

London, August 29. On Monday Night the Lords Carteret and Cadogan set out to imbark for Holland.

Whitehall, August 30. His Majesty has been pleased to appoint Grey Neville, John Plumtre, Leonard Smelt, Esqrs; Sir William Gordon, Sir Thomas Palmer, Barts. Robert Bristow, William Young, Esqrs; Commissioners for puting in Execution, An Act passed in the Third Year of His Majesty's Reign, Intituled, *An*

An Account of the Great Victory obtain'd over the Turks near Belgrade by the Imperial Army under the Command of Prince Eugene of Savoy, August 16. N. S. 1717.

THE Enemy having carry'd on their Approaches on our Right & Left, so near our Camp, that there was no room to doubt of their Design to attack us, and that in order thereunto they had taken Post on the Heighth near the Save and opened the Trenches against us, while they continually fir'd upon our Camp with 100 Pieces of Cannon and 20 Mortars, Prince Eugene resolved not to expect them in his Intrenchments, but with the Assistance of God to attack them in their own, whereupon all the Generals met the 15th about 3 in the Afternoon, in his Quarters, and made the following Dispositions, which were drawn in Writing.

The Mareschal General Count Palfi having under him Count Ebergeni, General of the Cavalry, the 3 Lieutenants General Counts Hauben, Lobkowitz, and Prince Frederick of Wirtembergh, and 4 Majors General, shall command the Right Wing of the Cavalry of the first Line, consisting of 6 Regiments of Horse; and the second Line of the said Wing, consisting of one Regiment of Dragoons & 4 of Horse, shall be commanded by the General Count Mercy, and under him by the Lieutenants General La Croix and Vehlen, and 3 Majors General.

The first Line of the Left Wing of the Cavalry, consisting of one Regiment of Dragoons and 5 of Horse, shall be commanded by Count Montecuculi, with the Lieutenants General Walmerode and Hautois and 4 Majors General; and the second Line of the said Wing consisting of 6 Regiments of Horse, is to be commanded by Count Martigny, General of the Cavalry, with the Lieutenants General Veterani and Contrecourt, and 3 Majors General.

The Infantry, which is to march out to attack the Intrenchments of the Enemy, shall be commanded by Prince Alexander of Wirtembergh General Velt-Mareschal. The first Line of the Right Wing thereof to be commanded by Count Maximilian of Staremberg, General of the Infantry, with the Lieutenants General Wagtendonck and Arembergh, and two Majors General, is to consist of 10 Batallions and 11 Companies of Grenadiers; and the Left Wing of the first Line, under the command of the Count de Harach, with the Lieutenants General Daun, Maffey and Bonneval, and 2 Majors General, is to consist of 15 Battallions and 15 Companies of Grenadies. The Prince de Bevern, General of the Artillery, is to command the second Line, with the Prince of Holstein, Wallis and Plischan, Lieutenants General, and 3 Majors General, which Line is to be compos'd of 17 Batallions and 17 Companies of Grenadiers. The Body of Reserve is to remain in the Line of Circumvallation, to be ready to march as occasion shall require, under the command of the Lieutenant General Seckendorf, and 2 Majors General, and is to consist of 9 Batallions & 6 Companies of Granadiers. They are particularly to guard the Plain before them. Some Regiments of Horse under the command of General Viard, and 8 Batallions with 4 Companies of Granadiers, commanded by Lieutenant General Brown, are to be left in the Intrenchments for the Security of the same against Sallies from the Town. The Troops posted in the Intrenchments on the other side of the Save, shall remain in the usual Posts, with a Colonel and 1000 Men, besides 300 more in the Island, 100 others in a Redoubt by the River, which shall be sustain'd in case of need by a Lieut. Colonel with 1300 Horse; the rest of the Troops that were posted about Semlin having already the Post assigned in the Army. Two Batallions besides are ordered to cover our Artillery, and the dismounted Troopers are to be divided and posted in the Lines of Circumvallation and Contravallation.

The Officers of the Artillery are to have their Train in a readiness; and at 10 of the Clock at Night, to have 30 Field Pieces with some Falconets before the two Wings. Some other Pieces are also to be posted on the Left, and 4 on the Right Wing of the Cavalry; care is to be taken likewise, that the necessary Number of Waggons with Ammunition attend at each Wing of the Infantry, and a General with other Officers of the Train, and Gunners are to be ready at hand to take Possession of the Batteries of the Enemy, as soon as they are taken. Some other Gunners and Officers are to continue in the Intrenchments, to make use of the Artillery that is to remain there.

These Dispositions having been given in Writing to the respective Generals, the Cavalry of the Right Wing march'd out about One after Midnight through the Interval between the Regiments or Harrach & Anspach, while the Cavalry of the Left did the like, close by the ...

Figure 21: *The Boston News-Letter issue dated Monday, October 28 to Monday, November 4, 1717.*

wrong. It seems like the editors were anxious to get the information out to the public. It reads:

> From Philadelphia, October 24. We are informed that a Pirate Sloop of 12 Guns 150 Men, Capt. Teach Commander took one Capt. Codd from Liverpool, two Snows outward Bound Seford for Ireland, and Budger for Oporto, and Peters from Madera, George from London, Farmer for New-York, a Sloop from Madery for Virginia, all of which met with most Barbarous inhumane Treatment from them.
>
> From New-York, Octob. 28. We are informed that Sipkins of that Place was taken by the Pirates, whose Sloop they have mounted with 12 Guns for a Pirate. Tis also said they've taken Capt. Rolland from Jamaica. A Sloop and a Snow were seen standing towards the Bar end of Long-Island, and thought to be the Pirates bound for Garders or Block-Island. We intend to be more particular about those Pirates in our Next.[10]

Notice the last sentence, "We intend to be more particular about those Pirates in our Next." They apparently knew this was a rush job and told their subscribers that they would do better in the following edition. An odd fact that I must point out is that *The Boston News-Letter*, for the most part, gives the name of the master rather than the name of the vessel. For example, as you will read in one of the upcoming excerpts, "Two Snows outward bound, Spofford loaden with Staves for Ireland and Budger of Bristol."[11] Spofford and Budger were the masters of those snows, not the names of the vessels. I'm sure the reader will quickly become accustomed to this oddity.

Master is a nautical term that is synonymous with *captain*. However, it generally applies to smaller vessels (i.e., *Captain* of a ship, but *master* of a sloop).

In *The Boston News-Letter*, November 4 to November 11 issue, the articles are identified as either coming from Philadelphia or New York. This helps in determining where the vessels were taken. The vessels listed under the heading "From Philadelphia" were probably taken within the area of Delaware Bay, the entrance to Philadelphia. Those vessels listed under the heading "From New York" were most likely taken somewhere between Cape May and Long Island. However, placing their location is easier than getting the proper sequence, that is to say, which one came first, then second, etc. Reading this rather archaic publication makes it difficult to place the events in any precise chronological order. Few dates are given and when one is provided, it is somewhat vague. However, after spending hours analyzing every detail within the text, I believe I have come up with the proper sequence of events.

Blackbeard approached Cape May as captain of the sloop *Revenge* and on October 12, 1717, he took his first vessel of those taken near the Delaware Capes. At least, it's the first one described in that edition of *The Boston News-Letter*. The unnamed vessel was from Dublin and was commanded by a man named Codd. Aboard were one hundred fifty passengers, mostly indentured servants. The pirate sloop was identified as the *Revenge* and Teach was named as the captain. Each pirate had about five pistols on their belts. After taking what they wanted, they threw the rest of the cargo overboard and let the vessel go. However, there is something far more stunning in the newspaper article. Teach, as they call him, was identified as a mate who formerly sailed out of Philadelphia. The article reads:

> Codd from Liverpool and Dublin with 150 Passengers, many whereof are Servants. He was taken about 12 days since off our Capes by a Pirate Sloop called the Revenge, of 12 Guns 150 Men, Commanded by one Teach, who formerly Sail'd Mate out of this Port: They have Arms to fire five rounds before they load again. They threw all Codds Cargo over board, excepting some small matters they fancied. One Merchant had a thousand Pounds Cargo on board, of which the greatest part went over board, he begg'd for Cloth to make him but one Suit of Cloth's, which they refus'd to grant him.[12]

We must pause for a moment to reflect on this striking revelation. Johnson certainly didn't mention the possibility of Blackbeard ever sailing out of Philadelphia. The idea that Blackbeard may have previously visited Philadelphia is supported in a letter from James Logan, an influential Philadelphia merchant, that is dated October 24, 1717.

Logan's original letter reads:

> Some of our Mastr Say they knew almost every man aboard most of them having been lately in this River, their Comandr is one Teach who was here a Mate from Jamca about 2 ys agoe[13]

My transcription into modern text reads:

> *Some of our ship's masters say they knew almost every man aboard, most of them having lately sailed in this river. Their commander is one Teach, who has sailed here as a mate onboard a vessel from Jamaica about 2 years ago.*

Later on, an entire chapter is devoted to Blackbeard's Philadelphia connection. However, for now, the intriguing possibilities are too complex to

The name Edward Teach (Blackbeard) appears in print for the first time in *The Boston News-Letter*.

Chapter Nine

explore at this point in our story. Therefore, we shall return to the action as I shed some light on the identity of Captain Codd's vessel from Dublin.

William Howard's charge sheet, which described the taking and destruction of the sloop *Betty*, also identifies two additional vessels taken in the Bay of Delaware. Those vessels are the sloop *Robert* and the ship *Good Intent*. Again, removing the repetitive legal language, the charges read:

> That the Said Wm Howard . . . on or about the 22d of Octor . . . in the Bay of Delaware . . . Pyratically take Seize and Rob the Sloop Robert of Philadelphia and the Ship Good Intent of Dublin.[14]

Considering the *Robert* was from Philadelphia, it is surprising that it wasn't mentioned at all in *The Boston News-Letter*. It makes one wonder how many other vessels taken by Blackbeard weren't reported. It's the "Ship *Good Intent* of Dublin" that proves to be the most interesting. As mentioned above, *The Boston News-Letter* describes Codd's vessel as "from Liverpool and Dublin with 150 Passengers, many whereof are Servants." Additionally, "One Merchant had a thousand Pounds Cargo on board." At the time, people were very specific with their descriptions of vessels. A ship has three masts, a snow has two, and a sloop only has one mast. Most of the vessels in *The Boston News-Letter* are identified as one of those types, but not Codd's vessel. Was this in fact the ship from Dublin named the *Good Intent*? It seems highly likely. The most obvious commonality is that it came from Dublin. But was this the "Ship *Good Intent*?" Considering that there were 150 passengers on board and a thousand pounds of cargo, it had to have been a large three-masted ship. There just wouldn't be enough room for all those passengers on a much smaller, one-masted sloop. In my opinion, the Good Intent is Captain Codd's ship, listed in this chapter as the first prize. That would make the sloop *Robert* prize number two.

Sailing on in the *Revenge*, he took his third and fourth prizes. They were both snows taken very close together as they sailed out from Philadelphia. One was commanded by a man named Spofford, loaded with staves for Ireland. The other was the *Sea Nymph*, of Bristol, commanded by Captain Budger, and loaded with wheat. On the *Sea Nymph*, the pirates roughed up a merchant named Richardson, perhaps to get him to reveal any hidden cash. They threw the cargo overboard, then loaded all the prisoners onto the first snow, and let them go. Blackbeard kept the *Sea Nymph* and put the crews from both prize snows onboard Spofford's vessel and let them go. With the addition of the *Sea Nymph*, Blackbeard had begun to build a fleet. The article from *The Boston News-Letter* reads:

The sloop Robert *and the ship* Good Intent *are taken.*

Sea Nymph and another snow are taken.

> The Pirate took Two Snows outward bound, Spofford loaden with Staves for Ireland and Budger of Bristol in the Sea Nymph loaden with Wheat for Oporto, which they threw over-board, and made a Pirate of the said Snow; And put all the Prisoners on board of Spofford, out of which they threw overboard about a Thousand Staves, and they very barberosly used Mr. Joseph Richardson Merchant of the Sea Nymph.[15]

Now Blackbeard had two vessels, the *Revenge* and the *Sea Nymph*. Their fifth prize was an inward bound sloop from Madera, commanded by Peter Peters. They kept twenty-seven Pipes of Wine, then cut the masts at the deck and pushed the sloop ashore, where it was beached. This was soon followed by their sixth, a sloop commanded by Captain Grigg carrying about thirty servants. Once it was looted, they cut the mast away and left it at anchor. Their seventh prize was a sloop from Madera, out of which they took two pipes of wine and then sunk it. Then they took their eighth prize, a sloop from Antigua. They put some of the captured servants aboard and let it go. The same article from *The Boston News-Letter* reads:

> They also took a Sloop Inwards Bound from Madera, Peter Peters Master out of which they took 27 Pipes of Wine, cut his Masts by the Board, after which She drove ashore and Stranded. They also took an other Sloop one Grigg Master, bound hither from London, with above 30 Servants, they took all out of her, cut away her Mast and left her at Anchor on the Sea. They also took another Sloop from Madera, bound to Virginia, out of which they took two Pipes of Wine, then Sunk her. It's also said they took a Sloop from Antigua, belonging to New-York, and put some of the London Servants and other things on board her.[16]

In reading these accounts, one can't help but notice the wanton destruction Blackbeard was causing. Merchandise that couldn't be loaded on the pirate vessels was thrown overboard. Unlike his old partner, Hornigold, who only took merchandise, Blackbeard kept everything of value, such as coins, jewelry, food, alcoholic beverages, and navigational instruments.[17]

So far, there were two vessels disabled by having their masts cut and left stranded and two vessels were sunk (including the *Betty*). There will be more. It may appear that Blackbeard was out for vengeance against the English. It also might be that as a new captain, perhaps he was trying to establish a reputation. Looking back over 300 years, it is impossible to say

Navigational instruments in 1717 included back staffs, magnetic compasses, hour glasses, and chip logs.

which is correct. One thing that is important to notice is that with all this destruction, none of his captives were killed or even injured.

Prize number nine was a ship under the command of Captain Tover, bound for Maryland. Information on this prize comes from a letter written by Jonathan Dickinson, who was a member of the Colonial Assembly and the Mayor of Philadelphia during Blackbeard's siege.[18] The letter was dated October 21, 1717, and still exists. It is stored at the Pennsylvania Historical Society and reads:

> My Son Jos went out of our Capes 29th last mo at wch time they were 3 Ships in Compa Jn Annis for London Capt Wells for N York & Capt Tover for Maryland. The Latter was taken by a Pyratt before others was out of Sight & Since have accot of Six more of our Vessells by sd pyratt taken who is yet at our Capes Plundering all that Comes Cuting away their maist and Leting them Dive a Shoar. Save a Ship wth Passengs he Spared & thus and thus is our River Blocked Up Until he goes hence.[19]

This account clearly describes the seriousness of the situation with the statement, "thus is our River Blocked Up." It was a major crisis for the merchants of that town. Not only for the cargo lost, but once Blackbeard began taking vessels, all shipping stopped for fear of being taken. The letter describes the vessel, most likely the sloop belonging to Peter Peters, whose mast was cut away and run aground. But this letter also mentions a ship belonging to "Capt Tover for Maryland" that "was taken by a Pyratt."

Figure 22: *Capes of Virginia and Delaware from a 1685 Map. The image is dedicated to the public domain under CC0.*

Tover's name isn't found in the *Boston News-Letter*. The date in the text of the letter, September 29, 1717, places the witness in the Philadelphia area toward the first phase of Blackbeard's piracies at Cape May. Logic then dictates that this prize is one not previously identified as one taken by Blackbeard would be his ninth prize.

After leaving Cape May, Blackbeard headed toward Long Island with his two vessels, the *Revenge* and the captured snow *Sea Nymph*. A line-item comment appeared in both pertinent issues of *The Boston News-Letter*, which reads:

A Sloop and a Snow were seen standing towards the Bar end of Long-Island, and thought to be the Pirates bound for Garders or Block-Island. We intend to be more particular about those Pirates in our Next.[20]

As they left Cape May on their way to New York, they took their tenth prize, a sloop commanded by Captain Farmer. This unlucky captain had been robbed by another pirate sloop earlier on his voyage. Blackbeard took his mast, anchors, cables, and all the money he had, then put some of the other captured servants onboard and set the sloop adrift. Fortunately, they managed to beach it at Sandy Hook. The eleventh was a great sloop commanded by Captain Sipkins. Blackbeard kept this sloop and added it to his fleet. He transferred 12 guns over to the sloop that he had taken from other vessels along the way and converted it into a pirate sloop. As its actual name is unknown, I shall refer to it as the *Great Sloop*. Blackbeard now had three vessels to command: the *Revenge*, the *Sea Nymph*, and the *Great Sloop*. His twelfth capture was a Jamaican sloop commanded by Captain Rolland. The article reads:

> Capt. Farmer from Jamaica, who was twice taken by the Pirates on his passage, the last off the Capes of Deleware by Capt. Teach, who took out his Mast, Anchors, Cables, what money was on board, and put some Servants on board him and then turned him a-drift he made a Shift to get into Sandy-hook, where he run his Sloop ashore, having no Anchors and came up for help to get her off. One Sipkins in a great Sloop of this place is taken by the Pirates, which Sloop they have mounted with 12 Guns and made a Pirate. It's also said that Capt. Rolland of this place in a large Sloop from Jamaica, is taken by the Pirates.[21]

Information on his thirteenth prize remains frustratingly elusive. Lucky number thirteen! The only copy of this November 4 to 11 issue of *The Boston News-Letter* that I can find has a large black blotch over the most valuable part of the page. Perhaps a better copy will emerge one day. For now, the article reads:

> Arriv'd from Barbadoes, who met Sandford, {unreadable} of our Pilates that was on board, Teach the {unreadable} they very much threaten New-England Men, in {unreadable} their fellow Pirates suffer there, that they will revenge of them.[22]

Apparently, Blackbeard captured a vessel from New England. As we shall see in several upcoming chapters, Blackbeard held a particular grudge

Chapter Nine

against all vessels from New England. His close friend, Sam Bellamy, and most of Bellamy's crew, died near Cape Cod when their ship capsized in a storm. Six of his pirates who survived the wreck were caught and executed. From this point on, every vessel from New England that Blackbeard caught, he burned.

A snow was a square-rigged merchant vessel with two masts with a loose-footed gaff trysail in the after-mast.

Captain Goelet's sloop was prize number fourteen. The pirates threw his cargo of cocoa overboard and added the sloop to their pirate fleet. However, Goelet and his crew had nothing to fear. Apparently, the *Sea Nymph* wasn't up to Blackbeard's standards, so he gave the snow to Captain Goelet and his prisoners and let them go. Blackbeard's fleet now consisted of the *Revenge*, the *Great Sloop*, and Goelet's sloop. As the vessels sailed apart on different courses, captain Goelet looked back and saw the pirate fleet attack and capture a ship and a brigantine. These were the fifteenth and sixteenth prizes attributed to Blackbeard during his piracies at the Delaware Capes. The article reads:

> On the 30th past arrived Capt. Goelet, who was lately taken by Teach the Pirate, coming hither in a Sloop from Curacoa, half loaden with Cocoa, which the Pirates threw overboard, and man'd the Sloop for a Pirate, and gave Goelet and his Crew the Sea Nymph Snow to bring them home in, Goelet saw the Pirate take a Ship & a Briganteen or Snow after parting with them.[23]

The name Blackbeard is mentioned for the first time in the historical record by Jonathan Dickinson.

Jonathan Dickinson, the Mayor of Philadelphia and Colonial Assemblyman, wrote another letter. One that is exceptionally valuable in understanding Blackbeard's life history. In a letter to merchant Joshua Crosby, dated October 23, 1717, Dickinson casually mentions one thing that leaps from the page as absolutely astounding. He mentions the name "Blackbeard."

The original letter reads:

> Thou mentons pyrating trade wth you, from the begining of this Mo until wthin this Week one Capn Tatch alls Blabeard in a Sloop whch they call Revengers Revenge About 130 Men, 12 or 14 Guns having layne of or Capes & Taken six or seven Vessels Inwd & outwd bound.[24]

My translation of this letter reads:

> *You mentioned that pirates were in the area from the beginning of the month until this week and that one, a captain named Thatch, alias Blackbeard, in a sloop which they call* Revenges Revenge *with about*

130 men and armed with 12 or 14 guns have been laying off our capes and have taken six or seven vessels, both inbound and outbound.

Before this letter was discovered, many historian authors, including Angus Konstam, attributed the source of his name, Blackbeard, to Henry Bostock, who described Blackbeard in his deposition in December 1718, over a year after this letter was written.[25] This remarkable source document pushes the date of the name *Blackbeard* back to October 1717. He was likely called Blackbeard much earlier than that, as he was apparently known as Blackbeard by those he encountered near Philadelphia. The first time he was called Blackbeard will most likely remain a mystery. Why he was called Blackbeard probably had something to do with the long, black beard he tied with ribbons. What is most interesting is the possibility that he was already known as *Blackbeard* by September 1717, very early on in his career as a pirate captain. Was he called Blackbeard earlier when he sailed with Hornigold? Or did he just make up that name as he approached Cape May and simply told all his captives that he was well-known as Blackbeard? Those are questions that can't be answered yet. Hopefully, some other documents will appear one day that may shed additional light on this enigmatic issue.

Meanwhile, where was Blackbeard's former partner, Ben Hornigold? There were several reports that Blackbeard was expecting his old friend to join him at any moment. Returning to *The Boston News-Letter* article mentioned earlier in this chapter concerning Captain Codd, the article also mentions that the pirates were expecting a consort ship to arrive at any moment. An excerpt from the article reads:

> He was taken about 12 days since off our Capes by a Pirate Sloop called the Revenge, of 12 Guns 150 Men, Commanded by one Teach, who formerly Sail'd Mate out of this Port:

> The Pirates told the Prisoners that they expected a Consort Ship of 30 Guns, and then they would go up into Philadelphia, others of them said they were bound to the Capes of Virginia in hopes to meet with a good ship there, which they much wanted.[26]

James Logan's October 24, 1717 letter also mentions this consort ship. Logan wrote:

> Some of our people having been Several dayes on Board them they had a great deal of free discourse wth them . . . The Sloop that came on our Coast had about 130 Men all Stout Fellows all English without any mixture & double armed they waited they Said for their Consort

a Ship of 26 Guns wth whom when joined they designed to Visit Philadia,[27]

Most historian authors suggest that this consort vessel was commanded by Ben Hornigold. This is supported by two articles that appear in *The Boston News-Letter*. The first article is from the Monday, November 18 to Monday, November 25, 1717 edition and states that Teach and Hornigold took a vessel at latitude 36 and 45, which in current navigational language, means latitude 36.45. This is the approximate latitude of the Virginia-North Carolina border. It also provides the date of October 18, 1717. The article reads:

> Pritchard from St. Lucie, who on the 18th of October in Lat. 36 and 45 was taken by Capt. Teach, in Compa, with whom was Capt. Hornygold, they took from him about 8 Cask Sugar and most of their cloaths at the same time, they took a Ship from London for Virginia, out of which they took something and let them go.[28]

The second article is from the Monday, December 30 to Monday, January 6, 1718 edition, and states that Teach and Hornigold took a ship from Maryland near the Virginia Capes. This event took place fourteen days before the arrival of the ship in Philadelphia, December 10, 1717, which would make November 26, 1717, the date the ship was taken. It reads:

> We are told from Maryland that a Ship from London was arrived there, who about fourteen days ago was taken off the Capes of Virginia by Teach & Hornigold, that took out of him a New Suit of Sailes and Rigging.[29]

These articles were addressed in Chapter Seven; however, for the sake of continuity, I shall offer my opinion once again. Earlier in this chapter, multiple reports appearing in *The Boston News-Letter* and letters mentioning Blackbeard that were written by officials in Philadelphia clearly place him in the Delaware River and along the shore of Long Island in October and early November 1717. The second is far more problematic for those who believe that Blackbeard and Hornigold were off Virginia in November. Indisputable French documents undoubtedly place Blackbeard on Bequia Island in the Grenadines on November 26, 1717.[30] That's about 1,900 miles away.

Shockingly, all historian authors haven't seen the problems with attributing these two attacks to Blackbeard simply because *The Boston News-Letter* said so. There are only two rational explanations for this conflict. Colin Woodard suggests that people were used to identifying the two pirates as

Figure 23: *Close-up Showing Blackbeard's Route from Long Island to Martinique from a Map of North America. The image is dedicated to the public domain under CC0.*

working together and when Hornigold boarded them, they just assumed the other sloop sailing with him belonged to Blackbeard.[31] A simpler explanation is that *Boston News-Letter* got the dates wrong.

As mentioned in the previous chapter, I suggest that these reports may have come in to *The Boston News-Letter* months earlier, but Teach wasn't well known and Hornigold was yesterday's news. After Teach's spectacular assault on the Delaware Capes, his actions became a much bigger story. After a moment or two of reflection, the editors may have intentionally changed the dates and added these stories to boost circulation and capitalize on the success of their earlier *editions. This must seem shocking to the readers. Imagine, a newspaper journalist changing the facts simply to sell more newspapers. No modern journalist would do such a thing, would they?*

When Blackbeard finally left the Delaware Capes, he had a fleet of three vessels: his sloop *Revenge*, the *Great Sloop* he had taken from Captain Sipkins, and Captain Goelet's sloop. Since this third vessel isn't mentioned in any later accounts, it obviously either sank or was abandoned by the pirates at some point. Blackbeard knew that if he wanted to continue to expand his fleet, he would need a large, fast ship comparable to Sam Bellamy's *Whydah*. Such coveted slave ships made the best pirate ships. They were built for speed and had plenty of space to berth 300 or more men.

Blackbeard must have known that the best place to hunt for slave ships was along the route these vessels often traversed from the African coast to the Caribbean, so that's where he headed next. One popular theory is that he sailed south through the Florida Straits, turned right at the Bahamas, and cruised along the north coast of Hispaniola toward the Lesser Antilles. But that route makes little sense. Why would Blackbeard go out of his way through the crowded shipping lanes of Florida and Hispaniola? The direct route would be much better. Some historian authors, like Colin Woodard, agree.[32] Blackbeard most likely sailed directly along the navigational lines found on all eighteenth-century charts that go from Cape May to the east side of the Lesser Antilles. From there, he headed south to the French waters east of Martinique.

On all English maps of the early eighteenth century, Martinique is spelled Martinico.

Ten
Getting the Dates Right

As Blackbeard entered French waters, he also entered French jurisdiction. Reports of the vessels he took there were made by French officials in Martinique. This presents a challenge for researchers, as the French calendar was eleven days ahead of the English calendar.

The difference in calendars is one aspect that most historian authors overlook when writing about the early eighteenth century. The British were still using the older Julian calendar while the French and Spanish were using the Gregorian Calendar. Therefore, dates recorded in documents by either the French or the Spanish are eleven days before the dates recorded in English documents.

The Julian calendar had been in standard use in Europe since the time of the Romans, but it was fairly flawed. By the mid-sixteenth century, it was ten days off. Pope Gregory XIII instituted a new calendar in 1582, which was named after him, the Gregorian calendar. Most Catholic countries, including France and Spain, switched over to it immediately. However, the English didn't follow suit, and by 1718, the English calendar was eleven days behind.[1]

In studying Blackbeard, this distinction becomes vitally important when using dates from original source documents to build timelines and accurately determine precisely where he was on any specific day. For example, according to the French documents, Blackbeard attacked and took *La Concorde* on November 28, 1717. This date is in the Gregorian calendar. At first glance, this date is problematic, because English records indicate that Blackbeard took the *Montserrat Merchant* on November 29, 1717. He couldn't possibly have taken those two vessels just one day apart, especially when you consider the time it would have taken Blackbeard to transform the captured slave ship into a pirate ship. But when you realize that the two dates are from different calendar systems, it makes sense. When you convert the French date to the Julian calendar, it shows that Blackbeard actually took *La Concorde* on November 17, 1718, in the English calendar.

In researching this book, I have used many source documents, including numerous copies of *The Boston News-Letter*, handwritten letters, deeds, and court proceedings. Occasionally, some of these source documents have been transcribed by modern historian authors. It occurred to me that some of the dates may have been converted over to match our modern calendar. In order to make sure that all the dates in those documents were indeed from the Julian calendar, I checked each one with an "easy to use" website that lists the days in both the Julian and the Gregorian calendars by year.[2] All one needs to verify the date is the day of the week associated with that date. For example, if the date on a document is given as Monday, December 2, 1717, one can look at the website and see that December 2, 1717, was indeed a Monday in the Julian calendar but was a Thursday in the Gregorian calendar. In that example, it is certain that the author was using the Julian calendar. I have found, as one would expect, that all English documents are in the Julian calendar and all French ones are in the Gregorian calendar.

Since almost all the source documents used as references in my research are English, all dates given in this book will be in accordance with the Julian calendar. To help eliminate any confusion, on the occasions when both French dates and English dates are used within the same paragraph, I'll remind the reader of this by specifying the date as being in the Gregorian Calendar, and when necessary, give the corresponding Julian calendar date, too.

But that's not the only issue! In England, the year didn't change on January 1, as it does in modern times. The specific date of *New Year's Day* changed many times over the centuries and wasn't consistent between nations. By the start of the eighteenth century, Catholic nations, including France, Spain, and even Scotland, were all using January 1 as the first day of the year, but England was using March 25. That means that in England, December 31, 1717 would be followed by January 1, 1717, and March 24, 1717 would be followed by March 25, 1718. This leaves us with an almost three-month period of immense confusion.

To help alleviate this confusion in the early eighteenth century, official documents such as court records were often dated during those problematic months, showing both years in the written date separated by a period. A good example of this is a Craven County appointment letter dated January 24, 1712, in which the date is written as "this 24th Day of Jan 11; Anno Domini 1711.12."[3] Notice that they first gave the old-style date of

"Jan 11" and then clarified it by writing "Anno Domini 1711.12." In modern times, historians often help alleviate confusion and get around this issue by giving both years in their dates but separated with a slash. The above-mentioned date would be written as January 11, 1711/12. That's a pretty good system, as long as the reader understands why this is being done. Within the text of this book, when writing a date during the problem months, I shall use the modern system. My December 31, 1717, will be followed by January 1, 1718.

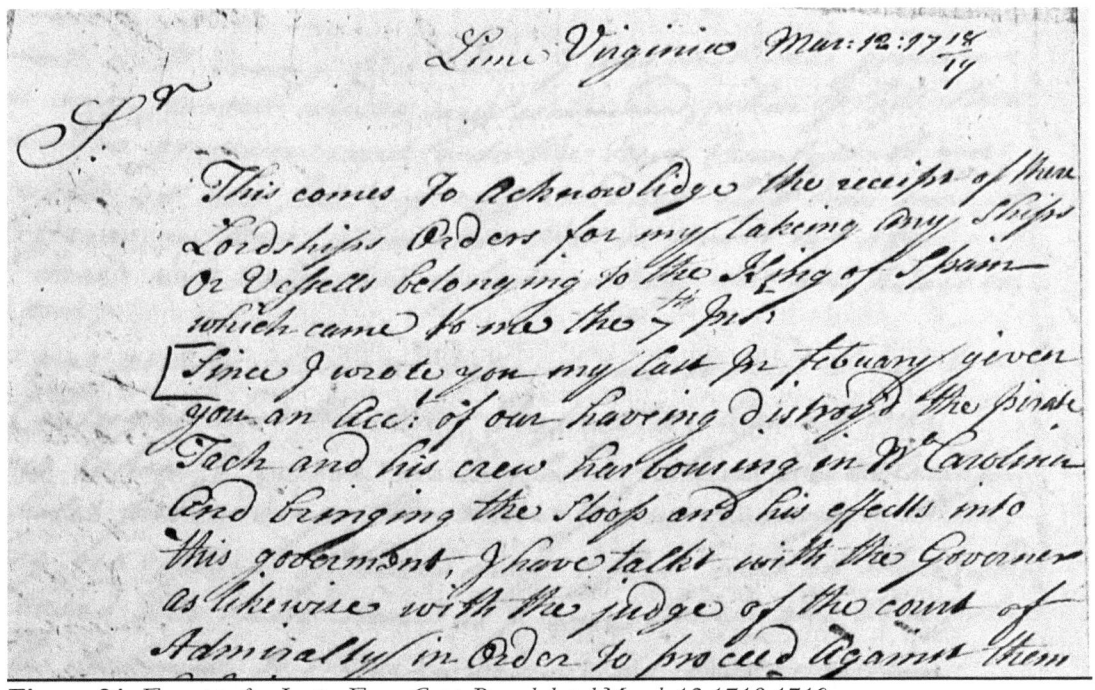

Figure 24: *Excerpt of a Letter From Capt. Brand dated March 12 1718-1719.*

England changed to the Gregorian calendar in 1752, following the British Calendar Act of 1751. Starting the new year on January 1 was easy, but adjusting for the eleven days was slightly more challenging. The British government solved this that year by simply skipping those days. This was done in September, when Wednesday, September 2, 1752, was followed by Thursday, September 14, 1752.[4] It would have sucked to have a birthday on one of those days.

A special note for those who like to recognize anniversaries of specific events in pirate history. Decide whether you wish to commemorate the recorded date or the actual anniversary. For example, Blackbeard's death is recorded as November 22, 1718, but that was in the Julian calendar. By our calendar, exactly 300 years later would have been December 3, 2018. Being aware of the differences in dates between the English and the French calendars will help us better comprehend the events that are about to transpire.

Eleven
The Queen Anne's Revenge

A large sloop was sighted in the distance by the dock workers at the port of Martinique. This sloop appeared to be pulling into port. Martinique was among the busiest ports in the Caribbean, so the sight of a sloop coming in wasn't unusual. However, this one was. From afar, the sloop could be identified as English-built. France and England had been at war and even though the war had ended years ago, ships from both nations remained cautious of each other and generally kept their distance. The approaching sloop began to attract attention, and a crowd of curious onlookers gathered on the dock. The surprised and perhaps even shocked dock workers looked on in wonder as the sloop grew nearer. They could see that the deck of this mysterious sloop was crowded with two hundred and forty-six blacks and thirty-two whites.[1] No one expected to see such a thing. Then, a frantic call for assistance came from the sloop, but this call was in French. Responding quickly, the men on the dock grabbed the sloop's lines and securely tied it to the mooring posts. A man stepped onto the dock and demanded to speak to the governor immediately. This man was Pierre Dosset, the captain of *La Concorde*.

Captain Dosset's ship was a slave ship that sailed from the port of Whydah on the west coast of Africa on October 2, 1717, with five hundred sixteen unfortunate Africans onboard. His destination was Martinique. He never made it. About sixty leagues before he made port, *La Concorde* was attacked by pirates.[2] To be more precise, *La Concorde* was attacked by Blackbeard. On November 28, 1717, (which was November 17, 1717, in the English calendar) at about 120 miles from their destination of Martinique, at 14 degrees 27 minutes north latitude, *La Concorde* was "attacked by two English bandit boats armed with two hundred and fifty-one men, one with twelve cannons, the other with 8 guns and both commandés par Edouard Titche anglaise."[3]

When Dosset reached Martinique, he provided a detailed description of all the circumstances surrounding the loss of his ship to Charles Mesnier, a government agent. His first account was dated December 10, 1717. Over the next two weeks, he made several additional statements which were all documented in the official reports. In addition to Dosset's statements, other reports were subsequently filed, including one highly detailed account from his lieutenant, François Ernaud.[4] All of those original documents still exist and are safely preserved in the French national archives. Fortunately, they were transcribed by Jaques Ducoin in 2001 as part of a research paper he prepared for the North Carolina Department of Cultural Resources.[5] Much of the information in this chapter comes from Ducoin's report.

La Concorde's voyage had been a difficult one. Since leaving France, disease had devastated the crew. Fifteen had already died, thirty-six men were down with "scurvy and blood flow," and one of the crewmen had drowned. Disease had also taken a large toll on their cargo. Of the 516 slaves put aboard at Whydah, only 455 remained.[6] However, this challenging trip was almost over. Captain Dosset and his crew must have taken comfort knowing that in just one day, they would reach their destination, Martinique.

Visibility was extremely poor. Even though the sun had just risen, there was a heavy blanket of fog resting on the water, completely enveloping *La Concorde* in a thick, gray mist. Lieutenant Ernaud was standing watch on deck, peering into the distance, which wasn't very far considering the awful conditions. With no warning, Ernaud saw two sloops emerge out of the fog and head directly toward his ship. From their course, he immediately realized that they were positioning for an attack. The feeling of apprehension that a ship's officer has while sailing in poor visibility quickly turned into the sense of fear one has before a battle. Ernaud called for the captain as the crewmen on deck rushed to the rail to see what was happening. By now, the sloops were upon them. Dosset and his crew could see that one sloop was armed with 12 guns and the other with 8 guns. These sloops were Blackbeard's *Revenge* and the 40-ton *Great Sloop* that he had taken from Captain Sipkins off the coast of New York.[7] The decks of both sloops were crowded with well-armed men. It was now clear to everyone that these were pirates and their ship was about to be attacked.

Both sloops let loose with a thunderous broadside as the pirates fired their muskets. The battle was on and Dosset faced a difficult decision. There were only twenty-one crewmen left onboard *La Concorde*. Even though the ship was armed with 16 guns, there simply weren't enough men to maneuver the ship and fire their guns at the same time. A shout came from one of the pirate sloops ordering them to lower a boat into the sea and to

Blackbeard's attack on *La Concorde*

send their captain over to the pirate sloop. Knowing that they were completely unable to defend themselves, Captain Dosset consented to the pirate's request. With all three vessels motionless in the water, the ship lowered their boat, and Dosset and a few of his crew boarded. The small boat pushed off and quietly rowed the short distance to the sloop that had called to him earlier. Once his row boat came alongside, Dosset hesitantly climbed aboard the *Revenge*. Standing on deck, Dosset got his first look at probably the most terrifying pirate he had ever seen: Blackbeard.

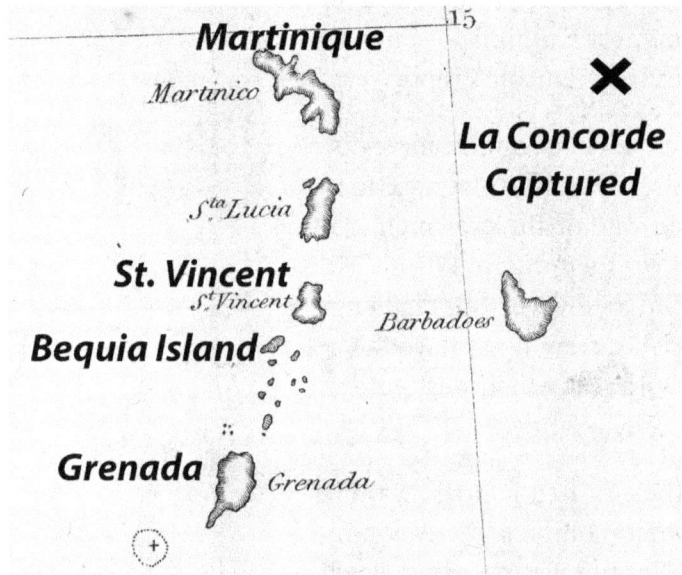

Figure 25: *Bequia to Martinique from Caribbean Map, Courtesy of Barry Lawrence Rudman.*

With the French captain onboard the *Revenge* as a hostage, the three vessels sailed to Bequia Island in the Grenadines. Once at anchor, the French crew, as well as all the slaves, were sent ashore. Dozens of pirates boarded *La Concorde* and began to loot the ship. Lieutenant François Ernaud provided a detailed account of the taking of his ship to the officials in Martinique. The following is an excerpt of the English translation of his statement:

The said vessel armed with 16 guns and 75 crewmen ... departed on the 18th of the said month to go to Whydah, where they anchored on the 8th of July, and where they made their trade and loaded onto the said ship the number of 516 heads of blacks of all sexes and ages and fourteen ounces of powdered gold. After which they left the said place on the 9th day of October following to go to Martinique and the French islands of America.

On the following 28 November, being 30 or 40 leagues from Martinique, by the latitude of 14° 30' North, they met around 8 o'clock in the morning in a foggy weather two pirate boats, one of which was armed with 12 guns and equipped with 120 men of crew and the other armed with 8 guns and equipped with 30 men. The declarant said that he had 16 men who died of disease including the one who had drowned, and in addition 36 men of their said crew sick of scurvy and blood flow so that they were only 21 men to make the maneuver and lead the said ship. So much so that the said two pirate boats having fired two volleys of cannon and musketry at them and shouted at them to put their canoe into the sea. The said captain and officers and crew seeing themselves unable to defend themselves from the said pirates, they came on board the said pirates which took them to Bicoya, islands of the Grenadines where the declarant and all the other members of his crew were searched and visited and plundered them and took the elite of their cargo and put the rest on the said island ashore.[8]

The Kingdon of Whydah, under local control of ruler King Haffon, was the center of the slave trade on the coast of West Africa. In modern times, it is the nation of Benin.

According to Charles Johnson, Ben Hornigold was with Blackbeard when he took *La Concorde*. Afterward, Hornigold left to accept the king's pardon. A week or so afterward, Blackbeard met Stede Bonnet and formed a partnership.

> After cleaning on the Coast of *Virginia*, they returned to the *West-Indies*, and in the Latitude of 24, made Prize of a large *French Guiney* Man, bound to *Martinico*, which by *Hornigold*'s Consent, *Teach* went aboard of as Captain, and took a Cruize in her; *Hornigold* returned with his Sloop to *Providence*, where, at the Arrival of Captain *Rogers*, the Governor, he surrendered to Mercy, pursuant to the King's Proclamation.
>
> In his Way he met with a Pyrate Sloop of ten Guns, commanded by one Major *Bonnet*, lately a Gentleman of good Reputation and Estate in the Island of *Barbadoes*, whom he joyned;[9]

This claim is absolutely absurd when compared to the facts. As we have seen, multiple sources have placed Bonnet and Blackbeard together since September of 1717. Additionally, as shall be described in the next chapter, when *La Concorde* was taken in November 1717, news of the king's pardon wasn't well known. Whereas Johnson's error is understandable, many other historian authors have subscribed to this notion, blindly following Johnson's work as if it were completely factual. An example of this is Robert E. Lee, author of the 1974 book *Blackbeard the Pirate*.[10]

Henry Bostock, the master of one of the vessels that fell victim to the pirates, described the ship as "Dutch built."[11] Bostock's deposition seems to be among the most quoted by pirate historian authors. Over the years, that comment has led many people to believe that *La Concorde* was originally either Dutch or English and that it had been captured by the French during the war. Investigating this possibility, Ducoin conducted a detailed and exhaustive search of original French, Dutch, and English documents which led him to conclude that *La Concorde* was constructed in France around 1710 and sailed as a French privateer in 1711.[12] *La Concorde* was

Chapter Eleven

specifically designed as a ship of war. It was originally a 280-ton privateer frigate, equipped with gun ports and 26 guns mounted.[13] The hull was designed for speed. The rounded shape of the hull gave the impression that it was a Dutch flute, but in the early eighteenth century, that hull shape was believed to make the ship easier to handle with all the sails up.[14] When the war was over, it was reconfigured to work in the slave trade and made voyages in 1713, 1715, and finally 1717.[15] These features made *La Concorde* the ideal pirate ship.

Blackbeard's vessels reached Bequia Island by November 19, 1717, where his pirates began to refit and reconfigure *La Concorde*. This was the painstaking process of converting a slave ship into a proper pirate ship. Unnecessary cargo needed to be offloaded to make room for treasure. Slave quarters were modified to accommodate the large pirate crew. The deck needed to be rearranged to make the ship more effective in battle. And finally, the guns from the *Great Sloop* were transferred to the ship to increase its armament. This entire process only took one week, as there are reports of the ship sailing off of St. Vincent seven days later. During that week, the pirates also searched for treasure.

According to Lieutenant Ernaud, there were over fourteen ounces of gold dust in the ship's cargo, with an additional thirteen ounces belonging to the ship's officers.[16] A 15-year-old cabin boy named Louis Arrot told one of Blackbeard's men that there was a significant amount of gold dust hidden onboard the ship. Some of the pirates threatened to cut the throat of Captain Dosset and the officers if they didn't tell them where it was. This was all part of Blackbeard's method of intimidation. Documentation shows that Blackbeard never killed any of his captives, so it is doubtful that he would have allowed any of the French officers to actually be harmed. But threatening them was something else. In this particular case, it worked, and the gold dust was handed over to him.[17]

Meanwhile, Stede Bonnet finally got his chance to command the *Revenge* once again. While Blackbeard's men were looting and refitting the ship, Stede Bonnet finally got the opportunity to once again be the captain of the *Revenge*. From the time Blackbeard had taken command two months earlier, Bonnet had remained in the background, recovering from his wounds. Once they reached Bequia Island, Bonnet had recovered and was fit for duty. This was his chance to prove his abilities to Blackbeard and to his crew. On November 23, 1717, Bonnet took the *Revenge* out to sea to hunt for prizes. This is supported by Captain Dossett's statement that "one of the two illegal boats had gone to cruise on the island of Saint Vincent."[18]

At the time, Bequia was considered too inaccessible for the French to colonize and was inhabited only by Kalinago and Arawaks, who used the island for fishing and farming.

Stede Bonnet, once again, assumes command of the Revenge.

Bonnet's first venture out was highly successful. He spotted two vessels anchored at Layou on the southwest tip of St. Vincent. One was an English ship. This would be the first prize taken by Bonnet since his partnership with Blackbeard. Bonnet's method of attack was to put about ten pirates into the sloop's boat and send them over to the prize vessel first. His sloop would be nearby, prepared to fire a broadside if the vessel refused to surrender. In this case, it was easy for his men to capture the English ship, because the captain had sent most of his crew ashore. However, the captain had sent most of his valuables ashore, too. Once in control of the ship, Bonnet's pirates confronted the captain. They repeatedly struck him and even threatened to hang him if he didn't send word to the shore for his crew to bring the treasure back to the ship. It worked, and Bonnet was given £6,000 worth of silver coins. Afterward, he burned the ship. This account was told to the French authorities in Martinique by an eyewitness, Pierre Raimond Olivier, the captain of Bonnet's next prize. Olivier's statement was included in the report of *La Concorde*:

> Here is what I learned about these two burnings from the testimony of Pierre Raimond Olivier . . . Half an hour before the privateer was taking an English ship anchored at Layou, another cove of Saint-Vincent and obliged the captain who had sent his crew ashore with all his money to go fourth on board. There, by dint of blows and threatening to have him hanged, he compelled him to send for his treasure which consisted of six thousand pounds sterling and then to hand it over to him if he wanted to guarantee himself from death.[19]

With the English ship burning, Bonnet quickly turned his attention to the French ship anchored nearby. It was the *Dauphin*, belonging to Sieur Simon.[20] The ship's master was Pierre Raimond Olivier. He had anchored there to take on firewood from the nearby tiny island of Mayougany. Olivier must have seen the pirate's attack on the English ship. Now they were heading for him. Bonnet sent his boat in first. These ten pirates attacked Olivier's vessel, firing small arms. Olivier and his crew of twelve abandoned ship and took refuge on the island. The pirate boat shifted course away from the vessel and rowed directly toward them.

The boat cut through the gentle surf and glided to a stop as the keel scraped the sand. The pirates jumped out, weapons in hand, and charged the French crew ashore. However, the French were armed too, and a sudden volley of musket fire came from their position. The pirates returned fire, and the battle was on. One can imagine this desperate struggle, each side maneuvering through the palm trees to gain a better position while cracking the muskets. However, a gun battle with flintlocks isn't so easy.

Chapter Eleven

It takes up to thirty seconds to reload, and if there is any dampness in the air, the weapon may not fire. Considering this, perhaps the fight even became hand-to-hand with cutlasses. The French crew managed to kill two of the pirates, then fled, running deep into the trees. Bonnet recovered his battered pirates and left the *Dauphin* stranded on Mayougany. Olivier mentions the 4th of this month as the date of the attacks. December 4, 1717, in the French calendar, would be November 23, 1717, in the English calendar. Olivier's statement was included in the report of *La Concorde* and reads:

> On the 4th of this month, Olivier being anchored at St Vincent to make firewood, a pirate boat and his canoe armed with ten men attacked him at a cove of the said island called Mayougany and forced him Olivier and his crew composed of twelve men, including him to abandon his boat and take refuge on land, Whence after fighting for some time and killing two pirates, they fled into the woods. Be that as it may, the pirates after seizing this ship sent for Simon's boat stranded in Mayougany.[21]

The next day, Bonnet took another French vessel that was owned by Sr. Henri St. Amour and commanded by his son, Charles. Bonnet sailed the *Revenge* back to Bequia Island with his captured prize in tow. While on shore, Blackbeard and Dosset discussed a ransom for the return of Charles St. Amour's vessel that consisted of "a pig, some poultry and several bunches of figs and bananas."[22] In his report, Dosset commented on that conversation and also mentioned that Blackbeard said, "he would also have given him twenty-five negroes."[23]

Bonnet takes Sr. Henri St. Amour's vessel.

Then, in a surprising turn of events, Blackbeard gave Captain Charles St. Amour one hundred six captured slaves and instructed him to set sail quickly and to pass behind the ship to prevent Bonnet from seeing his escape. It seems that Blackbeard had given Charles those slaves and released his vessel without Bonnet's knowledge or consent. These events could only have happened if Bonnet, Blackbeard, and Charles St. Amour were all together near the ship. When Captain Charles St. Amour finally returned to Martinique, he gave a statement that was recorded in a supplemental report dated December 21, 1717. The English translation of his statement reads:

> That the captain of the pirate ship having returned his boat to him and given one hundred and six negroes, he ordered him to pass behind his ship and to set sail promptly because the captain of the

pirate boat, his associate, not wishing to consent to the return of the boat or to the donation of the negroes.[24]

Note the use of the words "boat" and "ship." These were specific terms that mariners used to describe specific types of vessels. The "pirate ship" could only mean *La Concorde* and the "pirate boat" must have been Bonnet's sloop, *Revenge*. Unfortunately for Charles, he left in such a hurry that he didn't have the opportunity to take on food or water. He was becalmed in the waters between Bequia Island and St. Vincent and was forced to set out in his longboat to search for provisions. He was accompanied only by one of his friends and one of his slaves. Eventually, he made it to La Roche Percée and then to Martinique. His statement was included in the report and reads:

> Thus Charles Henri was obliged to force sails without food or water and, having found himself in full calm for three or four days four or five leagues off between St Vincent and Becouya, he decided to put his canoe into the sea to fetch food and water accompanied only by the son of an inhabitant of his neighborhood and one of the 106 negroes, with the help of which he arrived at last not without much difficulty at the place of St. Vincent called La Roche Percée, without having since any news of his boat in which he had left 105 negroes and a French-speaking Caribbean. Charles-Henri adds that he has every reason to fear that his boat has gone adrift, the currents being strong in this place and carrying all offshore.[25]

According to Lieutenant François Ernaud's follow-on report, Charles St. Amour's vessel drifted and ran aground on a sandbar off Grenada, or in his words, on "La Grenade" stranded on the "quay."[26] However, it appears that this vessel managed to make it to Grenada with sixty-four surviving blacks and one Caribbean native. This implies that forty-one must have died of thirst while stranded at sea. The lieutenant governor, de Pradines, wrote the governor of Martinique informing him of this and added that they would be sold on Grenada. His statement was included in the amended record made on December 21, 1717, and reads:

> I learn from the one I receive from M. de Pradines, lieutenant of the King at La Grenade, that the ship of Saint-Amour arrived there with sixty-four negros or negresses, and the Caribbean of which it has been spoken. He tells me that he will sell them to Grenada.[27]

This is reflected in the last report made on April 27, 1718, of the whole *La Concorde* mess, in which Lieutenant François Ernaud included the final disposition of all the slaves. The report begins with sixty-one slaves sold on Grenada. One surmises that three of the sixty-four blacks who arrived at Grenada on Charles St. Amour's vessel may have died. The statement ends with one hundred and five "blacks of loss," which is precisely the number that Charles St. Amour claimed were stranded aboard his vessel. It reads:

> The 61 negroes sold in Grenada produced 12,200 pounds, that of the number of blacks who were brought to Martinique, there are 56 exhausted (?) at 125 pounds each. The total of blacks . . . Claimed amounts to 376 of which it belongs six to the officers of the said ship and 20 who died coming from Bicoya and Martinique during the sail. Remains 350 blacks that have been sold. Deduct from 455. There are 105 blacks of loss since the capture of the said ship.[28]

At first glance, the numbers appear to add up nicely. However, after subsequent readings, one realizes that these numbers really don't make any sense. Perhaps something was lost in the translation. At any rate, it appears that Lieutenant François Ernaud believed that everyone was accounted for except the one hundred and five that left with Charles St. Amour. This leads one to believe that none of the Africans remained with the pirates.

The next day, November 25, 1717, Bonnet was back at sea again chasing vessels. He returned to Layou Bay, and sighted a French vessel sailing by. He sent his boat to attack with twenty-five pirates aboard. Two days earlier, Bonnet had sent ten men for his initial attack, resulting in two casualties. He had learned his lesson, so this time, his boat carried twenty-five pirates. They chased the French vessel while Bonnet looked on from the deck of the *Revenge*, but it was too fast for them to catch. Bonnet shifted course and pursued with the *Revenge*, but by that time, the French vessel was too far ahead. Bonnet eventually had to give up the chase. A statement in the official report reads:

> Since on the sixth of this month a boat coming from Grenada and passing through Layou, Bay of St. Vincent, was chased there by a canoe of twenty-five men which the pirates sent to him, and then by the boat of these same pirates.[29]

Before Blackbeard left Bequia Island, he decided to force ten of Dosset's crew to join the pirates. They all possessed sought-after skills: one pilot, three surgeons, two carpenters, a caulker, a cook, a gunsmith, and a master sailor. Four members of Dosset's crew joined willingly. This included Louis Arrot, the fifteen-year-old cabin boy who had alerted the pirates as to the hidden treasure.[30] Another, Julien Joseph Moisant, was also a cabin boy, according to the ship's passenger and crew list.[31]

Two days after Bonnet took Saint Amour's French vessel, the pirates were ready to set sail. Apparently, the *Great Sloop* was no longer of use to Blackbeard. It was a Bermuda design. Lt. Ernaud even commented on it, saying the "boat was of Bermudian construction."[32] Those sloops were notoriously slow, as they were built to carry lots of cargo, not for speed. After all, it was described as a "great sloop" but only mounted 8 guns. Such a vessel wasn't designed for speed or battle, and it just didn't fit into Blackbeard's plan. After transferring the guns over to his new ship, Blackbeard gave Captain Dosset the *Great Sloop* so he could sail to Martinique. He also gave them some barrels of beans.

Blackbeard's fleet sails from Bequia

> Before leaving Becoya the pirates released to Captain Dosset one of their two boats with about two hundred and fifty negroes or negresses and two or three barrels of beans. Two days before one of the two pirate boats, having gone to cruise on the island of Saint-Vincent, took the boat of Sieur Henri Saint Amour.[33]

Dossett sailed with thirty-two white crewmen and two hundred and forty-six blacks. They arrived at Martinique on December 7, 1717 (which is November 26, 1717 in the English calendar). This was just nine days after his ship was captured by Blackbeard.[34] The sloop that was given to Dosset wasn't large enough to carry everyone at once. He had to leave part of his crew and some of the slaves behind. Once safely at Martinique, he sent the sloop back to rescue the others. The report states that "part of his crew having been unable to contain them in this boat without risking losing many and which he will fetch in the same boat."[35]

La Concorde's refitting was complete; however, there is a question about how many guns it mounted. As mentioned earlier, *La Concorde* was originally built to carry 26 guns as a privateer in 1710,[36] but a slave ship didn't need that many, and some of them were taken off. There is a discrepancy between the French officers as to how many guns it carried at the time it was captured by Blackbeard. Captain Dosset reported 14 guns[37] while Lieutenant Ernaud mentioned 16 guns.[38] Perhaps two of them were swivel guns that the captain didn't count as part of the main armament. Regardless of which port is correct, with the 8 guns transferred from aboard from the *Great Sloop*, his ship now carried between 22 and 24 guns. But just two days after leaving Bequia Island, Captain Christopher Taylor reported that the ship had 32 guns.[39] It seems that Blackbeard had acquired up to 10 additional guns from the three vessels that Bonnet had taken at St. Vincent.

Blackbeard finally had the exact type of ship that most pirates dream of getting. It was even better than the ship his friend Sam Bellamy had. Even though *La Concorde* was built with only 26 gunports, it had ample space for placing additional guns on the quarterdeck and mounting swivel guns along the rails. This ship instantly made Blackbeard the most powerful pirate in the Caribbean, but it didn't yet have a name. Coming up with a name wouldn't be so easy. Such a grand ship needed a memorable name. A name that sounded impressive when told in sea stories by sailors in the taverns along the docks. A name that would inspire terror among its victims. Additionally, it needed to look good in print, for it was inevitable that his exploits were destined to be printed on broadsides and in newspaper articles in the colonies throughout the British Empire.

However, for a man like Blackbeard, that wasn't enough. He needed a socially significant name that linked him to a cause. His ship had to carry a name that demonstrated his support for a political movement dear to his heart. Many pirates were Jacobites, those who were in favor of placing James Edward Stuart on the throne. Stede Bonnet and Charles Vane certainly were. And so was Blackbeard. For him, there was no greater issue than that of the Jacobite revolution. Blackbeard chose to name his ship after James's sister, Queen Anne. She was the last of the Stuart monarchs. But simply naming it the *Queen Anne* wasn't enough. Jacobites wanted revenge against the English for exiling their king. That was it! His ship would be named *The Queen Anne's Revenge*.

Figure 26: *Queen Anne's Revenge Artist's Conception Painted by Sharon Glaze.*

Twelve
Cruising the Lesser Antilles

The *Queen Anne's Revenge* was the most powerful pirate ship in the Caribbean and was ready for action. With Blackbeard as the captain, it was armed with 32 guns. Stede Bonnet commanded the consort sloop *Revenge* with 12 guns. Together, they made an indomitable force. The two pirate vessels sailed from Bequia Island on November 26, 1717, and headed north along the western side of the Lesser Antilles. Shortly after sailing, they reached La Roche Percée near St. Vincent. As you may recall, this was the same location where just a day or two earlier, Bonnet had taken three vessels and chased another. Blackbeard could see the smoldering hull of the English ship that Bonnet burned as well as the beached ship, *Dauphin*, belonging to Sieur Simon, that Bonnet left on the tiny island of Mayougany.

Unfortunately, Blackbeard's first attack with his new ship would end in disappointment. The pirates sighted a small French sloop and shifted course to give chase. The menacing ship was rapidly closing on the small vessel when, unexpectedly, the wind completely stopped. Luckily for the French crew, their sloop was equipped with oars. They were able to row away as the two pirate vessels stood motionless in the calm sea. This account was included in the official report of *La Concorde* made at Martinique. It reads:

> The next day, another boat of this island also coming from Grenada was also chased in front of St. Vincent at the place called La Roche Percée and would have been infallibly caught, if in the full calm in which he found himself and the pirates, he had not used his oars to avoid them promptly. These two boatmasters assured me that the ship and the pirate boat were the same who had taken Dosset.[1]

Disappointment among the pirate crew over losing the French sloop disappeared when they sighted a large English ship lying at anchor nearby.

The *Great Allen* of Boston, Captain Christopher Taylor, taken by Blackbeard.

Chapter Twelve

The wind had freshened and the two pirate vessels were again on the move. As they grew nearer, they hoisted their pirate flags and positioned for an attack. When they were within range, Blackbeard fired two warning shots from the ship's guns. Sailing even closer, the pirate ship swung around and some of the pirates fired several muskets at the prize ship. For the crew, there was no chance of escape. They were still at anchor and their sails weren't even up. The large pirate ship bristled with 32 guns and had about three hundred well-armed men on deck. Such a vision was positively petrifying. As the *Queen Anne's Revenge* came alongside, the terrorized crew abandoned ship and fled to shore, taking all the ship's valuables and a young female passenger with them. The captain remained with his ship along with three of the crew. The English ship was the *Great Allen* out of Boston and the captain was Christopher Taylor. After Taylor was released, he made a detailed statement to the French authorities on Martinique. His statement was preserved in the French archives. Selected excerpts of Taylor's statement read:

> Bad weather and quantity of sick people ... forced him to go to St. Vincent ... and anchored there at three o'clock afternoon. Barely had it anchored when a ship and a boat carrying black and red flags, seemed to come right to him ... they had three hundred men and thirty-two mounted cannons, not to mention the help of their associates in the boat armed with twelve cannons with more than one hundred men ... within range of the cannon, he fired two shots, and he approached more closely and discharged several gunshots, from which his terrified crew fled to land and took away all the money that they were committed to guard including a young English girl, and that the Captain remained in his ship, he being one of four.[2]

Taylor watched with fear and trepidation as the pirates climbed aboard his ship. Within moments, his ship was completely under their control. It wasn't long before one of the three terrorized remaining crewmen told one of the pirates about the treasure that had been carried ashore. Taylor's hands were tied behind his back and the pirates threatened him with hanging if he didn't cooperate. Looking death in the face, Taylor told the pirates everything. Eventually, all the treasure, as well as the young girl, were brought on board. The treasure consisted of "silver and silverware, all for about eight thousand sterling livre."[3] That's worth about $152,000 in today's money.[4] They also took an exquisitely designed silver cup that was so unique and beautiful that one of Blackbeard's future captives felt compelled to mention it in his deposition. Henry Bostock described it as "one very fine cup they told deponent they had taken out of Capt. Taylor."[5]

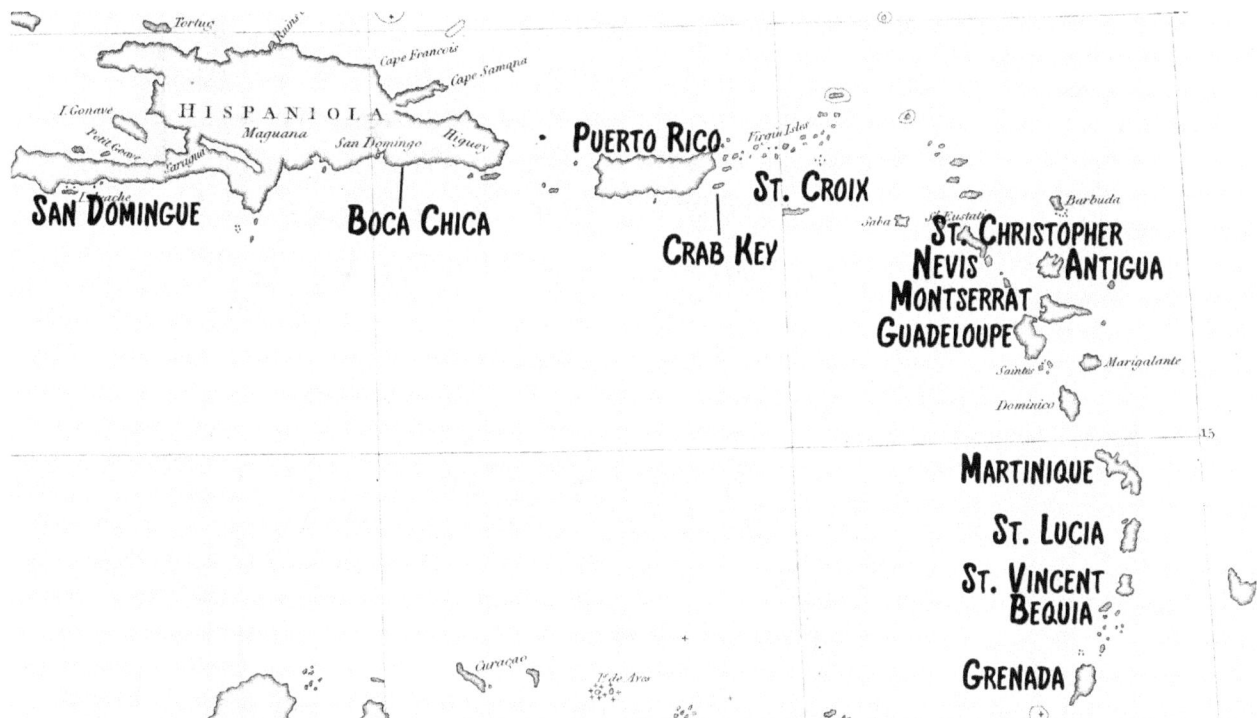

Figure 27: *Lesser Antilles to Hispaniola from Caribbean Map, Courtesy of Barry Lawrence Rudman.*

After detaining Taylor and his terrified crew for two days, Blackbeard ordered that the ship's longboat be prepared for the prize crew to use. Some provisions were placed onboard, consisting of three pieces of beef and some French biscuits. Blackbeard also released five slaves who had belonged to the French officers he forced to join his crew a few days earlier. The young girl was also allowed to leave. Four of Taylor's crew decided to remain with the pirates, including the sailor who had told the pirates about the hidden treasure two days before. On November 28, 1717, at about 9 p.m., they were permitted to leave. As the longboat slowly rowed away from the pirates, Taylor looked back and saw his ship burning, as well as Sieur Simon's ship *Dauphin*, the ship that Bonnet had left five days earlier.

The second half of Captain Christopher Taylor's statement to the French authorities on Martinique reads:

> The pirates having been informed that Captain Taylor had taken away his treasure to the shore, tied his hands close together and made threats of having him hanged, that they forced him to declare everything, to have them brought on board the money and the silverware, and to deliver the young girl to them ... and six other men of his crew who had brought the money, they returned to him the longboat of the English Merchant ship with three pieces of beef, French biscuit, and five negroes from Dosset's cargo who were given to the Pilot, the

Chapter Twelve

Surgeon, and the Cannoneer, the pirates didn't keep his crew except four sailors who surrendered of goodwill. One of these four deserters had declared the treasure of his Captain, and made him run great dangers of life . . . at 9 o'clock in the evening, the pirates before leaving, set fire to the English ship and Simon's boat which was nearby.[6]

The Boston News-Letter ran the story in their March 3 to March 10, 1718 edition. However, there is a discrepancy between their article and Captain Taylor's first-hand account. As shown below, the article mentions "Capt. Teach the Pirate in a French Ship of 32 Guns" and the "Sloop of 12 Guns" as expected. What is a surprise is that the article also mentions a "Briganteen of 20 Guns." Taylor didn't mention any such vessel. He stated that "they had three hundred men and thirty-two mounted cannons, not to mention the help of their associate in the boat armed with twelve cannons and mounted with more than hundred men."[7] Most likely, *The Boston News-Letter* simply made a mistake. *The Boston News-Letter* article reads:

> November Last Capt. Christopher Taylor in a great ship from Boston, was taken at or near St. Lucia or St. Vincent, by Capt. Teach the Pirate in a French Ship of 32 Guns, a Briganteen of 20 Guns, and a Sloop of 12 Guns, his Consorts: Capt. Taylor they put 24 hours in Irons, and whipt him, in order to make him confess what Money he had on board, burnt his Ship, put his Men on Shore at Martinico.[8]

Today, Basse-Terre is the capital of Guadeloupe.

This extremely harsh treatment was unusual for Blackbeard. But the *Great Allen* was from Boston, the place where Sam Bellamy's pirates had been hung. As you shall see in the rest of this chapter as well as the next, throughout Blackbeard's piratical rampage across the Caribbean, he burned every vessel from Boston that he encountered, occasionally commenting to the captives that it was in revenge for the way that part of the colony treated Sam Bellamy's men.

Blackbeard's first attack on a seaport was made in total darkness. After leaving the *Great Allen*, the pirate vessels sailed north to the French island colony of Guadeloupe, about one hundred ninety miles distance. Their target was Basse-Terre, on the southwest tip of the island. The pirates arrived very early in the morning on November 29, 1717, and the *Queen Anne's Revenge,* with the *Revenge* right beside, stealthily sailed into the harbor. The pirate vessels sailed past a single large ship lying at anchor. It was *La Ville de Nantes* and was loaded with sugar and ready to sail for France.[9] Few lights were shining. All was quiet in the port as the townsfolk slept.

The tranquility of the harbor was abruptly shattered when both pirate vessels began firing all their guns in multiple directions. Panic ensued among the startled French citizens as they awakened to the sound of broadside after broadside and the impact ashore of iron shot crashing through wooden structures. Some of the shots fired by Blackbeard's vessels were "hot shot." Heated until glowing red, hot shot is designed to set fire to whatever it hits. Within a very short time, half of the town was on fire. Then, the pirates turned their attention to a large French merchant ship.[10]

BASSE-TERRE, THE OFFICIAL CAPITAL OF GUADELOUPE

Figure 28: *Illustration of Basse-Terre. The image is dedicated to the public domain under CC0.*

By now, the crew of the ship were well aware of the pirate assault on their port. Everyone jumped overboard and swam ashore. Finding the ship deserted, it was an easy prize to take. Blackbeard added this ship to his fleet and sailed away, leaving as quickly as he had come. The ship's captain, Sieur Lemaître, made a report of the attack later on, which was included in the official report of the capture of *La Concorde* on Martinique. An excerpt of this report reads:

> The 9th of that month kidnapped by the ship and the pirate boat . . .
> The captain of this ship and all his crew had time to escape ashore

Chapter Twelve

before the privateers could make themselves masters of his ship which they took and in which only a small foam remained. There is reason to believe that they will have burned this ship, which is a very bad sailboat.[11]

Blackbeard had added another French ship to his fleet, but since he had taken it only a few hours earlier, the pirates didn't have time to make it battle-ready. It was now mid-morning, the same day as the Guadeloupe attack. Heading north, they were calmly sailing toward Montserrat when someone on the pirate vessels spotted a sail on the horizon. It was another prize ship just waiting to be taken. The pirate fleet altered course and sailed straight for it. As they closed the distance, everyone on the pirate vessels expected to see this ship attempt to escape. They were shocked when the potential prize ship steered into the wind and came to a stop. No guns had been fired. They had done nothing to reveal their intent to attack the ship. The pirates got another surprise as the ship lowered a longboat and began rowing over to Bonnet's sloop, *Revenge*. Blackbeard and his crew must have been wondering what was going on.

Montserrat was colonized by Britain in 1632 with many Irish settlers.

The longboat contained one of the ship's officers, Thomas Knight. As his boat came alongside the *Revenge*, Knight and Bonnet exchanged greetings. Finally, the amazed pirates realized what was happening. Captain Benjamin Hobhouse of the ship known as the *Montserrat Merchant*, didn't recognize them as pirates. It was considered good manners in those days for passing vessels to inquire if they had any letters to deliver. Captain Hobhouse mistook the pirates to be slave ships and had sent Knight over to politely ask if they had anything to deliver. In Knight's words, "seeing two ships and a sloop, and thinking one did belong to Bristol, and the other two to Guinea, he went in the long-boat to enquire for letters."[12] Thomas Knight gave a deposition after his release. The details of Knight's abduction, contained in this narrative, as well as the quotes, are all drawn from that deposition.[13]

Bonnet was the first to speak. He introduced himself as Captain Edwards and cordially invited Knight onboard. But just as Knight began reaching for the ladder, he casually glanced toward the sloop's stern and saw their pirate flag waving in the breeze. It had been raised earlier as they prepared to attack the ship. Apparently, no one thought to lower it after the *Montserrat Merchant* came to a stop. Realizing the whole truth of the matter, Knight refused to board. In Knight's words, "They desired us to come on board, but seeing Death Head in the stern we refused it." Bonnet's tone and manner quickly changed from polite to harsh and demanding. Bonnet ordered him onboard and Knight reluctantly complied. As soon as Knight stepped over the gunnel, Bonnet's demeanor changed back to a polite one.

In Knight's deposition, he said that "when we went on board, the first words they said to us were that we were welcome aboard the pirates."

Bonnet interrogated Knight as to the strength of the fort at Kinsale and the ships anchored at Plymouth. Both of those were located on the southwest side of Montserrat, which lay just ahead to the north of their current position. During the questioning, Bonnet remained very polite. But when Knight refused to answer, Bonnet pushed the issue saying that his crew would be angry and that "they would do mischief."[14]

Knight's prolonged absence onboard the *Revenge* caused Hobhouse to become suspicious. The *Montserrat Merchant* quickly made sail and escaped the clutches of Blackbeard and his men, leaving Knight and his oarsmen aboard the *Revenge* as captives. Knight was taken over to the *Queen Anne's Revenge* by longboat. When he arrived, he was welcomed with a similar cheerful greeting, "Welcome aboard the Pirate."[15]

Once onboard, Knight was denied the opportunity to actually meet Blackbeard. He was told that the captain was sick in his cabin. In an unexplainable twist of events, it appears that both Blackbeard and Bonnet were reluctant to reveal their identities. Bonnet maintained his alias as Captain Edwards, and Knight was told that the captain of the ship was named Kentish. In Knight's words, "They report the Captain of the pirates name is Kentish and Captain Edwards belonging to the sloop."[16] They didn't even tell Knight the names of their vessels.

This is an extremely odd aspect of this encounter. Everywhere else, Blackbeard seemed delighted to get his name out there. Everyone, even the French, reported that Teach, or some variation of that spelling, was the leader of the pirates. Several months earlier, as Blackbeard and Bonnet were taking ships along the Virginia and Delaware Capes, their names were well known by all of their captives. However, when dealing with Knight, it strangely seems that both men wanted to keep their identities a secret. To add to the mystery, Blackbeard uncharacteristically remained in his cabin as if he didn't want Knight to see him.

Sailing on, early the next morning, the pirate flotilla of the *Queen Anne's Revenge*, the *Revenge*, and *La Ville de Nantes* still in tow, passed Nevis Island. In the distance, a British ship of war was sighted, anchored in the harbor. Knight overheard the discussion onboard between some of the pirates and quartermaster Howard. They believed the ship to be HMS *Seaford*, which was a sixth-rate vessel and the smallest and least armed naval ship in the Caribbean. On top of that, it lay at anchor. Howard and many of the other pirates were eager to attack and try to capture it. They believed that they could row over in small boats, storm the deck, and easily

Chapter Twelve

New Division out of Antigua, Captain Richard Joy, taken by Blackbeard.

overwhelm the few crewmen who were left onboard as the ship's watch. Blackbeard remained tucked away in his cabin, apparently from his illness, and it was Howard's job to act as the go-between. The captain was strongly against the attack. His view was that this would be an unnecessary risk. For the time being, they should stick to easier prizes.[17] Blackbeard must have been very persuasive because he convinced Howard, who in turn convinced the crew. They sailed on.

It was late morning on November 30, 1717, when the pirates came upon the sloop *New Division* out of Antigua. The captain was Richard Joy, who gave a deposition after he was released. The record of his deposition begins with, "This morning he was taken by two pirate ships and a sloop."[18] As Blackbeard approached, his vessels were sailing under English colors. That way, they were able to sail among unsuspecting vessels without causing any concern. Bonnet's *Revenge* came alongside the *New Division* and, in his usual gentlemanly persona, politely invited the captain over for a visit. Something alerted Captain Joy that these might be pirates. Perhaps it was the weapons they all carried. However, it was too late. He was within pistol-shot range and knew that he could be wounded or killed if he attempted to go back to his sloop. Bonnet realized that Joy suspected them to be pirates and reassured him he would not be harmed. Once onboard, Bonnet and Captain Joy sat down for a meal and a few drinks. "When I went on board they asked me to eat and drink, and enquired what vessels were along the shoar."[19] With the captain securely on the *Revenge*, Bonnet's pirates calmly boarded the *New Division* and took control of the vessel.

During the course of their conversation, Bonnet told Joy that they would probably burn his sloop. Heart sickened, Joy pleaded with Bonnet to let him go, saying, "It was all I had to support my family." Joy's pleading must have been very convincing. Bonnet felt compassion for the captain and agreed to let him go. He even allowed Thomas Knight and his crewman to board the *New Division* and sail away. When Knight and Joy reached safety, they each made statements that were dated November 30, 1717.[20]

The next day, the pirate flotilla reached Saint Christopher and the harbor of Sandy Point. Not the usual seaport, Sandy Point looked more like a Caribbean beach resort. It was a calm, half-moon-shaped beach with gentle waves lapping on the warm sands. Overlooking the beach was a rather imposing rocky mountain about five hundred feet high and jutting up directly out of the sand. This mountain was called Brimstone-Hill and there was a fort constructed of a seaward-facing plateau about two-thirds of the way up the steep cliff face. However, the fort hadn't been well maintained and there was very little ammunition. This rendered the fort useless for the defense of this picturesque shoreline.

Figure 29: *Painting of Brimstone-Hill. The image is dedicated to the public domain under CC0.*

Unconcerned with the fort's guns, Blackbeard sailed directly into the harbor with his pirate flags flying. As a distraction, he set fire to the French ship, *La Ville de Nantes,* that they had been towing since capturing it at Guadeloupe, then headed the flaming ship directly toward the beach at the base of Brimstone Hill. With no shots ringing out from the fort, Blackbeard easily took several trading sloops that were at anchor in the harbor. Once he looted them, he set several of them on fire and sailed away.[21]

It is somewhat ironic that the French ship's captain predicted the ship's fate when he said, "There is reason to believe that they will have burned this ship, which is a very bad sailboat."[22] Leaving Saint Christopher, Blackbeard was back to having only two vessels, the *Queen Anne's Revenge* and the *Revenge*.

Approaching Saint Thomas on December 3, 1717, the pirates sighted the real HMS *Seaford*, which had been in the Virgin Islands, taking Governor Hamilton on a tour of his territory.[23] The vessel they had mistakenly identified as the Seaford four days earlier was an unknown merchant ship. Approaching the real *Seaford*, some of Blackbeard's crew may have considered attacking it. However, this attack would be far different from the one they planned four days earlier. That ship was at anchor. The *Seaford* they had just sighted was under sail. Even so, it was only armed with 24 guns and had a crew of only eighty-four.

1657

Saint Thomas is settled by the Dutch West India Company.

Chapter Twelve

It was conceivable that a combined attack by the *Queen Anne's Revenge* and the *Revenge* might be successful, although it would be extremely risky. The pirates weren't professional combat sailors. They used intimidation and an overwhelming show of strength to frighten their victims into surrendering. Engaging the *Seaford* while it was under sail would mean a running naval battle with a fully manned ship of war. This was a completely different matter.[24] Fighting a trained naval vessel would require skills and training beyond the pirate crew's capabilities. Blackbeard chose to sail past the *Seaford* without engaging, or even getting close. The logbook of the *Seaford* reveals that they saw the pirate vessels too, but thought they were merchantmen.

After passing the *Seaford*, Blackbeard sailed on to Saint Croix. En route, they took a Danish sloop and an English sloop from Antigua, which belonged to Robert McGill. In addition to the supplies they took, they also added to the number of guns aboard the *Queen Anne's Revenge*. All the captives from the English sloop were transferred aboard the Danish sloop unharmed. Afterward, they burned the English sloop and allowed the Danish sloop to sail on.[25]

Crab Key is a large island off the east end of Puerto Rico.

The *Queen Anne's Revenge* and the *Revenge* continued sailing in a westerly direction, passing south of Crab Key during the night. As the sun topped the horizon behind the ship's stern, the darkness of the night was gradually being replaced by the hazy light of morning. In the distance, Blackbeard sighted a solitary English sloop cutting across their bows to the north. It appeared to be unarmed and a very easy prize to take. The pirate vessels shifted course to intercept the unsuspecting sloop. The date was December 5, 1717, and this sloop's capture provided historians with a wealth of knowledge concerning Blackbeard, including the only first-hand description of this fierce and imposing captain.

Sloop Margaret, Captain Henry Bostock, taken by Blackbeard. Bostock was the first man to provide a physical description of Blackbeard.

The English sloop was the *Margaret*, whose master was Henry Bostock. From the quarterdeck of the small sloop, it was difficult for the crew to see the approaching vessels. The sun was behind them and the piercing light was blinding. As the pirate vessels grew nearer, their shadowy forms became clear. With less than one hundred yards between them, everyone on the *Margaret* finally realized that they were about to be taken by pirates. The captain watched anxiously as the two pirate vessels came alongside. Suddenly, the sound of a musket firing pierced the silence of the sea air. It came from the large ship. The *Margaret* offered no resistance and all three vessels turned into the wind and glided to a complete stop. Soon after, a man onboard the menacing ship shouted a greeting and ordered the captain of the *Margaret* to lower a boat and come over to the ship. Bostock fought to control his fear and maintain composure as he rowed over to the huge ship. As soon as he climbed over the gunnel, one of the pirates politely escorted him to the quarterdeck. A tall, well-built man stood patiently waiting next to the helm. He had a very long black beard that was striking in appearance. Bostock was led up to this impressive man, and the two of them stood face to face.

Blackbeard introduced himself as the captain and began a casual conversation with Bostock, inquiring about his cargo. Blackbeard's calm manner and polite tone must have put Bostock at ease. The *Margaret* carried a cargo of livestock, four cattle, and thirty-five hogs. The *Queen Anne's Revenge* launched a boat and sent it over to the *Margaret,* where the pirates took all the livestock, a barrel of gunpowder, five small arms, two cutlasses, all his books and navigational instruments, and some linen.

Eventually, the *Margaret* was released and sailed to St. Christopher, where its captain, Henry Bostock, gave a deposition. A copy of this deposition is in the British archives and a summarized version is included in the *Calendar of State Papers.*[26] Although slightly enhanced to increase interest and readability, all the details of the capture of the *Margaret* and Bostock's experiences while onboard the *Queen Anne's Revenge* come from that deposition.[27] I shall include portions of Bostock's deposition as the story unfolds. The deposition begins:

> This Deponent being duly Sworne on the Holy Gospel Says that in the Sloop Margaret of this Island whereof he was Master he on the 5th day of this Instant December at Break of Day turning up from Porto Rico met about ten Leagues to the Westward of Crab Island a large ship and a Sloop. The Ship fired a Small Arm at him and then Hal'd him ordering him to come on board which he with of his Men did in his Cause Being conducted to the Quarter Deck to the person that was called the Captain by the name (as he thinks) of Capt Tach, he asked this Deponent what he had on board to which this answered he had Cattle and Hogs then he said Capt Tach ordered his own boat to be hoisted out to go on board and fetch them which they did (they were four Beeves and about five and thirty Hogs) they took from him besides two thirds of a Bll of Gunpowder five small arms two Cutlaces, His Books and Instruments and Some Linnen.

Bostock spent eight hours onboard the *Queen Anne's Revenge,* during which time he was well treated. Later on, he described the ship as Dutch-built, but records prove that it was built in France. Originally built as a ship of war, the large hull was designed to hold a battery of guns. This may have given it the appearance of a Dutch merchant ship, whose hulls are larger

Chapter Twelve

to allow them to carry more cargo. Bostock noted that it carried thirty-six guns, and he estimated that there were about three hundred men in total. All of his crew were ferried over to the ship so the pirates could loot it more effectively. While Bostock quietly waited for the pirates to finish looting his sloop, Blackbeard forcibly retained Edward Salter, a cooper, and Martin Towler. A third crewman, Robert Bibby, volunteered to join the pirate crew.

> This deponent declares the Ship to be as he thinks Dutch Built, was a French Guinea man (he heard on board) that she had then thirty Six Guns mounted, that she was very full of Men he believes three Hundred that they told him they had taken her about Six or Seven Weeks before, that they did not seem to want provisions. That they kept him on board about Eight Hours did not abuse him or any or his Men except the forcing of two of To Stay With them whose Names were Edward Salter a Cooper formerly Sailed with Capt George Moulton and Martin Towler. But one of his men by name Robert Bibby . . . a Leverpool man voluntarily took on with them.

Feeling more at ease, Bostock was at liberty to wander about the pirate ship, where he noticed a large quantity of silver and gold bars onboard. One of Bostock's crewmen saw a uniquely designed silver cup and pointed it out to Bostock. It was so striking that Bostock asked Blackbeard about it and felt compelled to report his answer in his deposition. Always willing to boast about his successes, Blackbeard was delighted to tell Bostock of how he had taken the cup from Captain Taylor a few days earlier. He also told Bostock of how they had to beat him to get him to reveal where he had hidden his treasure and that they burned his ship. As long as he was on the subject, Blackbeard continued to describe all his recent burnings of vessels, which included two or three belonging to these Islands. He also told of the burning of a sloop just the day before that sailed from Antigua and belonged to a man named Mr. Gill. Some of the pirate crew mentioned that they had passed near an English Man-O-War, but that they carefully avoided it.

> He Saw a great deal of plate on board of them Tanka Cups &tc. particularly one his men took Notice of a very fine Cup which they told him they had taken out of one Capt Taylor whom they had taken going from Barbados to Jamaica which Captain Taylor they much abused, and as he this Deponent hears burnt his Ship that they told this Deponent they had burnt Several Vessels, among them two or three belonging to these Islands, particularly the day before they had

burnt a Sloop belonging to Antego one Mr Gill owner they owned they had met with the Man of Warr on this Station, but said they had no business with her, but if she had chased them they would have kept their Way.

News of an *Act of Grace*, also known as the *King's Proclamation*, had spread throughout the Caribbean. This proclamation offered a pardon for all pirates who turned themselves in to any colonial governor. Bostock optimistically told Blackbeard about the proclamation, perhaps hoping the pirates may change their minds and release his sloop at the prospect of being pardoned, but the pirates received the news with indifference. As they spoke, Bostock became fascinated with Blackbeard's long, black beard, so much so that he described it in his deposition. During their conversation, Blackbeard asked Bostock about other vessels that might be in the area. Not wanting to cause the capture of another vessel, Bostock wouldn't answer, but to Bostock's surprise, one of his crew eagerly told the pirates of a French sloop and a Dutch ship from St. Thomas that should be nearby. Upon hearing this, Blackbeard's eyes lighted with excitement. He rushed to the rail and shouted over to Stede Bonnet on the *Revenge*. He told Bonnet that there were two vessels expected soon and for him to go after them. Without a moment of hesitation, the *Revenge* raised sail and sped off to the east to look for them.

September 5, 1717

The *King's Proclamation* was signed on September 5, 1717, and gave every pirate exactly one year to surrender themselves and be granted a pardon. This document was also called the *Act of Grace* and the *King's Pardon*.

> This Deponent further Says that he told them an Act of Grace was expected out for them but they Seemed to Slight it. This Deponent Says the Captain was a tall Spare Man With a very black beard which he wore very long. That among the Men one Is a Nephew to Doctor Rowland of this Island that sailed in Captain Joseph Wood from this Island to London about three years ago, made himself known to this Deponent enquiring after his said Uncle &tc. They asked this Deponent whether there were any more Traders on the porto Rico Coast which he this Deponent would not give them an Account of but his men owned to them that there were two Traders on the Coast one a French Sloop and the other a Dunc from St. Thomas's that while he was on board they ordered their Consort Sloop (on board of which he heard there were about Seventy Men) to make Sail and Chace along Shoar to look for these Traders.

As with Hornigold's conversation with Timberlake a year earlier, Blackbeard was at ease discussing details of his ship's operations, as well as his plans with Bostock. He mentioned that they planned on sailing to Samana Bay on Hispaniola to careen and then to wait in ambush for a

Chapter Twelve

Spanish treasure fleet carrying the payroll for the garrison. These ships were expected to pass that way sometime just after Christmas. Blackbeard also asked if he knew a captain named Pinkethman, who had a privateer commission from St. Thomas allowing him to fish the wrecks.

> This Dept Says that by all he could guess from their Discourse they intended for Samana Bay in Hispaniola to Careen and thence to lye in Wait for the Spanish Armada that they expected should immediately after Christmas come out of the Havana for Hispaniola and & Porto Rico With ye money to pay the Garrisons this they declared to him the Captain Saying (they think We are gone but we will Soon be on the Backs of them unawares) they having been hereto-fore on these Coasts. This Deponent farther Says that they enquired of him where was Captain Pinkethman that he told them he heard he was at St Thomas's with a Commission from the King to go on the Wrecks.

Eight hours after the pirates first boarded the *Margaret*, they were finished looting it. They had taken everything they wanted. Bostock and his crew, minus the three men, were taken back to their sloop and Blackbeard let them sail on. The next day the *Margaret* put in to Tortola for supplies. As you may recall, the day before Blackbeard took the *Margaret*, he had taken two sloops, an English sloop from Antigua owned by Robert McGill and a Danish sloop, whose master was George Hanns. After both sloops were looted, Blackbeard burned McGill's sloop, but allowed the unharmed crew to board the Danish sloop, and he permitted them to leave. By coincidence, the Danish sloop and the Margaret both ended up on Tortola at the same time.

Tortola is one of the British Virgin Islands and is located about twelve miles east of St. Thomas.

The port of Tortola wasn't that big, and now there were three crews in town that had been taken by Blackbeard within the past several days. At some point, Bostock met with some of the men from the other two crews, along with the English sloop's owner, Robert McGill, who had arrived from St. Croix to rescue his stranded sailors. Bostock already knew of their misfortune, as Blackbeard had told him all about the English sloop he burned and the Danish sloop he looted and released. The men of all three crews shared a common experience and must have spent some time together discussing Blackbeard and his pirates. Henry Norton, one of McGill's sailors, told Bostock that their captain had hidden fifteen ounces of gold dust in the pockets of two of his crewmen, an Indian and a Negro. However, Blackbeard's men carefully searched them and found the gold dust, which they took. Henry Bostock added that he believed there was far more gold dust as well as other riches on board the *Queen Anne's Revenge*.

> This Deponent further saith that among other Riches he believed they had much Gold Dust on board for that an Indian and a Negro belonging to Bermuda (as this Defendent was informed at Tortola & by the above Said Robert McGill) having got from tha at Sta Cruix, got with the said McGills men on board a Danish Sloop and in her to Tortola. That one George Hanns an Englishman formerly belonging to this Island was then Master of the Said Danish Sloop, who when he had the Said Negro and Indian on board, Searched them and found on them as this Deponent heard about fifteen ounces of Gold Dust, which this deponent was told at Tortola by one Henry Norton the said Hanns Mc G———a Doctor on board shar'd between them. And further this Deponent Saith not.

A few days later, Bostock sailed his sloop to St. Christopher, where he made a deposition on December 19, 1717, to the Lieutenant Governor, William Mathew. This deposition has proven to be the most valuable of all depositions as far as Blackbeard historian authors are concerned. In addition to the details of the taking of his sloop, it includes details of other vessels that Blackbeard took and some of those that he burned. Bostock also mentions Taylor's silver cup. Although it may seem trivial, Bostock's comment about that cup will play a significant part in determining the truth of accusations toward Blackbeard in an incident described in a later chapter. Most importantly, Bostock's description of Blackbeard's appearance as "a tall Spare Man With a very black beard which he wore very long."[28] This description is one of only two in the historical record made by a person who actually saw Blackbeard. The other, which will be discussed in a later chapter, came from a man named Robert Maynard.

Finally, Bostock mentioned Edward Salter, a cooper, one of his crewmen who was forced to join Blackbeard's crew. Although he began his piratical career involuntarily, Edward Salter would become one of Blackbeard's most trusted crewmen. He will also play an important role in solving the mystery of the fate of seven of Blackbeard's crew, who were arrested and held in prison in Williamsburg months after Blackbeard's death.

After bidding farewell to Henry Bostock and the *Margaret*, Blackbeard continued westward toward Puerto Rico and remained in the waters of Hispaniola throughout the rest of the year. However, some historian authors disagree. To demonstrate how some more recent historian authors can confuse facts, Robert E. Lee, author of the 1974 publication, *Blackbeard the Pirate*, contends that Blackbeard sailed the *Queen Anne's Revenge* to Bath, North Carolina, where he accepted the king's pardon. After that, Lee contends that Blackbeard sailed to the Gulf of Honduras, where he first met Stede Bonnet and then formed their partnership.[29] As we have already seen, the historical record clearly established the partnership between Bonnet and Blackbeard in September 1717. As for the trip to Bath, there is no documentation suggesting such a trip, and there was no time for him to sail to Bath and return to the Caribbean in December 1717.

Blackbeard sailed westward along the south coast of Puerto Rico after December 6, 1717, where he may have taken the last prize of the year. The *Queen Anne's Revenge* and the *Revenge* sighted a French sloop and began the chase. This sloop was *La Volante*, which was commanded by Captain Jean Bleu Nesbayes. As the two pirate vessels bore down on the helpless French sloop, the captain could see their pirate flags waving in the wind. For three hours, *La Volante* desperately attempted to outsail the pirates, but they just didn't have the speed. As the *Queen Anne's Revenge* came within range, it fired two warning shots, and the French captain knew that escape was impossible. The pirates ordered him to sail to the nearby harbor of Boca Chica and drop anchor. Both pirate vessels came to anchor alongside. They were only about twenty miles east of Santo Domingo, the Spanish capital of the entire Caribbean. Blackbeard must have known that he was in danger of being spotted by a Spanish warship. Pressed for time, Blackbeard decided it was simpler to keep *La Volante* rather than remain there for hours looting it. The French captain and crew were quickly put ashore, and the anchor cable of *La Volante* was cut. The stranded Frenchmen stood helplessly on the beach and watched with great despair as Blackbeard sailed away with their vessel.

La Volante was taken by Blackbeard at Boca Chica Bay, which is on the south coast of Hispaniola (Dominican Republic) twenty miles to the east of Santo Domingo.

The owner of *La Volante* was Jean Morange. Months later, after the French crew finally reached friendly territory, Morange filed a letter of complaint with the French authorities. His letter is housed in the French National Archives. The following is an excerpt from that letter.

> Being at the coast of his boat at the Coast of Puerto Rico was attacked by a ship and a boat of pirates, it was La Concorde after having been hunted for three hours and having suffered two cannon shots under flag now this boat was obliged to anchor in the fairground harbor of Boca Chica, from which the pirate boat having approached and having anchored alongside, the French crew landed them, then the pirates took this boat from which they cut the cable.[30]

The stranded Frenchmen were left in unfriendly Spanish territory. Even though a state of war didn't exist between their two nations at the time, they weren't exactly on friendly terms. Spanish authorities arrested the French crew, and they were held and mistreated for over two months. Eventually, the French nationals were taken to Crab Island where they remained for five days and then were allowed to leave on two English vessels that took them to St. Thomas.

> The Spaniards of Boukachiquy guarded the French for three days at the seaside and took them by land to the port where they remained for two months . . . during the time the Governor of the Spanish coast . . . promising every time to give them a vessel to take them to St Thomas other the bad treatment given to the French crew he kept nothing of what he had promised . . . which forced them to return to crab island where there were still a few English and a few Negroes. They stayed five days at the end of which two English drummers having arrived there transported them to St Thomas.[31]

There is an interesting follow-on to this story. In his letter of complaint, Jean Morange added a report that he received from another French captain named Lasesre. At some point, Lasesre met an English captain who told him of how his sloop had been taken fifteen days earlier by a pirate ship of 40 guns accompanied by two pirate sloops. The English captain added that this was the same ship that had previously been called *La Concorde*.

> The Master of the Bateau de Mens. de la Teuche called Lasesre coming from St Alozie there met another English boat whose Captain told him that fifteen days earlier a new pirate of 40 guns accompanied by two boats also pirates had taken from him a boat of six years, and had told him that if it had been French, they would have seen it burned, it was the same pirates who previously captured la Concorde and then armed it with 40 cannon that they took from English vessels.[32]

Identifying this vessel is very problematic. The English captain didn't mention a specific date, just fifteen days earlier. Earlier from when! Several clues may help in determining which vessel this was. This mysterious English vessel was taken by a ship and two sloops. Blackbeard didn't have a second sloop until the very end of 1717. Additionally, the ship is described as having 40 guns. Blackbeard continually added guns to his ship as he took other vessels. Bostock mentioned 36 guns onboard. Reports of the *Queen Anne's Revenge* mounting 40 guns didn't occur until early 1718. The other clue is that this ship wasn't burned, but it was taken and kept by the pirates. As shown in the next chapter, during the first four months of 1718, the only known vessel that was taken by a ship and two sloops and that was kept by the pirates was the *Adventure*, Captain David Herriot. However, this couldn't be the same vessel mentioned by Lasesre because Captain

Chapter Twelve

Herriot also remained with the pirates until he was killed in Charles Town in November 1718.

It seems likely that this vessel was taken sometime in January or February of 1718. That date is based upon the fact that the English captain's report was included in Morange's letter. The French crew of *La Volante* weren't released until late February 1718. News of the loss of his vessel and the poor treatment of his captain and crew probably didn't reach the Morange until mid-March 1718. Outraged, Morange wrote a letter of complaint to the French officials. Such a letter would have been written shortly after Morange received the news. This places the date of his letter as late March or early April 1718. It would have taken at least a month for the English captain, whose vessel was taken by Blackbeard, to make his way to a port where Lasesre happened to be, and then for Lasesre to travel to Martinique, where he reported to Morange. If this is correct, this English vessel was among the vessels that Blackbeard took while cruising the Spanish waters near the Gulf of Mexico.

Thirteen
Death's Head

Blackbeard's iconic flag has come under a great deal of controversy in recent years. Traditionally, his flag is depicted as a black flag portraying a white skeleton of the devil with horns on his head. This skeleton is holding an hourglass in one hand as if to say, "Your time is running out." In the other hand, he is holding a spear which points to a bleeding heart. When I say "traditionally," I mean that it is the flag most recognized as Blackbeard's in many books written in the late twentieth and early twenty-first century, and which is sold as Blackbeard's flag in the thousands of gift shops that sell pirate flags.

So far, researchers haven't uncovered a detailed description of the flags used by Blackbeard or his vessels. There are a few vague reports like the one that appears in *The Boston News-Letter* dated June 9 to June 16, 1718. The article chronicles Blackbeard's capture of the *Protestant Caesar* and describes the flags used by his small fleet: "a large Ship and a Sloop with Black Flags and Deaths Heads in them and three more Sloops with Bloody Flags."[1] The "Black Flag and Deaths Head" refers to some sort of typical pirate flag and the "Bloody Flag" refers to a plain red flag which was used by all pirates in the seventeenth century as the signal for no quarter given. So, what is the origin of the flag with the devil, hourglass, spear, and bleeding heart?

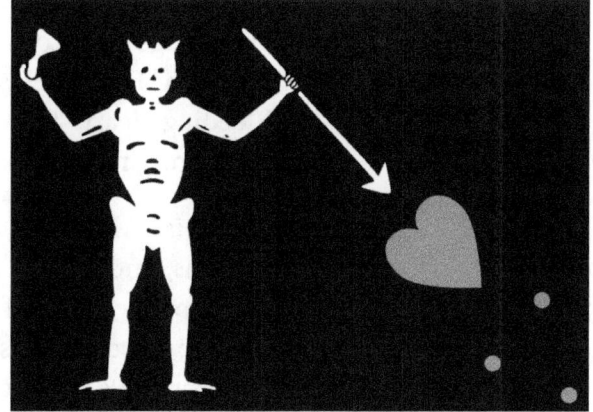

Figure 30: *Traditional Version of Blackbeard's Flag. The image is dedicated to the public domain under CC0.*

A description exactly matching such a flag appears in Patrick Pringle's 1953 book titled *Jolly Roger: The Story of the Great Age of* Piracy.[2] The book even includes illustrations of various pirate flags.[3] According to Pringle, the flag is described as belonging to pirate John Quelch in 1703, as he was

Chapter Thirteen

plundering Portuguese shipping along the South American coast. Quelch then returned to Marblehead in Massachusetts, trying to sell his captured goods. Pringle cites his source for this information as Philip Gosse's book, *The Pirates' Who's Who*, published in 1924. Gosse describes Quelch's flag as a black flag they called "Old Roger ... having in the middle of it an anatomy with an hourglass in one hand and a dart in the heart with three drops of blood proceeding from it in the other."[4] However, Gosse does not cite his source for this description, and no such description is contained in the transcript of Quelch's trial. In contrast, several witnesses in the trial testified that Quelch only flew the English colours.[5] This commonly retold story of Quelch's "Old Roger" flag is yet another example of historian authors incorrectly quoting undocumented sources as fact.

During the years when Blackbeard was operating as a pirate captain, there were other accounts of pirates who used similar flags. A most intriguing article was printed in *The Boston News-Letter*, issued in August of 1717. It reads, "Also arrived here the Snow Restoration, Nathaniel Brooker Master, from London, about 3 Months Passage, who meeting with Northerly Winds and wanting Water, he intended to seize Bermuda, but could not; at last he was taken by 2 Pirate Sloops (one of 12 Guns 100 Men, commanded by Capt. Napin, who had in his Flag a Death's Head and an Hour Glass, the other of 6 Guns 80 Men, commanded by one Nichols, who had in his Flag a Dart and Bleeding Heart) about 5 Leagues East of Harbour Island ... "[6]

This was not Blackbeard, of course. It was a team of two pirate captains, one named Napin (Napping) and the other named Nichols, whose combined flags seem to have inspired the design of the famous Blackbeard flag, one with "Death's Head and an Hour Glass" and the other with a "Dart and Bleeding Heart." However, there is another report of such a flag being used which might be the same pirates.

The *Saint-Michel de Nantes* was a French ship of 150 tons, armed with 12 guns, and crewed by 40 men. As she peacefully sailed about 20 miles north of the coast of Hispaniola in October 1717, she was attacked and taken by two pirate sloops. When the ship finally returned to France in January 1718, two reports were made to the Admiralty by the captain, Sieur Jean Dubois, one is dated January 3, 1718, and the other is dated January 6, 1718.[7] These dates are in the French Gregorian calendar and would be eleven days earlier in the English Julian calendar. Both reports still exist in the French archives, and even though they are very similar, they each contain slightly different information. By piecing them together, we can get a clear picture of the astonishing events surrounding this pirate attack. But more importantly, two aspects of these firsthand accounts by Captain Dubois render his reports incredibly valuable to pirate historian authors. The first is a detailed description of the attack. The second is his description of the pirates' flags.

It was about 10 o'clock in the morning on Wednesday, October 20, 1717, when Captain Dubois noticed two suspicious sloops approaching. One sloop was armed with 12 guns and the other was armed with 4 guns. The larger sloop was flying an English flag. Suspecting an attack, the French captain armed his crew as well as his 28 passengers and prepared for battle. After about two hours of pursuit, the larger pirate sloop came within range. Suddenly, the English flag was lowered, and both sloops raised their pirate flags. In the first account, Dubois wrote: "hoisted each of them a black flag with a figure of death holding in one hand a dart and in the other an hourglass at the top of their great mast with another flag of the same color having a figure of a man who holds another under his feet that crosses his throat." In his second report, Dubois describes the flag as "a black flag with a skeleton in the middle holding in one hand a dart and in the other a clock."[8]

The sight of these flags struck fear into the hearts of the passengers and crew. The larger sloop fired seven or eight volleys. Immediately after the flags were raised, the larger sloop fired seven or eight volleys. Captain Dubois wanted to return fire using the two guns that were mounted at the stern but the passengers pleaded with him to surrender without firing upon the pirates, fearing retaliation if they resisted.[9]

Captain Dubois signaled his surrender by lowering his colors and sent a small boat over to the pirate sloops to discuss surrender. The boat returned with the English captain and about 10 pirates who took Captain Dubois back to their sloop. He noted that there were about 135 men on the big sloop and about 35 men on the other. As he was held hostage, about 40 pirates rowed over to this ship and began taking everything they wanted. When Dubois told the pirates that another slave ship was due any day, they decided to wait. On Saturday, October 23, 1717, the pirate sloops took three more French ships. After taking what they wished and forcing a surgeon, a gunsmith, and four sailors to join their crew, the pirates let the French ships go.[10]

These accounts clearly show that a flag exactly like the traditional Blackbeard flag could have existed in 1717 and 1718. Blackbeard himself may have used that flag, and then again, he may not have. If one is a purist and only wants to attribute a flag to Blackbeard that is historically documented, then the only clue is a "black flag with deaths head." That type of flag normally has some sort of crossbones, but are they beneath the skull or behind it? Or is it a head with crossed swords beneath? And even if he did use one of those flags when he attacked the *Protestant Caesar*, he may have used other flags at other times. Many pirates, such as John Roberts, changed their flags often.[11] Until further documentation surfaces that gives a precise description of the flag used by the world's most famous pirate, no one can definitively describe precisely which flag Blackbeard used.

Figure 31: *Black Flag with Death's Head. The image is dedicated to the public domain under CC0.*

Fourteen
Cruising the Gulf of Honduras

When last seen, Blackbeard's fleet was near Samana Bay on the Island of Hispaniola. He had just taken the French sloop, *La Volante*, giving him a total of two sloops and a ship. Now, Blackbeard's fleet was entering French waters. The French colony of San Domingue, which would later become Haiti, encompassed the entire western tip of the island. Its principal seaport was Petit-Goave. Documents from that port dated January 1718 state that the officials were concerned about attacks from Blackbeard with his ship, *La Concorde*, and his other two vessels, the *Revenge* and the *Le Roy de Guillaume*.[1]

So far, detailed information on how Blackbeard took *Le Roy de Guillaume* or how he disposed of *La Volante* hasn't been found. Neither vessel was mentioned again by name. Subsequent reports only mention a ship accompanied by a constantly changing number of sloops. Keeping track of Blackbeard's consort vessels is challenging, especially as he entered the Gulf of Honduras. For example, on one day he is reported to have four sloops, and a day later, he only has two. There is one thing for certain when speaking about Blackbeard: he went through a lot of vessels.

From Hispaniola, it appears that Blackbeard sailed southwest toward Mexico. Spanish sailors putting into ports on Jamaica spoke of a pirate ship they called "The Great Devil" that was seen cruising near the Gulf of Mexico. It was also rumored to be "filled with much treasure." Those quotes came from a story that appeared in the March 28, 1718 edition of the *Jamaican Dispatch*.[2] There were several sightings of four sloops and a "ship of 42-guns" near Vera Cruiz, which were eventually reported in the June 1718 edition of the *Weekley Journal or British Gazette*.[3]

Around this time, Blackbeard took an unidentified English vessel and kept it. The captain and crew were released. Eventually, the captain met

Chapter Fourteen

a French captain named Lasesre, The English captain who told Lasesre of how his sloop was taken fifteen days earlier by a pirate ship of 40 guns accompanied by two pirate sloops. The English captain added that this was the same ship that had previously been called *La Concorde*. Upon reaching Martinique, Lasesre reported the incident to Jean Morange, who included the report in a letter he wrote to the French Admiralty. This letter and the taking of his English vessel were described at the end of an earlier chapter. The only information available on Blackbeard's capture of this vessel comes from that report.

> The Master of the Bateau de Mens. de la Teuche called Lasesre coming from St Alozie there met another English boat whose Captain told him that fifteen days earlier a new pirate of 40 guns accompanied by two boats also pirates had taken from him a boat of six years, and had told him that if it had been French, they would have seen it burned, it was the same pirates who previously captured la Concorde and then armed it with 40 cannon that they took from English vessels.[4]

The captain mentions that the *Queen Anne's Revenge* was accompanied by two sloops. One was certainly the *Revenge*. The second sloop could have been *Le Roy de Guillaume* or another sloop we know nothing about.

Even though these reports are most interesting, first-hand accounts of Blackbeard's movements and the vessels he took would be much more desirable. However, eighteenth-century prime source documents from the Spanish and French authorities are difficult to obtain. Fortunately for Blackbeard enthusiasts, he is about to return to the waters frequented by English shipping and the reports and articles that conveniently begin to appear in *The Boston News-Letter*. By March 1718, Blackbeard's fleet had rounded the Yucatan Peninsula and sailed south, deep into the tranquil waters of the Gulf of Honduras.

From the end of March to the middle of April, the pirate actions of Bonnet and Blackbeard are well documented in our favorite source, *The Boston News-Letter*. The Monday, June 9 to Monday, June 16, 1718 edition accurately chronicles several attacks.[5] Additionally, much of this information is corroborated by the deposition of David Herriot, a victim of one of those attacks. He remained with the pirates until June 1718, when he chose to join Stede Bonnet's crew.[6]

Blackbeard and Bonnet separate to cover more ocean in the hunt for prizes.

Around mid-March, Blackbeard separated from Bonnet in order to widen their search area and increase their chances of taking vessels. Blackbeard would not have done this if he didn't have complete trust and confidence in his partner to act independently. The *Queen Anne's Revenge* and at least one of their captured sloops sailed to the Turneffe Islands, which are located about twenty-two miles east of modern-day Belize City. Captain Bonnet and the *Revenge* headed toward the Bay Islands, which are located about 35 miles to the northeast of La Ceiba on the north coast of Honduras.

The Bay Islands were a lucrative spot for pirates to hunt for prizes. Today, the beaches of these islands are lined with luxury Caribbean resorts, and the capital city port of Roatán is crowded with cruise ships. Since the days of Henry Morgan in the 1660s, these islands were used by English buccaneers to raid Spanish shipping. By the early eighteenth century, a few tiny English settlements had been established on the main island, including the port of Roatán. These islands too were well known to all English merchantmen as a spot to take on fresh water and other supplies.[7]

On March 28, 1718, the *Revenge* quietly rounded the tip of the island and sailed along the south shore toward Roatán, where Bonnet spotted a large ship. He immediately closed the distance and positioned for an attack. The ship was the *Protestant Caesar* of Boston, commanded by Captain William Wyer. It was armed with 26 guns, more than four times the size of *Revenge*. To Bonnet's surprise, this ship prepared for battle. The last ship that had fired on Bonnet's *Revenge* ended in disaster. As you may recall, that engagement was the Spanish ship back in August 1717. In that battle, Bonnet's sloop was almost blasted to pieces, and Bonnet himself was severely wounded. Bonnet must have recalled that horrific event as he saw the gunports open on the *Protestant Caesar*. However, he was committed to the attack.

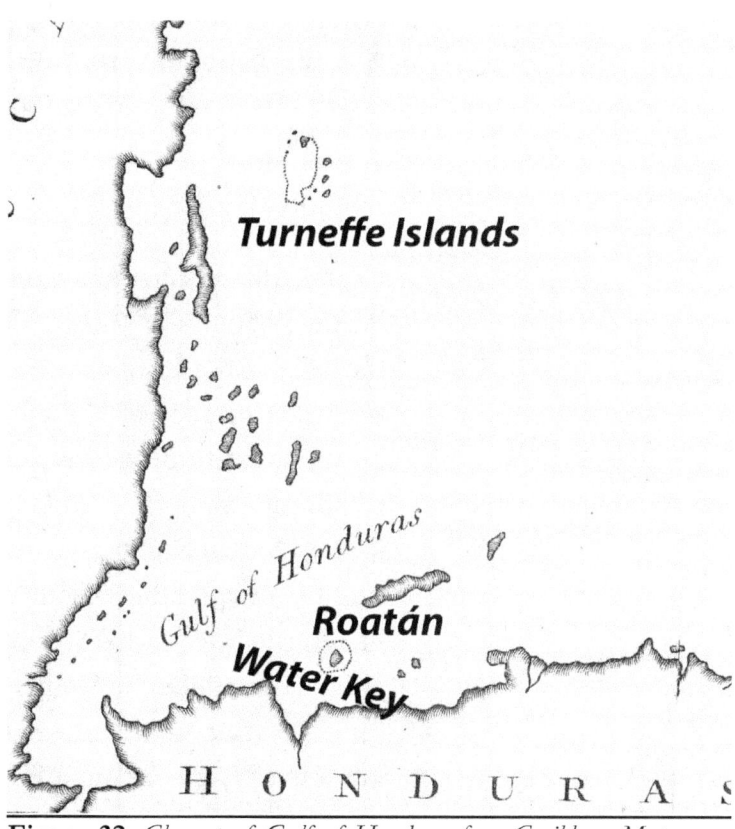

Figure 32: *Close-up of Gulf of Honduras from Caribbean Map, Courtesy of Barry Lawrence Rudman.*

1718

Roatán was officially vacated, and the safe harbor was often used by pirates. Today, the island's main port of Coxen Hole is a prime stop for cruise ships.

By nine o'clock, the *Revenge* had come up on the ship's stern quarter. That is an ideal position for an attack. Bonnet cut across the ship's stern and fired a 5-gun broadside supported with a hail of bullets from small arms fire. Solid shot must have torn through the stern of the big ship, but apparently did little actual damage. Undaunted, the *Protestant Caesar* returned fire with their two stern guns, also supported by a volley of small arms fire. Bonnet hailed the ship and ordered the captain to surrender immediately, or else they would all be killed. Hearing this challenge in English may have surprised Captain Wyer, who probably thought the sloop to be Spanish. However, that didn't matter to Wyer, and he continued the fight. For the next three hours, each vessel tried to outmaneuver the other in the darkness. Perhaps the only way to identify the position of the other vessel during this battle was from the occasional bright flashes of guns as they fired. Bonnet finally gave up and broke off the attack.

That encounter and battle are chronicled in *The Boston News-Letter* edition of Monday, June 9 to Monday, June 16, 1718, which reads:

> Capt. William Wyer late Commander of the Ship Protestant Caesar burthen about 400 Tuns, 26 Guns Navigated with 50 Men, who on the 28th of March last about 120 Leagues to the Westward of Jamaica, near the Latitude 16, off the Island Rattan, espied a large Sloop which he supposed to be a Pirate, and put his Ship in order to Fight her, which said Sloop had 10 Guns and upwards of 50 Men, and about nine a Clock at Night came under Capt. Wyers Stern, and fired several Cannon in upon the said Ship and a Volley of small Shot, unto which he returned two of his Stern Chase Guns, and a like Volley of small Shot, upon which the Sloop's Company hail'd him in English, telling him that if he fired another Gun they would give him no Quarter, but Capt. Wyer continued Fighting them till twelve a Clock at Night, when she left the Ship, and so he continued his Course to the Bay of Hundoras where he arrived the first of April last, and the eighth Day he had got on Board about 50 Tuns of Logwood, and the remaining part of the Ship's loaden lay ready cut to be taken on Board.[8]

Unsuccessful in his most recent attack, Bonnet sailed to the prearranged rendezvous spot with Blackbeard at the Turneffe Islands. These Islands are an atoll, which is a group of small islands forming a circle. Today, its crystal-clear waters and beautiful coral reefs make it a paradise for recreational divers. There are even a couple of resorts. In 1718, it was a well-known gathering place for English shipping. A small English settlement had been established on nearby Belize in the late seventeenth century, where they grew and exported valuable logwood. This profitable commodity always attracted English merchantmen anxious to turn a quick profit. Most importantly, the calm water of the five-mile-wide lagoon in the center of the atoll was an ideal spot for vessels to peacefully drop anchor, as the circular shape of the islands themselves offers moderate protection against rough weather. These geographic features made it a well-known rest spot for English merchantmen and it wasn't unusual to see vessels continually coming and going.⁹

The *Revenge* pulled alongside of the *Queen Anne's Revenge* on April 2, 1718. Stede Bonnet climbed aboard and reported his defeat to Blackbeard. Of course, there is no way of knowing exactly what was said, but Johnson's *A General History of the Pyrates* offers a very plausible version of this conversation.

Bonnet loses command of the Revenge, and Richards is put in his place.

> Teach, finding that Bonnet knew nothing of a maritime Life, with the Consent of his own Men, put in another Captain, one Richards, to Command Bonnet's Sloop, and took the Major on aboard his own Ship, telling him, that as he had not been used to the Fatigues and Care of such a Post, it would be better for him to decline it, and live easy and at his Pleasure, in such a Ship as his, where he should not be obliged to perform Duty, but follow his own Inclinations.¹⁰

Although this version may or may not be completely accurate, it most likely captures the gist of their conversation. At any rate, it is certain that Blackbeard had lost all confidence in Bonnet's ability to lead and replaced Bonnet with Richards. This is confirmed by the captain of their next prize, David Herriot. Later in his deposition, Herriot said that "the said Sloop Revenge, then commanded by one Richards . . . that the said Major Stede Bonnet was on board the said Thatch, but out of Command, being some time before turned out of his Command by the said Thatch and the Pirate Crew."¹¹

The sloop *Adventure*, master David Herriot, had left Jamaica on March 22, 1718, and reached the Turneffe Islands on April 4, 1718. Looking for a tranquil spot to anchor, the *Adventure* sailed into the lagoon. Herriot saw a ship and two sloops laying at anchor. At first, he thought the ship to be the *Protestant Caesar* under the command of Captain William Wyer, as he knew that ship was in the area. Thinking that this was a splendid opportunity to socialize with a fellow captain, the *Adventure* sailed close to the ship to anchor nearby. The dynamics of the situation quickly changed when, to Herriot's surprise, the gunports on the ship opened. Thinking the attacking ship was Spanish, Herriot quickly altered course in an attempt to avert danger.

Sloop Adventure, Captain David Herriot, taken by Blackbeard at Turneffe Islands.

The *Queen Anne's Revenge* fired a warning shot across the *Adventure's* bow as the crew of the *Revenge* cut their anchor cable and hoisted their pirate flag. A moment later, the *Revenge* raised sail and closed to block the *Adventure's* escape. A shout then came from the *Revenge* for the *Adventure* to heave to, lower a boat, and send their captain over to the pirate sloop.

Chapter Fourteen

After Herriot was aboard the *Revenge* as a hostage, Captain Richards sent five pirates over to the *Adventure* and anchored it under the stern of the *Queen Anne's Revenge*. Herriot described these events in his deposition, although he spells Captain Wyer's name differently. In Herriot's own words:

> ... and there saw a Ship and two Sloops, which this Deponent first apprehended to be Capt. Wyar, who came out of Jamaica with four other Sloops about a Week before this Deponent, and designed to come to an Anchor there. But soon after he perceiving the said Ship did not belong to the said Wyar, this Deponent took them for Spaniards, and then tacked about, and then the Ship fired a Gun at this Deponent's Sloop; and the said Sloop Revenge, then commanded by one Richards, a Pirate, slipped her Cable, and came up to this Deponent with a Black Flag hoisted, and ordered this Deponent to hoist out his Boat, and come on board them, which he did and then the said Sloop Revenge sent five of their Hands in this Deponent's Boat back again to this Deponent's Sloop, and brought this Deponent's Sloop to an Anchor under the Ship's Stern ... a Ship of forty Guns mounted, named the Queen Anne's Revenge, commanded by one Edward Thatch, a Pirate ... And this Deponent further says That at the time he was taken, as aforesaid, there was another Sloop in their Company, which the said Pirates called their Prize.[12]

Herriot also listed several members of Blackbeard's crew who remained with Bonnet after their partnership dissolved. All of these pirates were on trial in Charles Town seven months later, and David Herriot testified against them. They are:

> Edward Robinson Gunner, Neal Paterson, John Lopez, Job Beely alias Bayly, William Scot, Thomas Nichols, Zachariah Long, Matthew King William Livers alias Evis, Daniel Perry, Henry Virgin William Eddy alias Nedy, James Mullet alias Millet, Thomas Price, and James Wilson, but by reason of their frequent shifting from the said Ship the Queen Anne's Revenge ... cannot say properly to which of them they belonged.[13]

The prize sloop mentioned by Herriot is not mentioned again. It was either released or destroyed. Their next prize was taken a day later, on April 5, 1718. It was the sloop *Land of Promise* from Rhode Island under the command of Thomas Newton. Apparently, Blackbeard was still furious over Bonnet's failure to take the *Protestant Caesar* a week earlier. As Blackbeard

Land of Promise, Captain Thomas Newton, taken by Blackbeard.

looted the *Land of Promise,* he told Captain Newton that he intended to track down the *Protestant Caesar* and burn the ship because they had the audacity to fire upon the pirates. He also didn't want him to return to his home port and brag about defeating a pirate. In late May 1718, Captain Newton finally returned to Boston and gave his account of the incident to *The Boston News-Letter*. Their article reads:

> Arrived here the sloop Land of Promise Thomas Newton Master, who says that about the 5th of April last at the Island Turneff he was taken by Capt. Edward Teach Commander of a Pirate Ship of 40 Guns, and about 300 Men, and a Sloop of 10 Guns. Capt. Teach told Capt. Newton after he had took him, that he was, bound to the Bay of Huadoras to Burn the Ship Protestant Caesar, Commanded by Capt. Wyer who had lately fought the abovesaid Sloop, that Wyer might not brag when he went to New England that he had beat a Pirate.[14]

Meanwhile, the Boston ship that Blackbeard so desperately sought was still at the Bay of Honduras. With the issue of the *Protestant Caesar* pressing on Blackbeard's mind, he sailed from Turneffe on April 7, 1718 and headed south to take one of the few ships that had defied him. He sailed with the *Queen Anne's Revenge*, the *Revenge*, the *Adventure*, and the *Land of Promise*. On the way, they sighted a sloop in the distance. Blackbeard altered their course and gave chase. Shortly after the sun went down, the *Revenge* finally caught up to their next prize, the *Dolphin*, under command of Captain James Burchett.

As the two sloops came within hailing distance, Burchett shouted, "Whence came you?" Captain Richards replied, "From the sea." Then, two muskets fired and Richards ordered the *Dolphin* to lower a boat and to send their captain over. The pirates now had the sloop. A short time later, the fierce and imposing pirate ship, *Queen Anne's Revenge* came alongside, dwarfing the much smaller *Dolphin*. The prize crew were ordered to come aboard the ship, where they got their first sight of Blackbeard. The pirates took complete control of the *Dolphin* and added it to their fleet. After keeping the *Dolphin* for a few days, it was released. They sailed to Jamaica, where one of the passengers, Martin Preston, made a statement to Captain Thomas Jacob of HMS *Diamond*, which he notated in a letter to the Admiralty. It reads:

> There see in the Offing a Ship & two Sloops, they Chased and in the Evening one of the Sloops came up with us, Wee hailed and asked from whence they came, there reply was from Sea, then they fired two small Arms, and Orderd our boat to come on board, then the

Chapter Fourteen

men that had Possession of the Said Dolphin Sloop, Enterd themselves on board the Ship Comand'd by Edward Thatch, then the said Thatch sent several men in Lew, and carried the Dolphin Sloop along with them to the bay of Hinduras.[15]

Meanwhile, the *Protestant Caesar* was at the bay taking on a cargo of logwood. The pirates came to rest at Water Key, a small island about 20 miles to the southwest of the last reported position of the Boston ship that had eluded Bonnet. The next morning, they weighed anchor and sailed toward Honduras Bay. As they approached, they could see four sloops and a ship lying at anchor. Herriot commented on this saying, "Where there lay four Sloops, and a Ship named the Protestant Caesar."[16] Captain Burchett supports this in his statement saying, "There was lying there a New England Ship and four Sloops, from Jamaica."[17] As they came nearer, Blackbeard came to realize that his search had come to a quick end. He had found the *Protestant Caesar*, which would be the first vessel he took.

Protestant Caesar, Captain William Wyer, taken by Blackbeard.

The *Queen Anne's Revenge* and the *Revenge* moved in for the attack. They had already hoisted the black flags with death's heads. His other three sloops, *Adventure*, *Land of Promise*, and *Dolphin*, were flying red flags, making their fleet appear very imposing. As the pirates grew closer, the crew of the *Protestant Caesar* had to decide whether to risk death in a battle or abandon ship. However, Blackbeard's vessels were still too far away to be identified by Captain Wyer. The crew said they would fight if they were Spanish, but if they were pirates, they wouldn't. To help them decide the matter, Captain Wyer sent a pinnace out toward the approaching vessels to make a positive identification. To the horror of the men on the pinnace, they soon came to realize that not only were these pirate vessels, one of the sloops was the exact same sloop that they had fought twelve days earlier.

Panic ensued aboard the *Protestant Caesar* when the pinnace returned with the news. The crew believed Bonnet's threat to give no quarter when last they fought. They were certain that if they surrendered, that they would be murdered. As the *Queen Anne's Revenge* came within range, a warning shot was fired from one of the ship's guns, and every last man quickly jumped overboard and went ashore. This account also appeared in *The Boston News-Letter* edition of Monday, June 9 to Monday, June 16, 1718, which reads:

> Morning of the said Day, a large Ship and a Sloop with Black Flags and Deaths Heads in them and three more Sloops with Bloody Flags all bore down upon the said Ship Protestant Caesar, and Capt. Wyer judging them to be Pirates, call'd his Officers and Men upon Deck

asking them if they would stand by him and defend the Ship, they answered, if they were Spaniards they would stand by him as long as they had Life, but if they were Pirates they would not Fight, and thereupon Capt. Wyer sent out his second Mate with his Pinnace to discover who they were and finding the ship had 40 Guns 300 Men called the Queen Ann's Revenge, Commanded by Edward Teach a Pirate, and they found the Sloop was the same that they Fought the 28th of March last, Capt. Wyers Men all declared they would not Fight and quitted the Ship believing they would be Murthered by the Sloops company, and so all went on shore.[18]

Herriot confirms this in his deposition saying, "That one of the said Sloops came to descry what they were, and took said Thatch and Richards for Spaniards; but said Thatch fired a Gun, and hoisted his Black Flag. Whereupon Capt. Wyar and all his Men took to their Boat, and went ashore." He adds that after the ship was abandoned that "Thatch sent one Howard, his Quartermaster, and eight of his Crew, on board of Wyar's Ship,"[19] presumably to loot it.

What about the other four sloops that were laying at anchor nearby? Blackbeard took them, too. Three of those sloops belonged to Johnathan Bernard of Jamaica. He was actually the captain of one of them. In Herriot's depositions, he said, "That he knows not the Sloops Names; but three of them were commanded by Jonathan Bernard of Jamaica, Master of one of them, and Owner of three of the said four Sloops."[20]

One of Bernard's sloops was the *William and Samuel*, Captain William Wade. They had been at Roatán taking on logwood since February. They watched Blackbeard take the *Protestant Caesar* and then turn toward them. After the sloop's release, they sailed to Jamaica, where Captain William Wade made a statement to Captain Thomas Jacob of HMS *Diamond*, which he notated in a letter to the admiralty. It reads:

That he this deponent put into the Island of Rattan in his way from the bay of Hinduras in the beginning of Febuary past . . . Where came in a Ship of about 40 Guns & a Sloop of 10 Comand'd by Pyratts, the Masters name was Edward Thatch having in all about 250 Men (70 or thereabouts of which were Negroes) . . . they orderd this deponent to throw out his Logwood, and made use of his Sloop with the rest of the Vessells to creen their Ship and Sloop.[21]

On April 12, 1718, as Blackbeard was taking those sloops, another sloop named the *Revenge*, out of Jamaica and under the command of Captain

Chapter Fourteen

William Megerre, entered the bay. Captain Megerre had been there before, to take on logwood. Totally unaware that the bay was infested with pirates, he casually sailed toward his usual anchorage. In the distance, he could see eight sloops and two ships. He immediately recognized one of the ships to be the *Protestant Caesar*, as Captain Wyer and his Boston ship were well known in these waters. However, Captain Megerre was a bit puzzled as to the identity of the other vessels, as they weren't flying any flags. The two ships were anchored close to each other. Captain Megerre lowered a dory and sent it over to the *Protestant Caesar* to offer greetings, but as it came near the ships, the men in the dory were shocked to see a pirate flag being raised on the mast of the unidentified ship. The flag was accompanied by a musket shot.

Captain Megerre immediately assessed the truth of the situation and turned about to head for open water. As he raced out of the bay, he could see a fast-moving pirate sloop gliding through the water to intercept him. This was Captain Richards on the sloop *Revenge*. Both sloops were equally matched for speed. They were also both named *Revenge*. Richards chased Megerre for hours until he finally lost him in the darkness of night. After their escape, they sailed to Jamaica, where Captain William Megerre made a statement to Captain Thomas Jacob of HMS *Diamond*, which he notated in a letter to the Admiralty. It reads:

> He Arrived about the 12 April but Standing into the Customary place of Anchoring, found there riding 8 Sail of Sloops and 2 large Ships, one of which this deponent veryly belives, was belonging to New England Comanded by one Weir they not shewing any Collours he sent his dorey with three Men to acquaint them who he was ... but when he this deponent was got very near them, the other Ship hoisted Black Colours, and fired a Shot; he then (this deponent) stood out again for the Sea, but imediatly was followed by a Sloop of about 10 Guns who hoisted black Colours and fired several Guns at him; She Continued to chase till Night; when he lost her, having been obliged to cut away a large Canoe he had in Tow to sail the better, he then made the best of his way to this place.[22]

Meanwhile, Captain Wyer and his crew were still stranded on the island. Contrary to the evil way in which Blackbeard is usually portrayed, he sent a message to Wyer politely asking him to come aboard the *Queen Anne's Revenge* and talk. He followed the invitation with a promise that if the captain complies, that he will be treated civilly.[23]

Blackbeard's reputation was fairly well-established. He never went back on his word, and he had never killed anyone.[24] Captain Wyer agreed to comply with Blackbeard's offer and accompanied the crewman who brought him the message, probably William Howard, back to Blackbeard's ship. Their conversation was most polite. Blackbeard apologetically explained that he had to burn his ship because it was from Boston. Since some of Bellamy's pirates had been executed there, his sense of vengeance required all ships from Boston to be burned. David Herriot supports this in his deposition saying, "Richards set fire to Capt. Wyar's Ship, because she belonged to Boston, alleging the People of Boston had hanged some of the Pirates, and so burnt her."[25]

Blackbeard also promised not to harm any of Wyer's crew and that he would allow all of them to leave aboard the *Land of Promise* along with Captain Thomas Newton. Blackbeard kept his word. Captain Wyer and his crew sailed away aboard the *Land of Promise*, watching his ship burn as they departed. They safely arrived in Boston on May 31, 1718. Once again, *The Boston News-Letter* edition of Monday, June 9 to Monday, June 16, 1718, reads:

> And on the 11th of April three Days after Capt. Wyer's ship Protestant Caesar was taken, Capt. Teach the Pirate sent word on shore to Capt. Wyer that if he came on Board he would do him no hurt, accordingly he went on Board Teach's Ship, who told him he was glad that he left his ship, else his Men on Board his Sloop would have done him Damage for Fighting with them; and said he would burn his Ship because she belonged to Boston, adding he would burn all Vessels belonging to New England for Executing the six Pirates at Boston. And on the 12th of the said April Capt. Wyer saw the Pirates go on Board of his Ship, who set her on Fire and Burnt her with her Wood, and Capt. Wyer took his passage hither in Capt. Thomas Newton's Sloop, taken from him by Capt. Teach the Pirate unto whom he gave back his Sloop again, because she belonged to Rhode-Island.[26]

In addition to the *Protestant Caesar*, one of the captured sloops was burned. There is some confusion about the name of the captain. David Herriot said his name was Barnes, while Martin Preston said his name was James, from Port Royal. Since Preston was in Port Royal when he made that statement to Captain Jacob of HMS *Diamond*, most likely his name was James. Blackbeard kept the sloop *Adventure* and used the other sloops to help careen it, as well as *Queen Anne's Revenge* and the *Revenge*. The other sloops, *Land of Promise*, *William and Samuel*, *Dolphin*, and the other two sloops belonging to Jonathan Bernard, were released. In his deposition, David Herriot said:

> The said Thatch burnt one of the four Sloops, because she belonged to Capt. Barnes of Jamaica; which Barnes, as 'twas alleged, had said he would not employ those Sailors in his Service that had accepted of the King's Proclamation; and the other three Sloops, belonging to Bernard, they let go.[27]

As mentioned earlier in the last chapter, the king's proclamation offered a full pardon for all pirates. Those who accepted the pardon expected to be hired as honest seamen. Apparently, Captain Barnes still considered them to be pirates and refused to take them on as crewmen. That position angered Blackbeard and caused him to burn Barnes' sloop.

Captain William Wade of the *William and Samuel* said:

> [They] made use of his Sloop with the rest of the Vessels to creen their Ship and Sloop . . . they burnt the Ship and Sloop . . . them, and discharged this deponent & his Sloop.[28]

Captain James Burchett of the *Dolphin* said:

> Then they burnt the New England Ship, & a Sloop that belonged to Capt. James of Port Royall, then Sailed in company with them Two Sloops.[29]

Finished with the Gulf of Honduras, Blackbeard set sail for the Turks Islands aboard the *Queen Anne's Revenge*, with Captain Richards on the *Revenge*, and the *Adventure*. They headed west between Cuba and Jamaica to the Cayman Islands, perhaps hoping to pick up a few more prizes. As he passed the Caymans, he captured a small boat that was hunting for turtles. After taking on some fresh turtle meat, they continued around the west end of Cuba. Herriot's deposition says they "went to Turckcill Island, and from thence to the Grand Camania . . . where they took a small Turtler."[30]

Somewhere along the coast of Cuba or within the Florida Straits, Blackbeard took a Spanish sloop and kept it as a tender. According to historian author Kevin Duffus, it was 65 feet in length, probably Bermuda-built, and was "nimble, tight, and fair."[31] Prime source documentation for this capture can be found within the records of the trial of Stede Bonnet. The charges against Bonnet and his pirates describe the vessels that laid siege to Charles Town in May 1718 as "a large Ship mounted with forty Guns, their former

Sloop the *Revenge*, which was now called their *Privateer*, and two other Sloops, Prizes, which served them as Tenders."[32] Herriot also mentions this fourth vessel in that same account, saying that Blackbeard "set sail from Topsail-Inlet in the small Spanish Sloop."[33]

After such a long voyage, it was time for Blackbeard to put into a pirate-friendly port. There was only one such place, Nassau.

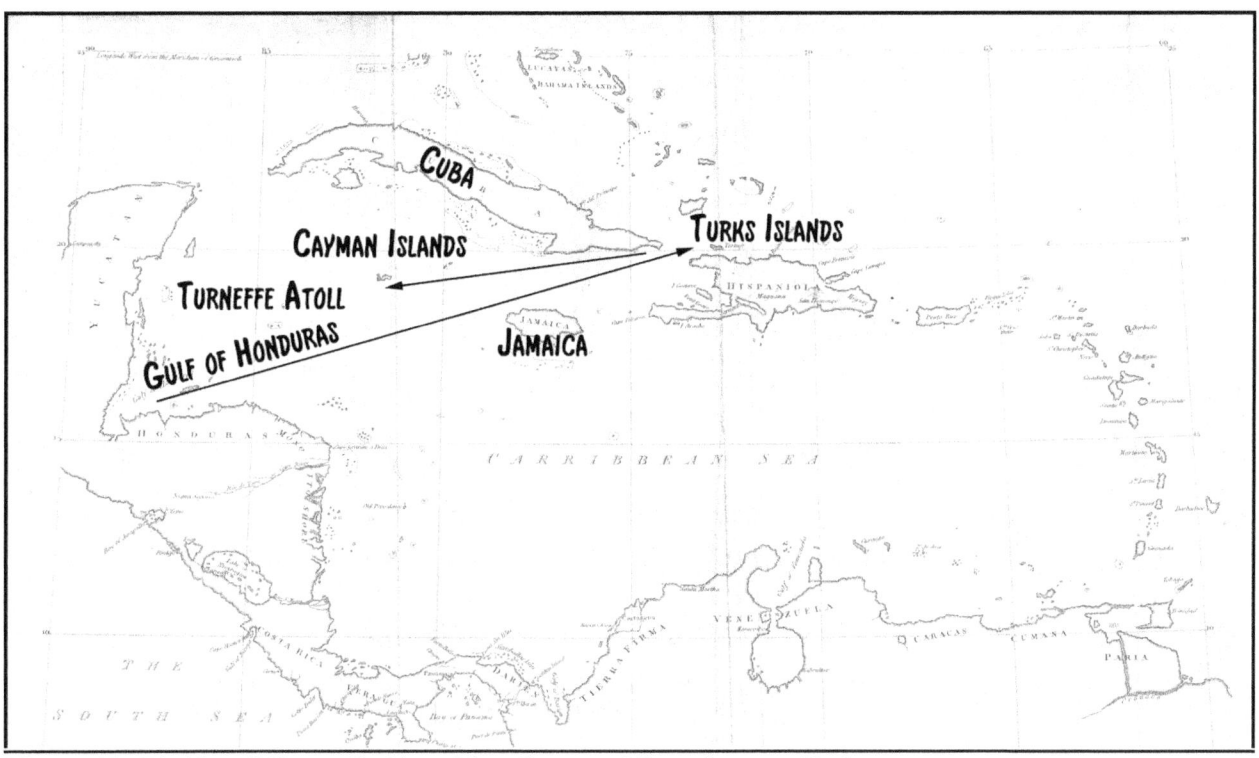

Figure 33: *Blackbeard's Route, Caribbean Map, Courtesy of Barry Lawrence Rudman*

Fifteen
Nassau and the King's Proclamation

"We came from the Bay of Honduras, and from thence to Providence."[1] This statement was made by Ignatius Pell at his trial in November 1718 at Charles Town. Pell was a former member of Blackbeard's crew who left Blackbeard to sail with Stede Bonnet and was subsequently arrested. Blackbeard's fleet of four vessels had just left the Caribbean. They sailed along the Florida coast for a short time, then turned east. Blackbeard was heading to his old home port of Nassau. However, Nassau was far different from the pirate town he had left nine months earlier.

King George I had recently issued his *Proclamation for Suppressing of Pirates*. It was often referred to as the king's proclamation, the king's pardon, or the act of grace. It was signed on September 5, 1717, but news of this proclamation wasn't widely circulated in North America until December, when *The Boston News-Letter* printed the entire proclamation in its December 2 to December 9 issue.[2]

With this one document, everything changed for English pirates throughout the world. The proclamation offered a full pardon for pirates who surrendered themselves to a governor before September 5, 1718, but only if they hadn't committed any acts of piracy after January 5, 1718. All pirates had to make a choice. They could either accept the king's pardon or risk being arrested as a pirate. The following excerpts are from this proclamation:[3]

> And We do hereby Promise and Declare, That in case any of the said Pirates shall, on or before the Fifth Day of September, in the Year of our Lord One thousand seven hundred and eighteen, 1 Surrender

him or themselves to One of Our Principal Secretaries of State in Great Britain or Ireland, or to any Governor or Deputy-Governor of any of Our Plantations or Dominions beyond the Seas, every such Pirate and Pirates, so Surrendring him or themselves, as aforesaid, shall have Our Gracious Pardon of and for such his or their Piracy or Piracies, by him or them Committed before the Fifth Day of January next.

Pirates who didn't accept the king's offer were to be arrested and tried for piracy. Civil, naval, and military officers were encouraged to arrest the pirates and would receive substantial rewards for their capture. Once the pirate was tried and convicted, the apprehender would receive a reward commensurate with the pirate's rank or status in the crew. The pardon specified the amounts as £100 for a captain, £40 for every senior officer, £30 for every junior officer, and £20 for every common sailor.

To Seize and Take such of the Pirates who shall refuse or neglect to Surrender themselves accordingly ... so as they may be brought to Justice, and Convicted of the said Offence... Such Person or Persons, so making such Discovery or Seizure, ... shall have and receive as a Reward for the same, viz. For every Commander of any Pirate-Ship or Vessel the Sum of One hundred Pounds; For every Lieutenant, Master, Boatswain, Carpenter, and Gunner, the Sum of Forty Pounds; For every Inferior Officer the Sum of Thirty Pounds; And for every Private Man the Sum of Twenty Pounds.

This delightful news quickly reached the governor of Bermuda, Benjamin Bennett, who had long been concerned about the pirate base at Nassau. Now it looked as though they would be effectively wiped out. Bennett immediately sent a fast sloop to Nassau with a copy of the proclamation.[4] Meanwhile, Ben Hornigold had been at sea, out of touch with the world, as news of the pardon circulated. He returned to Nassau in January 1718 with two Dutch merchant ships he had recently taken. One was armed with 26 guns and the other with 30 guns.[5]

As he dropped anchor, Hornigold was shocked to find a divided community where chaos and confusion abounded. Some pirates were in favor of accepting the pardon while others were bitterly opposed. Charles Johnson even mentions this in his book, *A General History of the Pyrates*, by writing that the pirates "called for a general Council, but there was so much Noise and Clamour, that nothing could be agreed on."[6]

Leigh Ashworth was captain of the *Mary*, sailing with Henry Jennings when Sam Bellamy escaped with 28,500 pieces of eight.

Chapter Fifteen

This was the Nassau that Hornigold had returned to. As he stepped ashore, Hornigold must have been immediately surrounded by frantic pirates and townsfolk alike, anxious to find out where he stood on this life-changing issue. The situation demanded that Hornigold choose quickly. Without hesitation, he sided with Henry Jennings, Richard Nolan, Josiah Burgess, Jean Bondavias, and Sam Bellamy's former quartermaster, Leigh Ashworth, who all supported the king and had agreed to accept the pardon. As a visible sign of this support, they went to the old fort that overlooked the entrance to the harbor and raised the British Union Jack atop the fort's tower.[7]

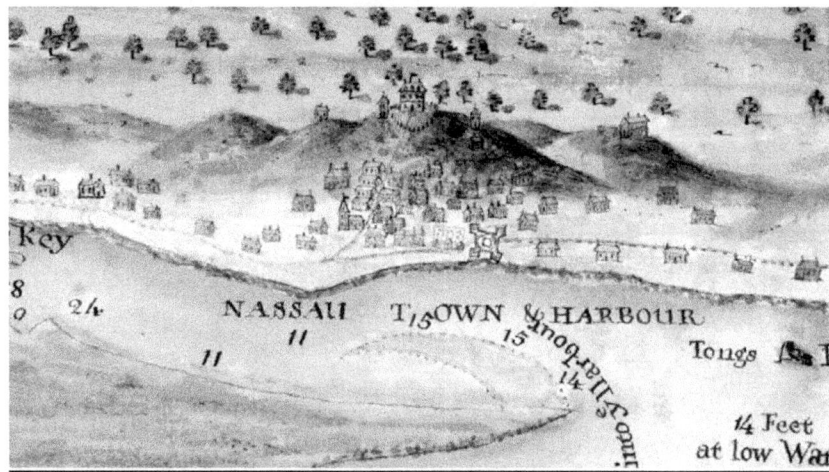

Figure 34: *Illustration of Nassau. The image is dedicated to the public domain under CC0.*

A group of Jacobite pirates, including Palsgrave Williams, Christopher Winter, Nicholas Brown, Edward England, Edmund Condent, Jack Rackham, and Charles Vane, objected to the king's proclamation. They were naturally opposed to King George I on every issue. For them, it was less about piracy and more about politics. Charles Vane assumed the role of leader and organized about one hundred armed pirates into an angry mob right in the middle of the town square. When they were ready, Vane led the irate pirates in an assault on the fort. After overpowering the minimal resistance offered by the fort's defenders, Vane and his followers took control, pulled down the British flag, and hoisted a "Black Flag with the Death's Head."[8]

Jacobite Charles Vane opposes the British navy

The situation in Nassau rapidly deteriorated into a total breakdown of any sense of order and began to take on the appearance of a civil war. Hornigold himself, fearing that the situation would erupt into a full-fledged battle, sent word to the naval authorities at Jamaica and requested that they send a warship immediately. However, this was short-lived. The anti-pardon pirates soon realized that it wouldn't be long before the royal navy arrived and that they couldn't fight the British government and their old comrades at the same time. Many Jacobite pirates fled, including Edmund Condent aboard his sloop *Dragon*. However, they weren't the only ones to leave. Many colonists left for their protection sailing to ports throughout North America.[9] Even Henry Jennings fled the chaos. Boarding his sloop *Bersheba*, he sailed directly to Bermuda to accept the pardon without having to fight his friends or wait for the arrival of a naval vessel.[10]

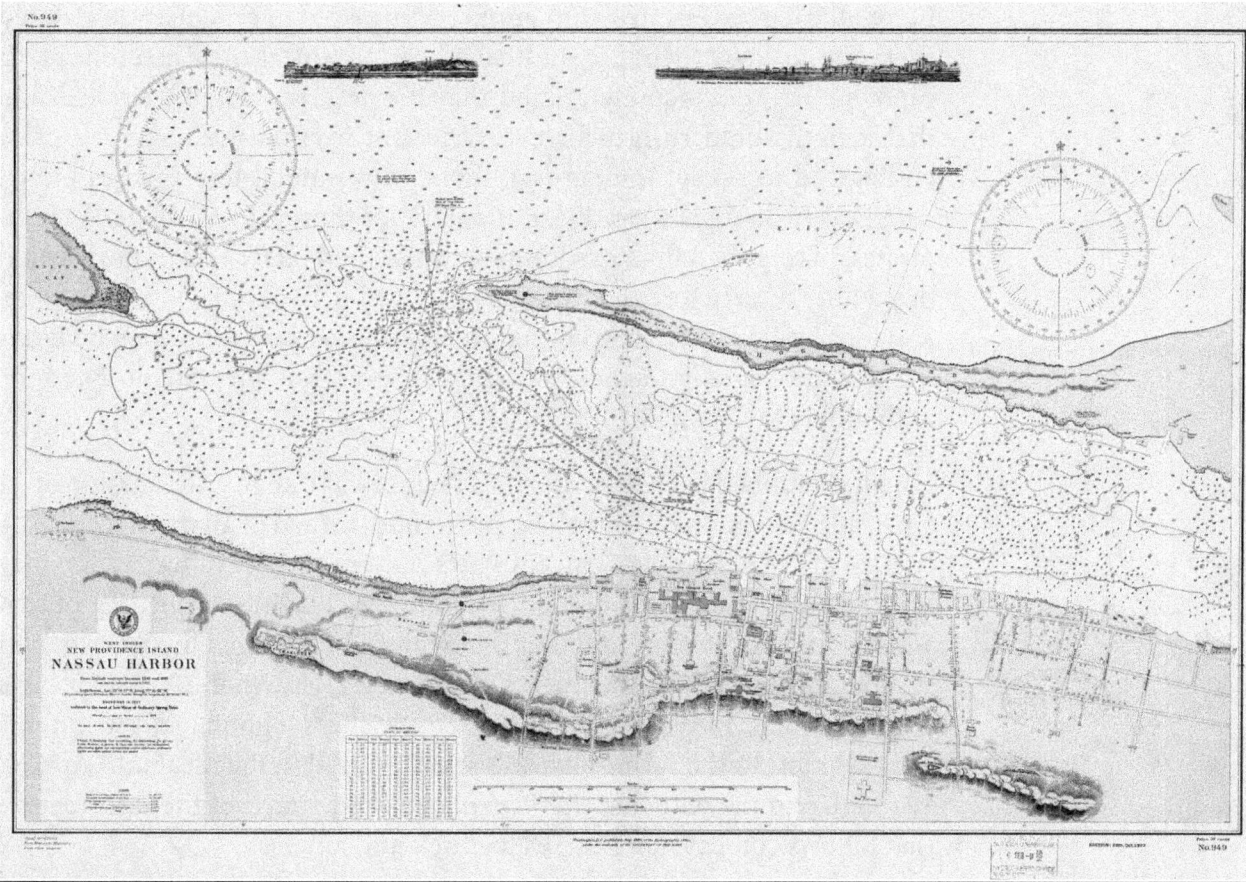

Figure 35: *Nassau Harbour New Providence Bahamas US Navy chart town plan 1885 (1924) map. Image shot 1924. Exact date unknown. Licensed from Almay.*

Charles Vane didn't have any vessel of his own at the time. He had made so much money raiding the 1715 Spanish wrecks with Jennings that he had been living in Nassau for the past two years without the need to go to sea.[11] Now, the need caught up to him. Vane, with the assistance of sixteen pirate friends, seized the merchant sloop that happened to be lying at anchor in the harbor. They quickly converted it into a pirate vessel.[12]

Nassau was a tinderbox waiting to explode into a violent conflict. The fort overlooking the harbor was still flying the pirate flags that Vane and his men had raised. There were fourteen vessels in the harbor, some flying British flags and others flying black or red pirate flags. The two Dutch ships that Hornigold had brought into the harbor over one month earlier had been seized by the anti-government faction and were now flying either black or red pirate flags atop their masts. To add to the danger, the guns on the Dutch ships had been run out and were ready to fire.[13] There were also two other ships that the pirates had seized. One from Bristol, and the other, a French ship of 30 guns.[14] Suddenly, without warning, on February 23, 1718, a single ship quietly entered the harbor that commanded everyone's attention.[15] It was HMS *Phenix*. The Royal Navy had arrived.

Chapter Fifteen

Before the harbor erupted with gunfire, Hornigold intervened. He ordered everyone to hold their fire and called for an emergency meeting of all captains. He convincingly argued that if even a single gun opened fire, that action would ruin it for everyone, but if everyone remained calm and agreed to accept the pardon, those who wanted to return to piracy could easily do so afterward. Additionally, for those who wished to remain pirates, taking the king's pardon now while the *Phenix* was in port would buy a little time before the royal navy arrived with a much stronger force. As Hornigold worked to keep the situation under control, a long boat flying a white flag approached. Aboard this boat was Lieutenant Symonds with orders to negotiate.[16]

Usually Stationed at New York, HMS Phenix was a sixth-rate warship of 24 guns.[17] This is among the smallest and lightest armed of all naval ships. The captain of the *Phenix* was Vincent Pearse, a very capable officer.[18] One may have already noted the spelling. Generally, the traditional spelling of that word would be *Phoenix,* and that is how the name of this ship is written in several contemporary documents and articles. However, Captain Pearse spells the name of his ship as *Phenix* on the ship's logbook and within his official letters to the Admiralty. The *Phenix* stayed in the area until April 6, 1718.[19] Up to the date that the *Phenix* departed, much of the information that follows comes from Pearse's logbook and letters.

Vincent Pearse (1679–1745) was promoted captain and assigned HMS Phenix on May 10, 1715. He served with distinction until 1741.

Lt. Symonds was met by Hornigold, and in the ship's logbook, Pearse wrote that he "was received by a great number of Pirates with much Civility." As he read the proclamation, the pirates took it with a "great deal of joy." Hornigold warned Lt. Symonds that Charles Vane would be the greatest problem and told Symonds where to find him. Vane had taken his recently captured sloop to Bushes Cay.[20]

As soon as Symonds returned to the *Phenix* and made his report, Captain Pearse ordered his ship to set sail. A short time later, Pearse found Vane's sloop laying at anchor right where Hornigold said it was and carefully positioned the *Phenix* to block Vane's sloop from escaping. Without waiting for any explanation from the pirates as to why they were hiding there, Pearse ordered his gun crews to open fire. A moment later, solid shot rounds splashed all around Vane's helpless sloop. Vane raised a white flag and sent his quartermaster, Edward England, over to the *Phenix* aboard a rowboat. The pirates desperately explained that they were just about to set sail for Nassau to accept the pardon when the *Phenix* arrived and opened fire. Captain Pearse didn't believe a word of it and arrested Vane and his crew on the spot.[21]

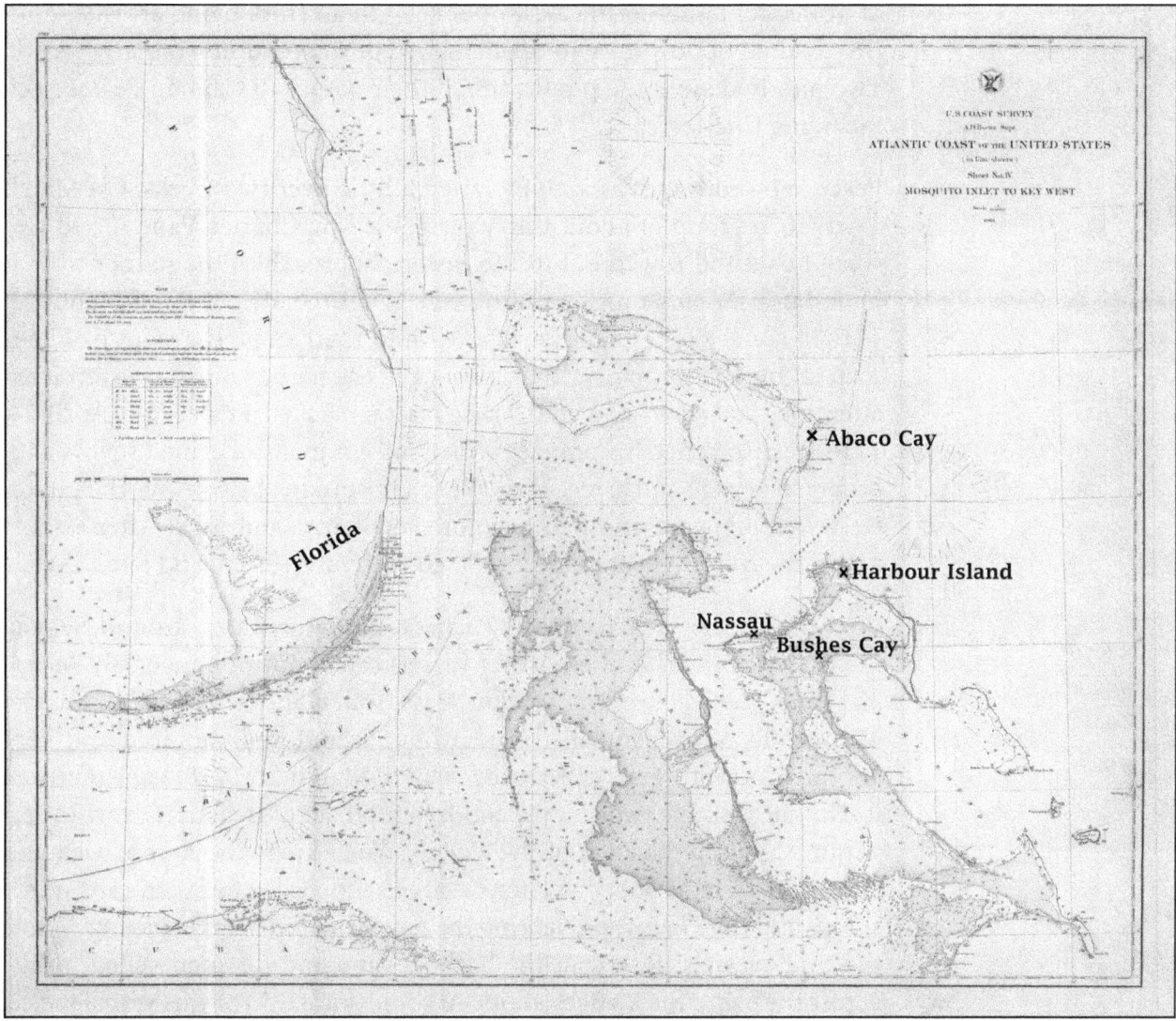

Figure 36: *Bahamas and South Florida, Courtesy of Barry Lawrence Rudman*

Hornigold once again proved himself to be the leader of the community. When the *Phenix* arrived back at Nassau with Vane's sloop in tow, many of the pirates feared that Vane and his men would be executed and that they might be next. Hornigold led a delegation that rowed over to the *Phenix* and explained their fears to Captain Pearse. A logbook entry from the *Phenix* made by Pearse reads that they "informed me that my taking of the sloop had very much alarm all, the Pyrates in general believing that men taken in her would be executed."[22]

Hornigold convinced Pearse that if he allowed Vane and his men to accept the pardon, just about everyone else would, too. Between February 26 and February 28, 1718, all the former pirates of Nassau rowed over to the *Phenix* in small groups to accept the pardon. I'm sure that some of them went willingly while others went reluctantly. As each man climbed aboard, they signed Captain Pearse's list and were then given a certificate

Chapter Fifteen

of protection. This certificate would keep them from being arrested until they had an opportunity to officially accept the pardon from a governor. Two hundred and nine pirates added their names to Captain Pearse's list, including Charles Vane.[23]

Peace and order in Nassau only lasted a little over two weeks. On March 18, 1718, a group of about thirty pirates led by Charles Vane seized two small boats and rowed out to the eastern approach of the harbor to wait in ambush for some unsuspecting vessel. On the afternoon of March 22, 1718, a small sloop from Jamaica fell into their trap. The sight of Vane's men taking the sloop near the entrance of the harbor must have attracted attention, and news of Vane's latest piracy reached Pearse. On April 24, 1718, he dispatched his pinnace with a full company of armed sailors with orders to engage and arrest Vane. But that wouldn't be so easy. The pirates were well prepared and had far more small arms onboard. After a short battle, the royal navy pinnace retreated and Vane set sail and left Nassau.

Over the next several weeks, the *Phenix* cruised the waters around Nassau, looking for pirates and escorting vessels safely out of danger. By March 30, 1718, the *Phenix* was low on fresh water and returned to Nassau. Meanwhile, Vane had sailed to Harbour Island and captured and kept another sloop of 9 guns called the *Lark*. On March 31, 1718, Vane returned to Nassau with his two sloops, and in a total act of defiance, came to rest within shouting distance of the *Phenix*. Vane knew the harbor well and chose the perfect spot to anchor. A large sandbar lay between the pirates' sloops and the *Phenix*, preventing the navy warship from engaging. While the three vessels floated in the harbor, Vane and his men yelled insults across the bar to the British sailors. At some point during this verbal melee, three of Captain Pearse's men decided that life as a pirate was better than life as a British sailor and quietly slipped over the side and joined Vane's crew. Finally, on April 3, 1718, Vane became tired of the game, sailed out of Nassau's harbor, and then sailed southward. The *Phenix* left three days later, on April 6, 1718. The following are excerpts of Captain Pearse's reports to the Admiralty.[24]

> For on the 18th & 19th March last in the Night, two Gangs of them being upwards of thirty in Number, went away with two Boats from the Towne of Nassau and on the 22nd in the afternoon they took a small sloop from Jamaica that was comeing into the Eastward upon information thereof I that Night sent my Pinnace Man'd & arm'd to attack them on board the said sloop, but they being too strong she was oblig'd after exchanging several small shot with them, to returne aboard.

Nassau and the King's Proclamation

> Cruis'd off the harbour . . . But I being then short of water I put into Providence, the day following the said Pirate Sloop who had been out since the 24th Inst. March came to an anchor at the Easternmost part of the harbour, with 'an other called the Lark, which they had taken three days before off Harbour Island, onboard of which they had Removed themselves, she being a good Sailer & Equipt with 9 Guns & 8 Patraroes, there was a Barr between me & them on which was no more than eight foot water, which made them very Insolent.
>
> The said Pirate Sloop was Commanded by one Cha. Veine & Man'd with 45 Men, three of which was of my own Ships Company that deserted me & took on with them. She sail'd from the Backside of providence on the 3rd April, & stood to the Southward, I remain'd there till the 6th and then Sail'd.

One month after HMS *Phenix* left Nassau, the population was still reeling from the recent events. Everyone's future was now uncertain. Every tavern was filled with discussions about what to do next. Many of the pirates had willingly accepted the pardon in hopes of becoming a privateer in the upcoming war with Spain, while others had accepted knowing that they fully intended to return to piracy at the first opportunity. Each pirate was confronted with the reality that they had to make a serious career choice. However, this news not only affected the pirates, it was of great concern to the many business owners who had greatly profited from those pirates over the past few years. Tavern owners faced the prospect of losing patrons as the vast number of pirates were replaced by a modest number of honest sailors. Brothel owners faced the prospect of being completely shut down. To make matters worse, everyone knew that Nassau's newly appointed governor, Woodes Rogers, was due to arrive sometime in the summer with enough ships to enforce the will of King George I. Their pirate kingdom was finished!

Amidst the uncertainty and confusion, a familiar face quietly sailed into port. One can only imagine the excitement among the population of the town of Nassau when Blackbeard's large ship and three sloops came to a stop in the harbor and dropped anchor. Everyone recognized him immediately. After all, a tall and imposing man with a long black beard stands out among his peers. Additionally, Blackbeard had called Nassau his home port for over two years and was a regular in Nassau's taverns and shops.[25] He had only been gone for nine months. But it wasn't only Blackbeard; he had brought about seven hundred men into the community.[26] Imagine what that did for the town's economy.

Woodes Rogers was a highly successful privateer during the War of the Spanish Succession.

149

Chapter Fifteen

There is no contemporary documentation of a meeting between Blackbeard and his old partner, Ben Hornigold, but they must have met. They were both present in a town that was perhaps only one mile wide, and Hornigold had assumed the role of the town's leader prior to the arrival of the new governor, Woodes Rogers. The news of the arrival of Blackbeard's vessels with seven hundred pirates would have commanded Hornigold's attention. I like to think that their meeting was not only friendly but somewhat touching. They had been close friends and partners. They must have had a few drinks at their favorite tavern and reminisced about old times. At some point, Hornigold would have tactfully altered their conversation from the past to the future. He must have asked Blackbeard what his intentions were. After HMS *Phenix* landed the previous February, Hornigold had become a pirate hunter. He must have wondered if he would be pitted against his old friend.

Thinking back to December 5, 1717, Blackbeard recalled his captive, Henry Bostock, captain of the *Margaret*, mentioning the new proclamation offering a pardon for pirates.[27] Now he was able to get the full details. Hornigold and Blackbeard must have discussed this proclamation and what Blackbeard was going to do about it. Hornigold's new quartermaster, John Martin, was certainly in on the discussion, as well as Blackbeard's current quartermaster, William Howard. As previously mentioned, Howard had served as Hornigold's quartermaster until joining Blackbeard's crew in September 1717. Obviously, these four men had known each other for a long time.[28]

A few days after Blackbeard dropped anchor at Nassau, Charles Vane returned to port. As we have already seen, Vane and Hornigold were on opposite sides of the debate over accepting the king's pardon. Vane had already returned to piracy, while Hornigold was destined to become a pirate hunter. However, Blackbeard and Vane were old friends, and they were delighted to see each other.[29] They both shared Jacobite sympathies.[30]

There is no contemporary documentation of a meeting between Blackbeard and Vane, but it must have happened. However, one must wonder what Hornigold's reaction was as he watched his chief adversary greet his former partner. The historical record remains silent.

There is no way of telling how many of Blackbeard's pirates remained in Nassau or if he picked up a few new ones. Reports of the total size of his force widely differ from one account to another. This is understandable, as it would be impossible to estimate the number of pirates on board two or three vessels accurately, especially from a distance. A report made by

Captain Thomas Jacob of HMS *Diamond* noted that had about 250 to 270 men or thereabouts near the end of his Gulf of Honduras cruise.[31] After leaving Nassau, Captain Ellis Brand of the Royal Navy estimated his total strength to be about three hundred and twenty men. In a letter dated July 12, 1717, Brand wrote, "When they first came on the coast there number consisted of three hundred and twentie."[32] If these two accounts are even remotely accurate, it would appear that he picked up a few new crewmen in Nassau.

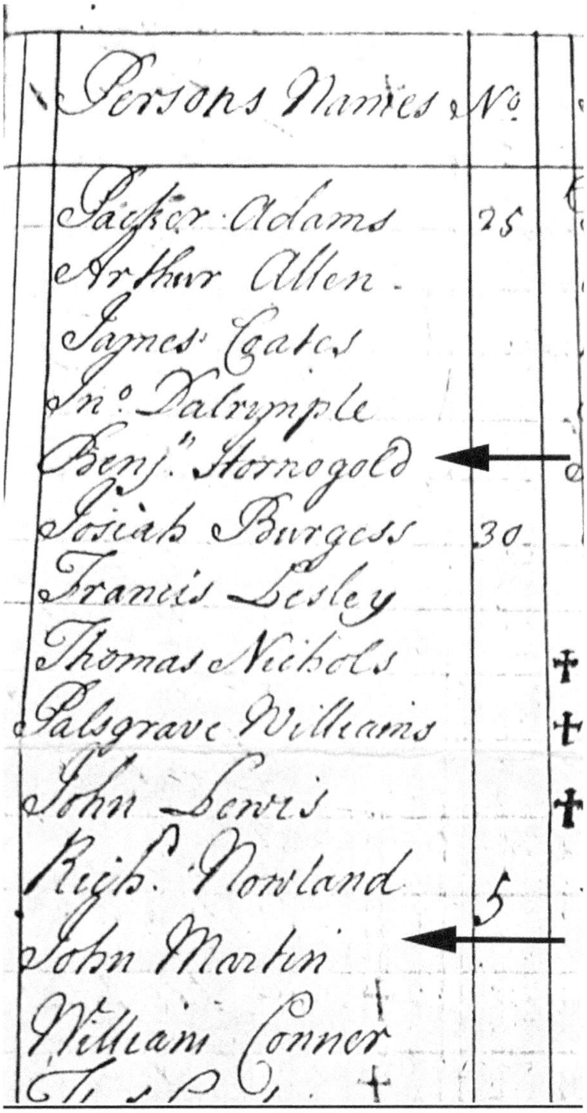
Persons Names	No
Packer Adams	25
Arthur Allen	
James Coates	
Jn.o Dalrymple	
Benj.n Hornogold	
Josiah Burgess	30
Francis Lesley	
Thomas Nichols	
Palsgrave Williams	
John Lewis	
Rich.d Nowland	5
John Martin	
William Connor	

Figure 37: *Excerpt from Phenix Log Showing Hornigold and Martin.*

Some historian authors do not believe that Blackbeard stopped at Nassau after leaving the Caribbean. They contend that his four-ship fleet sailed north along the Florida coast and directly to Charles Town. However, they have overlooked the testimony of Ignatius Pell, mentioned at the start of this chapter. Additionally, there is strong circumstantial evidence that is quite compelling. This evidence comes in the form of Hornigold's former quartermaster, John Martin. When Blackbeard partnered with Stede Bonnet and left Nassau for the Jersey Capes, John Martin stayed with the captain, Ben Hornigold. Martin was still with his captain at Nassau in February of 1718 when the *Phenix* arrived. This is evident from Capt. Pearse's list of those who accepted the pardon. Benjamin Hornigold is the fifth name on the list and John Martin is the twelfth.[33] It is thought-provoking to visualize Martin standing in line just seven places behind his captain.

John Martin was among Blackbeard's pirates who were arrested in Bath in November of 1718.[34] As you will read in a future chapter, when Blackbeard sailed to Ocracoke in November of 1718, six of his pirates remained in Bath. After Blackbeard was killed, those six pirates, including John Martin, were arrested. How did Martin get from Nassau to Bath? He might have traveled to Bath on his own sometime in the summer of 1718; but if that were the case, he wouldn't have been identified as one of Blackbeard's men and wouldn't have been arrested. Blackbeard and his pirates had been in Bath since July of 1718 and were fairly well known to the town's folk. Because John Martin had arrived at Bath along with Blackbeard, he was identified as a member of his crew. The only explanation that makes any sense is that

Chapter Fifteen

Martin boarded one of Blackbeard's vessels when they stopped at Nassau in May of 1718 and then accompanied Blackbeard to Bath.

But their relationships go far deeper than just shipmates sailing aboard Hornigold's vessels. John Martin and William Howard knew Blackbeard long before they became pirates. As you shall read later, all three were originally from Bath.[35] John Martin was the son of Joel Martin, who owned a plantation on the west side of Bath since 1702. Joel Martin died in 1715 and left 220 acres to his son, John.[36] So when John Martin joined Blackbeard's crew at Nassau in May, he wasn't going out pirating again; he was going home.

Sixteen
Siege of Charles Town

Blackbeard's fleet rounded the Bahama Bank, peacefully entered the Gulf Stream current, and sailed northward along the Florida coast. The water was a deep royal blue, in contrast to the turquoise color of the Bahamas. He had passed by this shore just eight months earlier when he was on his way to begin his career as a pirate captain aboard the *Revenge* and had sailed the same way several other times. Once past the Spanish territory of Saint Augustine, the pirate fleet entered the busy English shipping lanes of Charles Town. These were very familiar waters to Blackbeard. As they approached the entrance to Charles Town harbor, the crew could see dozens of sailing vessels in the distance.

Taking vessels wouldn't be as easy as most other ports. Charles Town was an extremely difficult port to enter. There was a large sandbar that traversed the entrance to the harbor. The only way to safely navigate over this bar was through one of the few channels. To make matters worse, those channels were constantly shifting positions. The only safe way for a vessel to enter Charles Town was for them to hire a local pilot. Consequently, pilots were often beyond the bar, sailing about in small boats looking for potential clients.

There were four pirate vessels: The *Queen Anne's Revenge*, a ship of 40 guns with Blackbeard in command, the *Revenge*, a sloop of 12 guns with Richards in command, the *Adventure*, David Herriot's captured sloop with Hands in command, and a small Spanish sloop they captured on their way to Nassau. Their total pirate force was estimated between three hundred and four hundred men. In his deposition, David Herriot described their vessels as:

> Having then under their Command the said Ship Queen Anne's Revenge, the Sloop commanded by Richards, this Deponent's Sloop,

A network of shifting sandbars at the mouth of the harbor complicated the passage of all large vessels. An experienced pilot was necessary to help them "get over the bar."

commanded by one Capt. Hands, one of the said Pirate Crew and a small empty Sloop which they found near Havana.[1]

Governor Johnson of South Carolina described a similar force in his letter to the Proprietors, in which he wrote, "This Company is Comanded by One Teach alias Blackbeard has a Ship of 40 od Guns under him and 3 Sloopes Tenders besides & are in all above 400 Men;"[2]

Stephen Godin, a prominent merchant in Charles Town, wrote several letters to the Lords Proprietors of Carolina complaining about the pirates. Extracts of these letters are in the Calendar of State Papers. Godin wrote:

> One a large French ship mounted with 40 guns and the other a sloop mounted with 12 guns with two other sloopes for their tenders having in all about 300 men all English the ship is commanded by one Theach and the sloop by one Richards.[3]

Robert Johnson was commissioned Governor of South Carolina by the Lords Proprietors and served from 1717 to 1719.

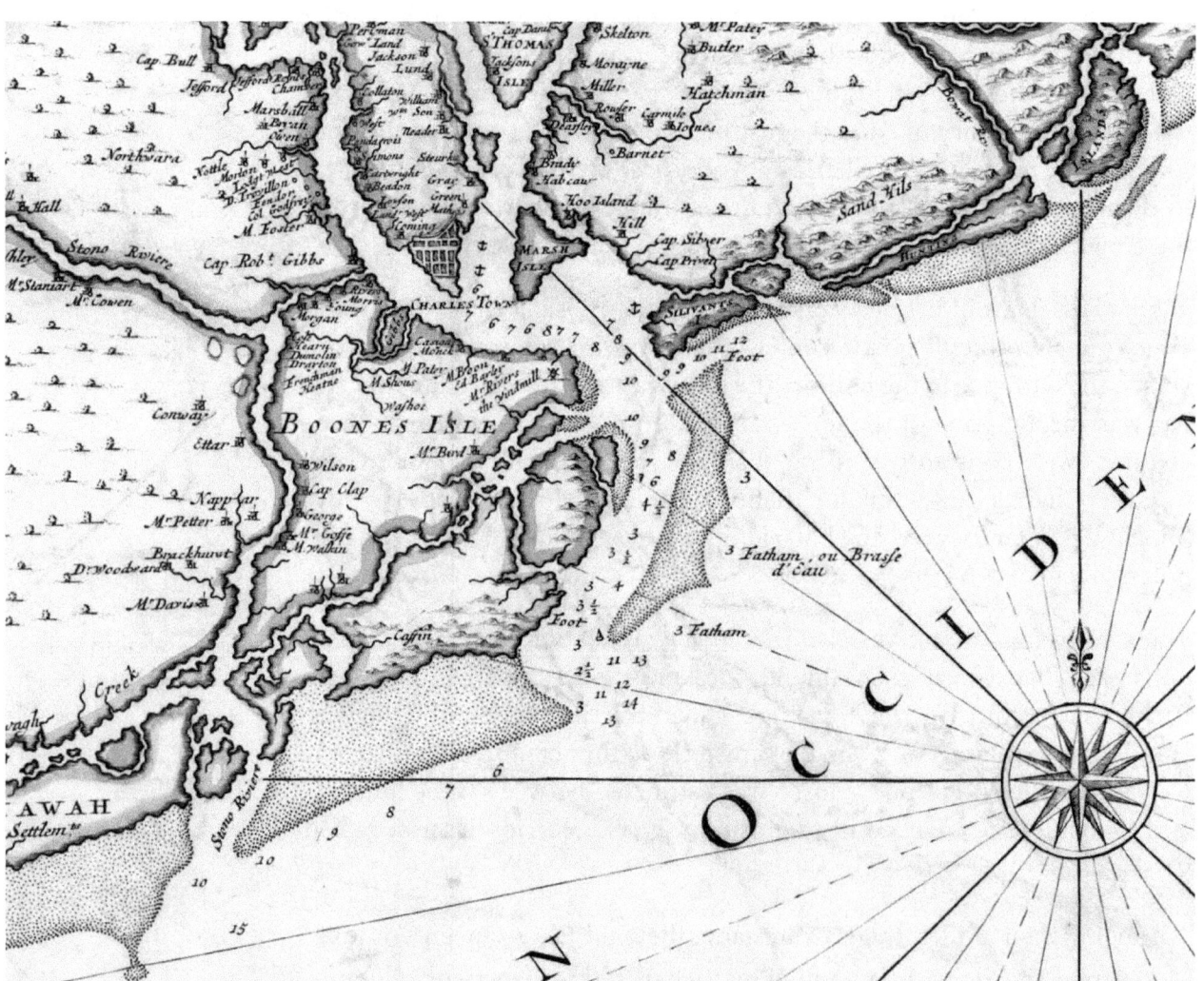

Figure 38: *Close-up of Charles Town Harbor from a 1690 Map. The image is dedicated to the public domain under CC0.*

As soon as Blackbeard's fleet arrived at the Charles Town bar, they began taking prizes. Many details on the piracies of Blackbeard at Charles Town in May 1718 entered the historical record during the trial of Stede Bonnet and his crew in November 1718. A few of Bonnet's pirates provided testimony or depositions, specifically mentioning the vessels taken at Charles Town. The entire trial transcript was published in London in 1719 under the title, *The Tryals of Major Stede Bonnet,* and was used by Charles Johnson as one of his primary sources for his book, *A General History of the Pyrates*. A summary of Blackbeard's and Bonnet's piracies at Charles Town was included at the beginning of the trial and states, "They were no sooner off the Bar, but they took five Prizes viz. two Ships bound in from London, two bound out to the same Place, and a small coasting sloop."[4] In his letter, Governor Johnson mentioned that they took eight or nine vessels, but this seems to be an exaggeration.[5]

The records show that Blackbeard took five vessels while at Charles Town, as mentioned above, and two more vessels shortly after leaving harbor. Identifying the names and captains of those vessels is far more complex. There isn't a single source that explains everything. By scrutinizing several primary source documents, I am confident that I have been able to carefully piece the clues together and arrive at an accurate summation of the vessels Blackbeard took at Charles Town. *The Tryals of Major Stede Bonnet* was a good start.

The two most important documents I used were Governor Robert Johnson's June 18, 1718 letter to the Proprietors and Stephen Godin's letter dated June 13, 1718. Both of these letters are included in Headlam's *Calendar of State Papers*. A copy of Johnson's original hand-written letter is housed in the Charleston County Library and is included in its entirety in Headlam's *Calendar of State Papers*. Governor Johnson's letter was used as the primary source for the trial of Stede Bonnet, *The Boston News-Letter* article, and for *A General History of the Pyrates*. The other valuable primary source that filled in some of the details was a document that lists all the vessels and captains entering and leaving Charles Town in 1718. This document is called the *Shipping Lists for South Carolina 1717–1721,* and it is available on microfilm in the Charleston Public Library.

It was May 22, 1718, and the *Crowley,* under command of Captain Robert Clarke, had just cleared the bar. The captain bid their pilot adieu and turned his ship toward the open sea. No sooner had the crew settled down for a peaceful voyage back to their home port of London when a lookout noticed a ship and three sloops sailing on a course to intercept them. This seemed very odd to Captain Clarke. He wondered why they weren't headed for the entrance of the bar. He and the crew watched curiously as

MAY 22, 1718

Blackbeard arrives in Charles Town.

the mysterious vessels grew closer. Their curiosity turned to fear as pirate flags atop their masts came into focus. Within a few moments, the *Queen Anne's Revenge* and the *Revenge* positioned themselves near the *Crowley*, eliminating any possibility of escape. It was all over without a battle. The *Crowley* had become Blackbeard's first prize at Charles Town. David Herriot's deposition reads, "where the said Thatch and Richards took a Ship commanded by one Robert Clark, bound from Charles-Town aforesaid to London."[6]

Stephen Godin's letter reads:

> South Carolina, 13th June, 1718. Capt. Clarck in the Crowley ... the 22nd as he was just proceeding from the barr was unfortunately taken by two pirates, one a large French ship mounted with 40 guns and the other a sloop mounted with 12 guns with two other sloopes.[7]

For Blackbeard, the *Crowley* turned out to be the jackpot. The ship was carrying passengers in addition to the cargo. Among them was Mr. Samuel Wragg, a member of the Council of South Carolina. His young son was with him and was also taken prisoner. Wragg was a VIP, and Blackbeard immediately realized that such a prominent citizen could be used as a hostage to bargain for anything he wanted. Blackbeard's demands have puzzled historian authors for centuries. Instead of asking for gold dust or silver coins, Blackbeard wanted a chest of medicines that valued no more than four hundred pounds. There are no descriptions of this in the historical record of the types of medicines in the chest. Theories and opinions on the reason for this unusual request will be discussed later in this chapter.

Obviously, a message had to be taken to the authorities at Charles Town with Blackbeard's demands. One of the passengers, Mr. Marks, was chosen to deliver the message. He was a gentleman and well-known in the community. Blackbeard stressed the seriousness of the situation to Mr. Marks and made a horrific threat. Blackbeard told Marks that if he didn't return in a day or so with the medicines as requested, everyone would be killed, their heads would be sent ashore, and the vessels already captured or taken later would be burned. Blackbeard also threatened to sail over the bar and burn every ship in the harbor. Two of Blackbeard's most senior officers, Captain Richards of the *Revenge* and Captain Hands of the *Adventure* were selected to go with Marks and ensure he delivered the message. A few additional pirates probably accompanied the shore party.

The type of boat used to transport Mr. Marks across the bar to deliver his message wasn't specified anywhere in the historical record. Governor Johnson does offer a clue, however. In his letter, he mentioned that the pirates "in Sight of the town tooke our Pilot Boat."[8] Therefore, it is logical to assume that the pirates used this pilot boat to shuttle Mr. Marks ashore.

None of Blackbeard's vessels ever crossed the bar during their siege of Charles Town. They all remained outside. He would have had no use for a pilot boat unless he needed it to transport Marks and his pirates across the bar and to the harbor docks. It is likely that Blackbeard's men hailed a passing pilot boat and forced the pilot to take Marks, Richards, Hands, and a few others onboard. There are a few anecdotes about the nature of this boat and that it capsized on route. This will be discussed later in this chapter.

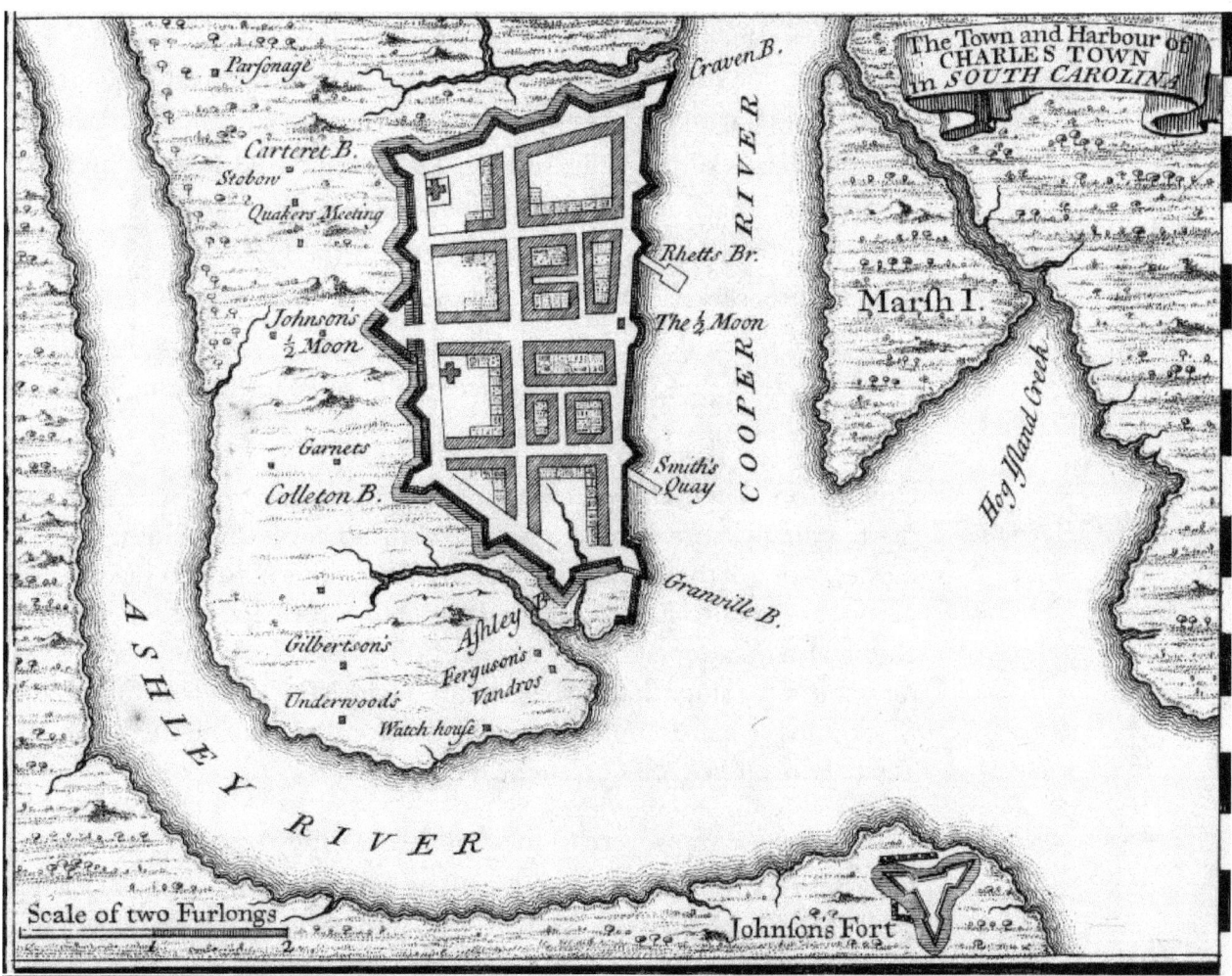

Figure 39: *Inset of Charles Town from a 1733 Map by Herman Moll. The image is dedicated to the public domain under CC0.*

While ashore, Richards, Hands, and possibly one or two others boldly walked the streets, threatening or at least intimidating everyone they encountered. Their presence made an enormous impression on the people of Charles Town, as their lawless presence in the community was mentioned in every account of Blackbeard's visit. Although details of their carousing while on their brief shore leave weren't included in these reports, it is reasonable to assume that they entered the taverns and took whatever drink they fancied. The most informative source documentation includes a statement made by Richard Allein, Attorney-General at Stede Bonnet's trial, and in Stephen Godin's letter. Richard Allein's statement reads:

> And after having taken Mr. Samuel Wragg, one of the Council of this Province, bound out from this Place to London, as also one Mr. Marks, and several other Vessels going out and coming into this Harbour, they plundered those Vessels going home to England from hence of about fifteen hundred Pounds Sterling, in Gold and Pieces

of Eight. And after that, they had the most unheard-of Impudence to send up one Richards, and two or three more Pirates, with the said Mr. Marks, with a Message to the Government, to demand a Chest of Medicines of the Value of three or four hundred Pounds, and to send them back with the Medicines, without offering any Violence to them, or otherwise they would send in the Heads of Mr. Wragg and all those Prisoners they had on board; and Richards, and two or three more of the Pirates, walked upon the Bay, and in our publick Streets, to and fro in the Face of all the People, waiting for the Governor's Answer.[9]

Godin's comments add some very interesting information. When discussing the pirate shore party, he wrote, "with Richards and another person master of one of their tenders." The only other ship's master in Blackbeard's party was Captain Hands, master of one of the other sloops, the *Adventure*. Godin also mentioned that Blackbeard threatened "to come over the barr for to burn the ships that lay before the Towne." This could only mean that Blackbeard remained outside of the bar the entire time he was at Charles Town. In his June 13, 1718 letter, Godin wrote:

Whilst these ships were in their possession they sent one of Clark's passengers with Richards and another person master of one of their tenders to towne with a message to send them a chest of medecines which if was refused by the Government they would imediately put to death all the persons that were in their possession and burn their ships etc. and threatn'd to come over the barr for to burn the ships that lay before the Towne and to beat it about our ears, as the Town is at present in a very indifferent condition of making much resistance if them or any other enemye should attempt it and that we were very desirious to gett them off our coast by fair means which we could not doe otherwise for want of such helps as other Governments are supply'd with from the Crown, the chest of medecines was sent etc.[10]

While waiting for the medicines, Blackbeard took one more ship as it left Charles Town, and two more ships as they approached the entrance of the harbor. Additionally, he took a small coasting sloop. As mentioned earlier in this chapter, Blackbeard took "two Ships bound in from London, two bound out to the same Place, and a small coasting sloop."[11] The other outbound ship was the *Ruby* from Charles Town and bound for London. Its captain was Jas. Craigh. In his deposition, David Herriot said "that said Thatch and Richards while they lay off the Bar of Charles-Town, took another Vessel coming out from Charles-Town."[12] Godin mentions that Blackbeard took "inward bound they now took besides Capt. Clarck, Capt. Craigh in a small ship belonging to this place as he went over the barr bound for London."[13] The shipping records show that on May 24, the ship *Ruby* left for London, commanded by "Capt. James Craigh."[14] That day, May 24, 1718, was just two days after Blackbeard took the *Crowley*.

The first inbound vessel taken was the pink *William* from Weymouth, England, under command of Captain Hewes. Bonnet's trial record states that Blackbeard took two inbound ships from London[15] and David Herriot mentioned that Blackbeard "Took two Pinks coming into Charles-town from England."[16] A pink is a type of ship, and the two terms were often used to describe the same vessel. Godin added to the statement above by writing, "and the William Capt. Hewes from Weymouth."[17] The second inbound ship was the *Artimesia*, inbound from London under command of Captain John Dornford. In addition, the shipping records show that the ship *Artimesia* of London, Captain John Dornford arrived around the same day as the ship *William*.[18] The small coasting sloop was the sloop *William*, of Philadelphia under command of Captain Thomas Hurst. Shipping records show that it left Charles Town on May 24, 1718, the same day that the *Ruby* left port.[19]

A pink is a type of sailing ship with a very narrow stern.

Eventually, the chest of medicine was delivered to Blackbeard and all of his prisoners were released. Governor Johnson mentioned that all their clothes had been taken, along with their valuables. He wrote, "After Plundering them of all they had were Sent a Shore almost Naked;"[20] After they left, they sailed to the north. Godin wrote:

> Soon after they dismissed our people and their ships having first taken from the two vessells that were homeward bound what little money they had on board and all their provisions and from the two others the same and distroy'd most of their cargoes etc. all for pure mischief sake and to keep their hands in. They made no farther stay (thanks to God) but are gone to the Northward.[21]

Artemisia I was the queen of Halicarnassus, a Greco-Carian city during the reign of the Persian king Xerxes. She commanded five ships in a major naval battle during the invasion of Greece from 480–479. It is likely the ship, *Artimesia*, was named for her or the Greek goddess Artemis.

Their length of stay at Charles Town is uncertain. The only reference came from David Herriot's deposition. He said that they stayed there "for the space of five or six Days."[22] As far as the amount of loot Blackbeard took, in addition to the chest of medicine, the only contemporary source is *The Boston News-Letter*, in the Monday, June 30 to Monday, July 7, 1718 edition. This article is the only story *The Boston News-Letter* ran on Blackbeard in Charles Town. It reads:

Chapter Sixteen

> Capt. Teach in a Ship of 40 Guns with two Sloops and about 300 Men, came to our Barr and took two Outward Bound Ships for London, one Clark and Craig, with two Inward Bound Ships from England: Clark had several Gentlemen Passengers, as Mr. Mark and MR. Wragg, &c. the former they sent up with two of the Pirates to demand a Chest of Medicines (which was sent them down) on Penalty of Burning all the Vessels and Men, which they afterwards plundred, and took all their Provisions, and some Rice, and about 4000 Pieces of Eight, and stript them of all their Cloaths and dismist the Ships and Men, which were forced to come in to refit; they staid here about ten dayes, & we are afraid they are not far off this Coast. We hear that they are bound to the Northward and Sware Revenge upon New-England Men & Vessels.[23]

About a week after Blackbeard finally left Charles Town, two vessels, which had also been taken by Blackbeard and his pirates as they sailed north along the coast, entered port. Stephen Godin wrote:

> Since they are gone severall vessells are come in amongst which is a brigantine from Angola with 86 negroes which was mett with by the pirates they took from her 14 of their best negroes, she belongs to Bristol, a ship from Boston is also come in which was likewise plunder'd by them.[24]

The first of these vessels was the brigantine *Princess* of Bristol, Captain John Redford commanding. It was taken almost immediately after Blackbeard left the entrance to the bar. After looting the vessel, fourteen slaves were retained by the pirates. Details of this brigantine come from the shipping records which state that on June 16, a "Bridgg. Princess of Bristol" arrived with captain "John Redford," carrying a cargo of "86 Negroes from Guinea."[25] Apparently, the capture of this vessel was pivotal in the prosecution of both Bonnet's men and William Howard, Blackbeard's Quartermaster. Howard's trial will be covered in detail in Chapter Twenty-One. A portion of Howard's charge sheet reads:

> In Company with the afforsaid Edwd Tach . . . in the Afforsaid ship . . . by the name of Queen Anns Revenge . . . did detain and upon the ____day of May in the year of our Lord 1718 the Briganteen _____ of London bound on a Voyage from Guinea to South Carolina.[26]

Even though the clerk who prepared the charge sheet didn't write the day in the blank, it does indicate May as the month the brigantine was taken, confirming them leaving Charles Town at the end of that month. At Bonnet's trial, William Scot testified "That we took a Brigantine, out of which we took fourteen Negroes. After we had discharged the Brigantine, we set sail and went to Topsail Inlet at North Carolina."[27] Also at that trial, Ignatius Pell stated:

> That said Thatch took out of the Brigantine he took off the Bar of Charles-Town fourteen Negroes and that he heard Thatch tell the Commander of the said Brigantine, That he had got a Baker's Dozen.[28]

The other vessel that Blackbeard's fleet took after leaving Charles Town was the ship *William* of Boston, Captain Nathaniel Mason, commanding. Those details can be found in the Carolina shipping records which show that it arrived on June 16, the same day as the brigantine *Princess*.[29] This ship was captured by Richards on the *Revenge*, probably while Blackbeard was busy with the brigantine *Princess*. This ship was from Boston, and Blackbeard's standing order was to destroy all vessels from that port. Apparently, Captain Richards chose to ignore that order and let the ship sail on unharmed. Blackbeard was upset with his captain for letting the *William* go. In Herriot's deposition he said that he "heard they took one Mason, and heard Thatch afterwards blame Richards for not burning said Mason's Vessel, because she belonged to Boston."[30]

Overall, Governor Johnson wasn't too far off when he wrote that Blackbeard took "8 or 9 Sail." He took five vessels outside of Charles Town proper and two vessels shortly after he left. That makes seven. Considering that Johnson wrote his letter only a few days after the *Princess* and the *William* made port, it is very likely that he expected a few more vessels to arrive with tales of Blackbeard's treacheries. Also, his letter was focused on requesting more military support from the proprietors writing that they should "be induced to Afford us the Assistance of a Frigate or two to cruse hereabouts."[31] Exaggeration is standard procedure in that type of letter. From there, Blackbeard sailed for Topsail Inlet and to the next phase of his career. But this isn't the end of our Charles Town saga. There are a few more points that require careful examination and clarification.

Perhaps no segment of Blackbeard's life has spawned as many unsubstantiated and fantastic stories as his Charles Town adventure. This chapter is a perfect example of how historian authors have embellished and twisted the truth over the years. Throughout the rest of this chapter, we shall abandon all prime source documentation and enter the writer's world of speculation and imagination. We shall discuss some of the unsubstantiated events about Blackbeard at Charles Town that have worked their way into the mainstream media over the years. It began with Charles Johnson.

Chapter Sixteen

He asserted that Blackbeard took a brigantine and two sloops while sailing between the Bahamas and the Carolina coast when he wrote, "and from the *Bahama* Wrecks, they sailed to *Carolina*, taking a Brigantine and two Sloops in their Way."[32] I have not been able to find a single source document that mentions any pirates taking two sloops and a brigantine between the Bahamas and the Carolinas in May of 1718. With no other sources to quote, most authors have simply repeated Johnson's words and cited his book as their source.

Recently, Colin Woodard in his 2007 book, *The Republic of Pirates*, went a bit further and actually identified those two sloops. The first was the *Providence*, captained by Blackbeard's old friend and former pirate, Josiah Burgess. The second was the *Ann*, commanded by Leigh Ashworth, another former pirate. Woodard paints an interesting story of a pirate reunion at sea. He contends that Burgess sold his cargo to Blackbeard and then returned to Charles Town to gather intelligence for Blackbeard. In Ashworth's case, Woodard explains that after a brief and pleasant meeting, Ashworth told Blackbeard that he planned on retiring to his Jamaican manor, and both sloops were released without being looted.[33]

The only source Woodard cites for these claims are the shipping records from South Carolina. The records show both sloops clearing Charles Town harbor in May 1718, the *Providence* with Captain Burgess on May 1, 1718, and the *Ann* with Captain Ashworth on May 15, 1718.[34] There is nothing else to indicate that they ever encountered Blackbeard. These records also show dozens of other sloops leaving at about the same time. It appears that Woodard chose these two sloops as the ones mentioned by Johnson simply because their captains were former pirates and may have had an association with Blackbeard previously. However, if this were true, the sloops weren't actually captured by Blackbeard and they wouldn't have been reported as such. Woodard's claim makes for an interesting story, but it is solely based on highly circumstantial evidence. I believe that Charles Johnson was in error when he mentioned two sloops and a brigantine being taken by Blackbeard before Charles Town. Perhaps he confused these three vessels with some of the other vessels taken at Charles Town.

Another mythical account of Blackbeard's Charles Town siege is the tale of the boat capsizing in a squall as it took Mr. Marks, Richards, Hands, and a few other pirates ashore. In this account, when the boat didn't return by the appointed deadline, Blackbeard told Wragg to prepare for death. Fortunately, Blackbeard was convinced to hold off for a while. Marks was eventually able to send a message to Blackbeard explaining the delay, and

Wragg and the other prisoners were spared. This tale first appeared in Shirley Hughson's 1894 book, *Blackbeard & the Carolina Pirates*. He wrote:

> Marks was given two days in which to accomplish his mission, and the prisoners, who had been acquainted with the demands and the attached condition, awaited with the most intense anxiety the return of the embassy. Two days passed and the party did not return. Thatch suspected that his men had been seized by the authorities, and notified Wragg that his entire party could prepare for immediate death. He was persuaded, however, to stay his bloody order for at least a day, and while awaiting the expiration of that time a message was received from Marks that their boat had been overturned by a squall, and that after many difficulties and much delay, they had succeeded in reaching Charles Town. Begging for a further reprieve, they agreed to pilot the pirate fleet into the harbor, and assist Thatch in battering down the defenseless town, in the event that proof was brought of the detention of the messengers by the authorities.[35]

Hughson was an honest author. He stated in the footnotes that his account was made up of a combination of information contained in Governor Johnson's letter and Charles Johnson's book. But this disclaimer didn't dissuade other authors from using this dramatic and exciting story. In 1927, another edition of Charles Johnson's *A General History of the Pyrates* was released. This time, it contained additional content, which was added by its editor, Arthur L. Hayward. The story of the capsized boat appears in this edition with a few additional and exciting details. In this version, the men survive by using a floating hatch. Of the many editions of Johnson's book, this is the only one that contains this account. It reads:

> This boat, it seems, was sent off by Mr. Marks very discreetly, lest a misconstruction should be put upon the stay that an unfortunate accident had occasioned, and which the men that belonged to her was cast away, being overset by a sudden squall of wind, and the men with great trouble had got ashore at an uninhabited island three or four leagues from the Main; that having stayed there some time till reduced to extremity, there being no provision of any kind, and fearing what disaster might befall the prisoners aboard, the persons belonging to their company set Mr. Marks upon a hatch and floated it upon the sea, after which they stripped and flung themselves in, and swimming after it, thrust the float forward, endeavouring by that means to get to town. This proved a very tedious voiture and in all

likelihood they had perished, had not this fishing boat sailed by in the morning, and perceived something in the water, made to it, and took them in, when they were near spent with their labour. When they were thus providentially preserved Mr. Marks hired a boat which carried them to Charleston. In the meantime he had sent this boat to give them an account of the accident.[36]

Patrick Pringle's 1953 book, *Jolly Roger: The Story of the Great Age of Piracy*, repeats this story citing Hughson.[37] Robert Lee repeats this account in his 1974 book *Blackbeard the Pirate*, citing the Arthur L. Haywood edition of Johnson's book shown above.[38] Finally, Colin Woodard also includes this story in his 2007 book, *The Republic of Pirates*.[39] This event isn't mentioned in any account prior to Hughson's 1894 book. Considering that Hughson says he made it up, it most likely never occurred.

The final piece of exuberant author exaggeration concerns the mysterious chest of medicine that Blackbeard demanded. Historian authors have puzzled over this for over three hundred years. What makes this so intriguing is that it was the only thing Blackbeard demanded. He had many prisoners, including Mr. Wragg, a member of the council. He could have demanded anything. Instead, he demanded a chest of medicine only worth about three or four hundred pounds. To complicate the issue, the historical record makes no mention of the type of medicines sent to Blackbeard. In 1718, there were lots of medicines in common use that could have been included in this chest. They included purgatives, emetics, opium, cinchona bark, camphor, potassium nitrate, and mercury. Treatment for many diseases usually entailed some sort of induced vomiting through the use of a purgative. Other treatments included the application of mercury or potassium nitrate to treat infections. This included mercury injections to treat venereal disease. Pain was often treated with opium.[40]

The enigmatic nature of these medicines has compelled historians and authors to speculate on the nature of the crisis that triggered Blackbeard's unusual demand. It could have been some sort of disease that was widespread among the crew, or it could have been an illness that affected only one or two of the men. It also could have been a traumatic physical injury that required pain relief. Until the early twenty-first century, authors had avoided any conclusion. That changed when a pewter syringe was discovered at the wreck site of the *Queen Anne's Revenge*. Throughout the late seventeenth and eighteenth centuries, syringes like the one found were used to inject mercury into a patient to treat syphilis as well as other venereal diseases. For some twenty-first century authors, the conclusion was obvious. Blackbeard and/or some of his men had syphilis and required immediate treatment.

In *The Republic of Pirates,* Colin Woodard speculates that many of Blackbeard's crew contracted syphilis while visiting the prostitutes in Nassau a few weeks earlier.[41] Apparently, he overlooked the fact that it takes at least three weeks for symptoms to develop. Author Kevin Duffus speculates in his book, *The Last Days of Black Beard the Pirate,* that Blackbeard had been suffering from the effects of syphilis for some time and that he was in the second stages of the dreaded disease when he was off the Charles Town bar.[42] Whereas syphilis among the crew is certainly a possibility, there is absolutely no historical evidence to support that claim.

Pewter syringes like the one found on the *Queen Anne's Revenge* were common on sailing vessels in the early eighteenth century. Just about every vessel had one. Even so, there is no proof that the syringe they found at the wreck site was used by Blackbeard's crew. It may have been onboard *La Concorde* long before Blackbeard ever took the ship, since syringes were usually included as part of the standard medical supplies. Another facet to consider is that the treatment of venereal disease among sailors wasn't uncommon. It was a normal part of life at sea. If all of Blackbeard's vessels were out of their supply of mercury, why didn't they stock up when they were in Nassau? A pirate port with prostitutes like Nassau must have had a good supply on hand. When I consider these facts, the immediate need for mercury to treat syphilis just doesn't seem feasible. There had to be some other reason for Blackbeard's demand for this chest of medicines.

With Charles Town behind them, and after looting the brigantine *Princess* and the ship *William*, Blackbeard's fleet—the *Queen Anne's Revenge*, the *Revenge*, the *Adventure*, and the Spanish sloop sailed northward toward Topsail Inlet in North Carolina. Upon arrival, the events that transpired had a profound impact on the lives of Blackbeard, Bonnet, and all their crewmen. Those events forced dramatic changes that split the partnership and sent both captains on a new course with destiny.

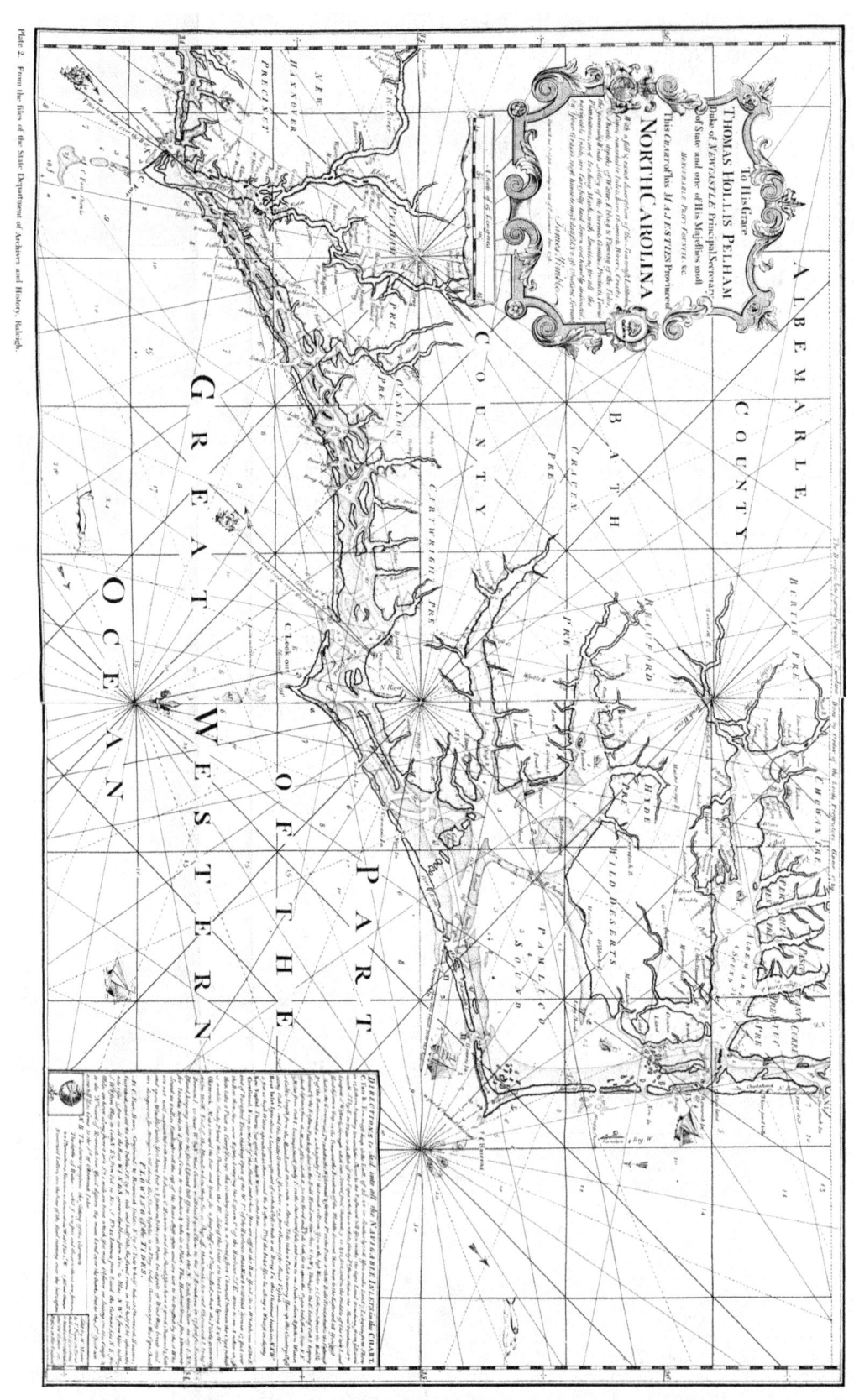

Figure 40: *1738 James Wimble Map of North Carolina Courtesy of East Carolina University.*

Seventeen
Topsail Inlet

The *Queen Anne's Revenge* slowly made its way toward an endless green shoreline of barrier islands that were covered by tall beach grass and dense scrub bushes, with a solid wall of trees behind the low-lying vegetation. The ship rolled back and forth from the groundswell of the increasingly shallow water. From the quarterdeck of their ship, the pirates strained to see some indication of an inlet. However, they were still quite a distance away. Their consort sloops, the *Revenge*, the *Adventure*, and the Spanish sloop were out in front. The pirates noticed a few small fishing boats scattered about, working the shoals and oyster beds. The air was heavy with heat and humidity. As they neared the inlet, the pleasant smell of the sea was replaced by the musty smell of the shoreline, which resembled a combination of seaweed, fish, and wet wood.

Topsail Inlet was among the most challenging entrances in North Carolina, even for experienced pilots. There was only one narrow channel that zigzagged between two barrier islands and then around numerous sand bars that lay hidden just below the surface of the water. The singular reference ashore by which to navigate this channel was a small white house surrounded by moss-draped oak trees. Approaching Topsail Inlet from the sea, this house only became visible when the vessel was at the entrance of the channel. On the *Queen Anne's Revenge*, a lookout shouted down from the crow's nest and pointed in the direction of the house that was now just a tiny white speck among the lush green background. The three sloops, sailing ahead, formed a line and safely entered the channel, with the large ship following at a suitable distance behind. However, the light was fading rapidly as the sun set. The *Queen Anne's Revenge* anchored to wait for the morning light.

The next day, just before the sun broke the horizon, Blackbeard stepped onto the deck and peered toward the inlet. Through the early morning

Chapter Seventeen

mist, he could barely see his three sloops safely waiting beyond the inlet at Core Sound. It was time to get underway. His crew hauled up the anchor and set the sails, and the ship gradually picked up speed as it moved toward the channel. With no warning, the *Queen Anne's Revenge* came to a crashing halt. The entire ship shuttered violently as lanterns and other unsecured equipment flew forward and fell to the deck. Everyone felt the deck rise and heard the gut-wrenching sound of snapping timbre as the hull scraped along a bed of sand and shell. Blackbeard's pirate ship had run aground.

The 1738 James Wimble chart clearly shows this channel and the white house used as a navigational aid. Built around 1700, the house was known as "Hammock House".

Topsail Inlet was the entrance to a small and sparsely populated fishing village called Fish Town. It really wasn't a town, it was a cluster of shacks, with only a few permanent structures scattered among them. The white house was there, of course. It was built around the year 1700.[1] A few years after the *Queen Anne's Revenge* sank, Fish Town developed into the town of Beaufort. However, when Blackbeard and his crew were there, it was a remote and isolated village with very limited supplies and nothing to offer a stranded crew.

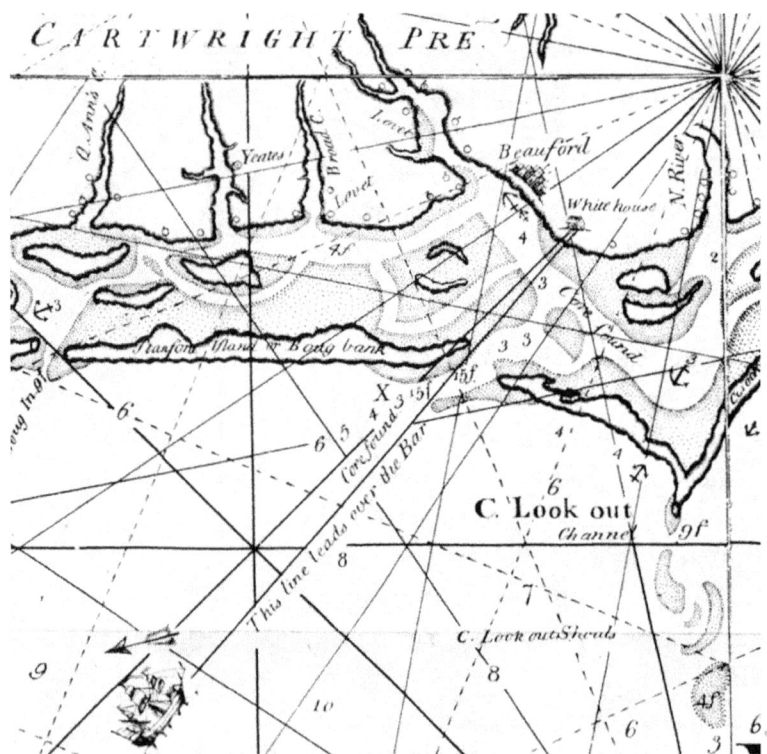

Figure 41: *Close-up of Topsail Inlet showing White House used for navigation from the 1738 James Wimble Map, Courtesy of East Carolina University.*

The pirates on the three sloops looked on in shock and disbelief. Most of them certainly didn't expect to see their ship run aground, but it is likely that this was exactly what a few of them expected.

William Howard, Blackbeard's quartermaster, rowed over to the sloop *Adventure* and took command. He sailed the *Adventure* back through the channel and attempted to assist the grounded ship. He was unsuccessful. With only about fifty yards to go before reaching the ship, the *Adventure* ran aground, too. Both the *Queen Anne's Revenge* and the *Adventure* were hopelessly wrecked.

All the details on the wrecking of both vessels and the events that transpired afterward come from the eyewitness accounts provided by several of Stede Bonnet's pirates. Four months after losing the *Queen Anne's Revenge* and *Adventure*, Bonnet and his men were arrested and tried for piracy. The transcript of their testimony provides us with many answers; however, it also leaves many questions unresolved. David Herriot, who had been the captain of the *Adventure* before Blackbeard took it at Turneffe Atoll, made the following statement:

> That about six Days after they left the Bar of Charleston, they arrived at Topsail-Inlet in North Carolina, having then under their Command the said Ship Queen Anne's Revenge, the Sloop commanded by Richards, this Deponent's Sloop, commanded by one Capt. Hands, one of the said Pirate Crew and a small empty Sloop which they found near Havana ... That the next Morning after they had all got safe into Topsail-Inlet, except Thatch, the said Thatch's Ship Queen Anne's Revenge ran aground off of the Bar of Topsail-Inlet, and the said Thatch sent his Quartermaster to command this Deponent's Sloop to come to his Assistance; but she ran aground likewise about Gunshot from the said Thatch, before his said Sloop could come to their Assistance, and both the said Thatch's Ship and this Deponent's Sloop were wrecked.[2]

The day the vessels wrecked was challenging for everyone. The damaged sloop presumably took on water and gradually slipped off of the sandbar. It would take months or even years for storms and erosion to push the ship into deeper water.[3] The bewildered crew had plenty of time to remove their personal possessions as well as provisions. For the rest of that day, everyone was focused on making it safely ashore with everything they could carry. The remaining two sloops, as well as some local fishermen, were available to help. Once on shore, their bewilderment turned into confusion, apprehension, and rage as the stranded crew pondered what they would do next.

The overwhelming feeling among the crew was that Blackbeard intentionally ran his ship aground to somehow cheat them out of their share. If true, it seems likely that Blackbeard's most trusted confidant, William Howard, was an accomplice. When Howard sailed the *Adventure* supposedly to help the *Queen Anne's Revenge* work its way off the sandbar, he was actually there to ensure that it didn't. In the process, Howard also intentionally sank the sloop. This is reflected in the testimony of Bonnet's men who were there. An excerpt from the prosecuting attorney's opening statement reads:

> In about a Month after we had Advice from *North Carolina* that these Pirates having some Difference among themselves, ran their Ship and one of the Sloops on shore on that Coast; and afterwards the Captain

Chapter Seventeen

(*Blackbeard*) cheated most of his Crew of their Share of the Riches they had taken.[4]

David Herriot's statement reads:

> It was generally believed the said Thatch run his Vessel a-ground on purpose to break up the Companies, and to secure what Moneys and Effects he had got for himself and such other of them as he had most Value for.[5]

Ignatius Pell testified that:

> We set sail and went to Topsail Inlet at North Carolina, where the Ship was run ashore and lost, which Thatch caused to be done.[6]

William Scot said exactly the same thing in his testimony, saying that:

> We set sail and went to Topsail Inlet at North Carolina, where the Ship was run ashore and lost, which Thatch caused to be done.[7]

The crew's belief is effectively summarized in the closing statement of the trial of Bonnet's men. Richard Alien and Thomas Hepworth, two of His Majesty's Justices of the Peace, stated:

> He heard by the Pirate Crew aboard Thatch, that Thatch took out of the Vessels that were taken off of the Bar of South Carolina, in Gold and Silver, to the Value of one thousand Pounds Sterling Money; and by others of them, to the Value of fifteen hundred Pounds Sterling Money: But that when Thatch broke up the Company, and before they came to any Share of what was taken by Thatch, Thatch took all away with him.[8]

Stede Bonnet leaves for Bath to accept the King's Pardon.

Amidst the confusion on the ship and onshore, Stede Bonnet took the opportunity to travel to Bath, North Carolina, to turn himself in to the governor and receive the king's pardon. At that time, Bath was the capital of the province and the home of Governor Charles Eden. Bonnet obtained a small boat and chose four of his men to accompany him and handle the craft. At his trial, Judge Trott mentioned that "Bonnet had not above five Hands" on his boat.[9]

The journey by water was about ninety miles one way. They could make it there in one day if they were lucky. William Morrison, one of Bonnet's pirates, testified that "Capt. Thatch had run the Sloop ashore, and Maj. Bonnet went up to the Governor for the Act of Grace."[10]

While Bonnet was away at Bath, thirty-nine pirates were onboard the *Revenge*, most likely performing maintenance and preparing it to sail. In the distance, they could see Blackbeard and a few of his men casually taking possession of the small Spanish sloop they had taken near the Florida Cape. That sloop, which had been used as a tender, now mounted 6 guns aboard. The small sloop's mainsail was raised, and it began to sail directly toward the *Revenge*. There were hard feelings toward Blackbeard among some of the pirates on the *Revenge*, because he had "cheated most of his Crew of their Share of the Riches." Their resentment was about to get much worse. As the sloop came alongside, Blackbeard's men raised their small arms and captured the *Revenge*.

John Lopez, one of Bonnet's pirates on trial, stated that Blackbeard only had five men when he took the *Revenge*. It's hard to imagine thirty-nine pirates being overpowered by only five men, even if they were armed. Once the *Revenge* was under Blackbeard's control, he ordered fourteen of the men on the *Revenge* to join him. Perhaps those fourteen were secretly on Blackbeard's side all along. He also took all of their money, small arms, and provisions. Afterward, everyone was put on board the sloop and the *Revenge* was abandoned. They sailed away from Topsail Inlet through the shallow channel that runs northeast between the barrier islands and the main shore. They came to a small island where Blackbeard put the remaining twenty-five pirates ashore and marooned them. Blackbeard sailed away and out of sight.

In the trial testimony, many of Bonnet's men mentioned this marooning. Neal Paterson said, "Thatch came on board and carried away fourteen of our best Hand, and marooned twenty-five of us on an Island." When asked, "Did not Thatch carry away your Money and what you had besides of Goods?" he said, "Yes." Ignatius Pell said, "After we had been there some time, Captain Thatch came aboard, and demanded all our arms, and took our best Hands, and all our Provision, and all that we had, and left us." John Lopez said, "I was at the Bay of Honduras, and was taken by Thatche, and carried to Topsail-Inlet, and there he marooned me on an Island, and came with five Hands, and carried all away that we had, and left us." Others who said that they were also marooned included James Mullet, Thomas Price, Edward Robinson, William Morrison, Robert Tucker, and Matthew King.[11]

Blackbeard maroons some pirates on Harbor Island.

David Herriot's deposition gives us more details about the island but differs in the number of men marooned. Judge Trott mentioned twenty-five men along with Neal Patterson.[12] As you shall see, Herriot sets the total number at seventeen. Herriot's deposition reads:

Chapter Seventeen

But instead thereof, ordered this Deponent, with about sixteen more, to be put on shore on a small Sandy Hill or Bank, a League distant from the Main; on which Place there was no Inhabitant, nor Provisions. Where this Deponent and the rest remained two Nights and one Day, and expected to perish; for that said Thatch took away their Boat. That said Thatch having taken what Number of Men he thought fit along with him, he set sail from Topsail-Inlet in the small Spanish Sloop, about eight Guns mounted, forty White Men, and sixty Negroes, and left the Revenge belonging to Bonnet there, who sent for this Deponent and Company from the said Sandy Bank. And then said Major Stede Bonnet resumed the Command of his Vessel.[13]

Two and a half days later, Bonnet returned from Bath on his boat. He had been successful and had a certificate given to him by Governor Eden, pardoning him for his piratical crimes. His men called to him from their island and he put his boat ashore. They were all desperate to be rescued. They told Bonnet of Blackbeard's unexpected attack and that the *Revenge* was still at Topsail Inlet. As they climbed aboard Bonnet's boat, he explained to them that he intended to sail to St. Thomas and obtain letters of marque which would legally allow him to take Spanish Vessels. This is consistent throughout the testimony. John Carman said, "I would not have went on board, but Maj. Bonnet shewed me the Act of Grace."[14] This "Act of Grace" was in reference to Bonnet's certificate. William Morrison said, "Capt. Thatch had run the Sloop ashore, and Maj. Bonnet went up to the Governor for the Act of Grace; and when he returned, he told me I might go to St. Thomas's."[15] Ignatius Pell stated it best by saying:

> Maj. Bonnet came with the Boat, and told us, as we were on a Maroon Island, that he was going to St. Thomas's to get a Commission from the Emperor to go against the Spaniards a Privateering, and we might go with him, or continue there: so we having nothing left, was willing to go with him.[16]

During the trial, Judge Trott asked the pirates who said that they went with Bonnet why they didn't resist. He asked, "Bonnet had not above five Hands, and there was of you twenty-five; why would you be all commanded by them?"[17] This statement confirms the number of twenty-five men marooned on the island as well as the four men that accompanied Bonnet to Bath. Bonnet would be the fifth man. They explained that at the time, they believed that he was going to become a legal privateer and that no one thought that they were returning to a life of piracy. Of course, what would you expect a man to say when on trial for piracy?

1718–1720

The War of the Quadruple Alliance was fought against Spain by Britain, France, Austria, and The Dutch Republic. In 1718, the Dutch colony on St. Thomas was issuing letters of marque, while the English were not.

After rescuing the twenty-five marooned men, Bonnet sailed to Topsail Inlet and reclaimed the *Revenge*. All of Blackbeard's men were gone. For the first time in months, Bonnet was once again in command of the sloop he had purchased over one year ago. Once everyone was onboard, Robert Tucker was chosen as quartermaster.[18] Before sailing, Bonnet renamed his sloop the *Royal James,* in direct reference to James Edward Stuart, and David Herriot's deposition reads:

> The said Thatch and all the other Sloop's Companies went on board the Revenge, afterwards called the Royal James, and on board the other Sloop they found empty off Havana.[19]

As Bonnet left Topsail Inlet, he had about fifty men onboard his sloop, the *Royal James*. Apparently, he picked up an additional twenty men from those who were still on shore in Fish Town. Dozens of other pirates stranded in Fish Town simply wandered through North Carolina into Virginia. As Captain Brand stated, "severall came into Virginia I am told by a man that left them."[20] After Marooning the twenty-five pirates on a deserted island, Blackbeard had continued on through the channel and arrived at Ocracoke Island. This was an ideal spot, as Ocracoke is located at the main entrance to Pamlico Sound, which is the only water route to all of North Carolina's major seaports. Blackbeard also renamed his sloop the *Adventure*, not to be confused with the sloop *Adventure* that sank next to the *Queen Anne's Revenge*. Aboard his 8-gun sloop was a crew of loyal pirates. David Herriot puts that number at "forty White Men, and sixty Negroes,"[21] however, the preliminary statement of Bonnet's trial puts that number at about 30. That statement reads:

> Upon this they separated. *Bonnet* got his own Sloop the *Revenge*, and with about fifty of the oldest Pirates went to the old Trade. *Blackbeard*, with about thirty more, sailed from thence on the same account in the other Sloop.[22]

As Bonnet left Topsail Inlet, some of the local fishermen told him where Blackbeard was. Anxious to avenge the man wrongs that Blackbeard had done him, Bonnet set his course for Ocracoke, too. However, when he arrived, Blackbeard was already gone. David Herriot's deposition reads:

> That the said Major Bonnet being informed by a Bomb-Boat that brought Apples and Cider, that Thatch lay at Ocracoke-Inlet with only eighteen or twenty Hands, he resolved to pursue him, and cruised after him for four Days.[23]

Chapter Seventeen

Of course, I would be remiss if I didn't include Charles Johnson's comment on the subject. It is certain that Johnson read a copy of *The Tryals of Major Stede Bonnet,* as his words echo those in David Herriot's statement. In his second edition of *A General History of the Pyrates*, Johnson wrote:

> Major *Bonnet* told all his Company, that he would take a Commission to go against the *Spaniards,* and to that End, was going to St. *Thomas's* therefore if they would go with him, they should be welcome; whereupon they all consented, but as the Sloop was preparing to sail, a Bom-Boat, that brought Apples and Sider to sell to the Sloop's Men, informed them, that Captain *Teach* lay at *Ocricock* Inlet, with only 18 or 20 Hands. *Bonnet,* who bore him a mortal Hatred for some Insults offered him, went immediately in pursuit of *Black-beard,* but it happened too late, for he missed of him there.[24]

These events were also documented in two reports made by officers in the Royal Navy, Captain Vincent Pearse of HMS *Phenix* and Captain Ellis Brand of HMS *Lyme*. Captain Pearse's letter, dated September 5, 1718, reads:

> I presume e're this comes to hand their Lordships will hear that one Teech Commander of a pirate, has lost his ship at Topsell inlet in North Carolina, & that himself and the greatest part of his Company has Surrender'd themselves to the Governor there, and accepted of his Masjsties most Gracious Pardon. Signed, Vinct: Pearse.[25]

Captain Brand's letter is dated July 12, 1718, and reads:

> This comes by the globe of Maryland which I am seeing of the coast, I am to acquaint you for there Lordships information that on the 10 June or thereabouts a large pyrate ship of forty guns with three sloops in her company came upon the coast of North Carolina ware they endeavour'd to goe in to a harbor, called Topsail Inlett, the ship wreck upon the bar at the entrance of the harbour and is lost, as is one of the sloops the other two sloops being still in there posesion with two hundred and, 30 of the pyrats they continue together given out they design for Currico and other of the islands, when they first came on the coast there number consisted of three hundred and twentie, whites and negroes, the rest having been to surrender, some to the governers of N. Carolina and severall came into Virginia I am told by a man that left them, above seventeen days since that the two sloops crew are fallen out and it was expected they would engage

each other if there disputes, are not soon reconsil'd amongst them; I have enquir'd of severall peoples that are acquainted with the place.[26]

Several unanswered questions remain. The first is the location of the island where the pirates were marooned. In recent years, some historian authors have identified as the Bogue Banks.[27] This island is at the entrance of Topsail Inlet and is near the spot where the *Queen Anne's Revenge* sank. The problem is that the eastern tip of this island is within shouting distance of Fish Town, and every fisherman in the area would sail right past the island, close enough to where they could be flagged down. Additionally, Herriot said that the island was "a League distant from the Main," which is a little over three miles. Bogue Banks is about one and a half miles from the main shore.

Captain Ellis Brand (1681–1759) assumed command of HMS *Lyme* on February 19, 1717.

The most likely spot is Harbor Island, which is about four miles from the mainland, too far to swim, but directly along the route from Topsail Inlet to Ocracoke. But more importantly, it is directly along the route that Bonnet must have taken from Bath back to Topsail Inlet. It is an Island where Bonnet would have easily seen his men hailing him from the beach. A close-up view of James Wimble's 1738 map shows the location of Harbor Island along Bonnet's route.

This leaves us with the strong possibility that Blackbeard intentionally marooned those men on an island along the route to Bath, knowing that Stede Bonnet would see them upon his return to Topsail Inlet. If this is true, it establishes that Blackbeard wanted to dissolve his partnership with Bonnet and most of his men, but he didn't want to kill anyone.

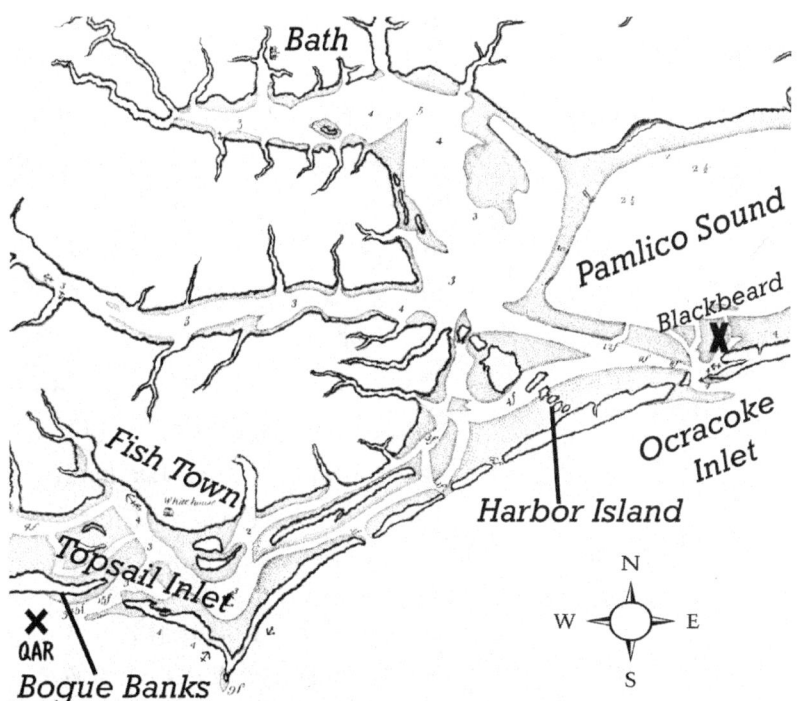

Figure 42: *Close-up of Route to Bath from the 1738 James Wimble Map, Courtesy of East Carolina University.*

Another issue is Blackbeard's motivation. Cheating his partners out of their share of the profits is a pretty good one, but what does Charles Johnson have to say on the matter? After Blackbeard's death, a handwritten journal belonging to Blackbeard was supposedly found. Johnson included a small excerpt from that journal in his chapter on Blackbeard. There is no way of telling if Johnson actually saw a transcript of that journal or simply invented the entire passage, but it is worth mentioning. According to Johnson's second edition, Blackbeard's journal

> Such a Day, Rum all out:—Our Company somewhat sober:—A damn'd Confusion amongst us!—Rogues a plotting;—great Talk of Separation.—So I look'd sharp for a Prize;—such a Day took one, with a great deal of Liquor on Board, so kept the Company hot, damned hot, then all Things went well again.[28]

Johnson also wrote a summary of all of these events, but he changed one important fact. He identifies Hands as the man who commanded the sloop that came to the assistance of the *Queen Anne's Revenge*. Herriot's testimony from the trial identified Hands as the captain of the sloop *Adventure* up to the point when the *Queen Anne's Revenge* ran aground, but also states that "Thatch sent his Quartermaster to command this Deponent's Sloop to come to his Assistance;" meaning William Howard. Perhaps Johnson didn't want to complicate the story with the addition of a new character. William Howard isn't mentioned anywhere in *A General History of the Pyrates*. Johnson wrote:

> *Teach* began now to think of breaking up the Company, and securing the Money and the best of the Effects for himself, and some others of his Companions he had most Friendship for, and to cheat the rest: Accordingly, on Pretence of running into *Topsail* Inlet to clean, he grounded his Ship, and then, as if it had been done undesignedly, and by Accident; he orders *Hands*'s Sloop to come to his Assistance, and get him off again, which he endeavouring to do, ran the Sloop on Shore near the other, and so were both lost. This done, *Teach* goes into the Tender Sloop, with forty Hands, and leaves the *Revenge* there; then takes seventeen others and Marroons them upon a small sandy Island, about a League from the Main, where there was neither Bird, Beast or Herb for their Subsistance, and where they must have perished if Major *Bonnet* had not two Days after taken them off.[29]

The biggest question that has spawned from these events is whether the sinking of the *Queen Anne's Revenge* was intentional or not. This has been the subject of many debates among historian authors and pirate enthusiasts for decades. Fortunately, substantial evidence has come to light that goes far beyond the speculative theories expressed by Charles Johnson. This evidence is the *Queen Anne's Revenge* itself.

The actual shipwreck was discovered in 1996 by the treasure-salvage company, Intersal LLC.[30] It lies about 1,500 feet west of the entrance to the current channel in twenty-three feet of water.[31,32] It has been identified through the examination of a multitude of evidence. Although they have found nothing that specifically says "La Concorde," nothing on the ship dates after 1718, and many construction features have been proven common to early eighteenth-century France.[33] Additionally, the ship's bell is dated 1705 and has IHS MARIA ANO 1705 written on it. The design was a common type of bell made by traveling French bell makers.[34]

Underwater archaeologists found a large pile of anchors and guns in the midsection of the ship. That has led some of them to speculate that they were moved from the bow to lighten that part of the ship, making it easier to get off the bar. Additionally, they found an anchor about 450 feet away from the ship, which also caused some to speculate that it was there because some of the crewmen were trying to pull the ship free.[35]

If this is true, it does nothing to dispel the notion that the grounding was intentional. Perhaps others tried to free the ship after Blackbeard left. Perhaps Blackbeard went through the motions of trying to free the ship to reinforce his deception. More likely, since the ship was there for months or years, the possibility that local fishermen attempted to salvage the ship is very real.

In light of all the eyewitnesses believing that it was intentional, one must ask, "Why go to Topsail Inlet in the first place?" The channel beyond was too shallow for a ship to navigate. Once inside, the only place for them to go would be to sail right back out again. The inlet was essentially a dead-end. Unless Blackbeard had an overwhelming craving for fresh oysters, Fish Town had absolutely nothing to offer. An experienced sea captain like Blackbeard would have known all this, but his crew wouldn't. That includes Stede Bonnet, who wasn't a sea captain and probably couldn't navigate from a chart.

Blackbeard had pulled it off nicely. He downsized from hundreds to about twenty-five or so, without having to hand out hundreds of shares. Blackbeard had ended his unlikely partnership with a man that he could no longer trust. It appears that he got away with everything of value onboard his ship. The most intriguing aspect of all the artifacts recovered from the wreck of the *Queen Anne's Revenge* was the treasure. They found precisely four silver coins and some loose bits of gold dust.[36] Not exactly the treasure one would expect. That leads us to the last unresolved question pertaining to the shipwreck. What happened to the treasure?

Artifacts from the wreck can be seen at the North Carolina Maritime Museum in Beaufort, NC, which sits at the entrance of Topsail Inlet.

Eighteen
Life in Bath

Blackbeard sailed away from the small island where he had just marooned twenty-five of Bonnet's pirates. The sound of their angry cries and shouts pleading for him to come back and pick them up faded in the distance. He was aboard his new sloop *Adventure* with 6 guns mounted and about twenty-five men. They sailed to Ocracoke Island, at the main entrance to Pamlico Sound. He briefly scouted the area and found it to be an ideal spot for a base. He remained there for about a day. Before he could go about establishing a base, he first needed to get a pardon from the governor of the province. Until then, he was in danger of being taken by pirate hunters, or at the very least, of being arrested. The *Adventure* set sail and headed northwest, up the Pamlico River.

His timing was perfect, and his plan was working. Blackbeard was certain that as Stede Bonnet was returning from Bath, he would stop at Harbor Island to pick up the marooned men. He also knew that as soon as Bonnet reclaimed the *Revenge*, he would be heading directly toward Ocracoke to even the score. Blackbeard left Ocracoke and sailed for Bath as Bonnet was leaving Topsail Inlet. By the time Bonnet finally caught sight of Ocracoke, Blackbeard had already gone.

In good weather, the Pamlico River is very scenic, with smooth, calm water and tree-lined banks. It's a fairly wide river at the entrance; however, as serene as the water may appear, it can be a tricky river to navigate. Parts of it are just too shallow for large vessels, and several dangerous shoals lie unseen beneath the surface of the water. About twenty miles up the river, Bath Creek becomes visible on the right. As the *Adventure* approached, it hugged the right bank to avoid a shoal, then gently swung over and entered Bath Creek. Picturesque and peaceful, the creek was about half a mile wide with plantations every so often along both banks. About two miles later, Blackbeard finally reached the docks.

Bath was a sleepy little port. There were only a few taverns in town and perhaps a dozen other buildings. Even so, it was one of the largest ports in North Carolina, and the docks could easily accommodate the *Adventure*. Bath's deeds list many landowners as "Mariner," indicating many captains and sailors were local residents.¹ Additionally, it was the home of the governor, and as such, the capital of North Carolina. When court was in session, the number of people in town dramatically increased, overwhelming the tavern keepers who didn't have any beds in their taverns.²

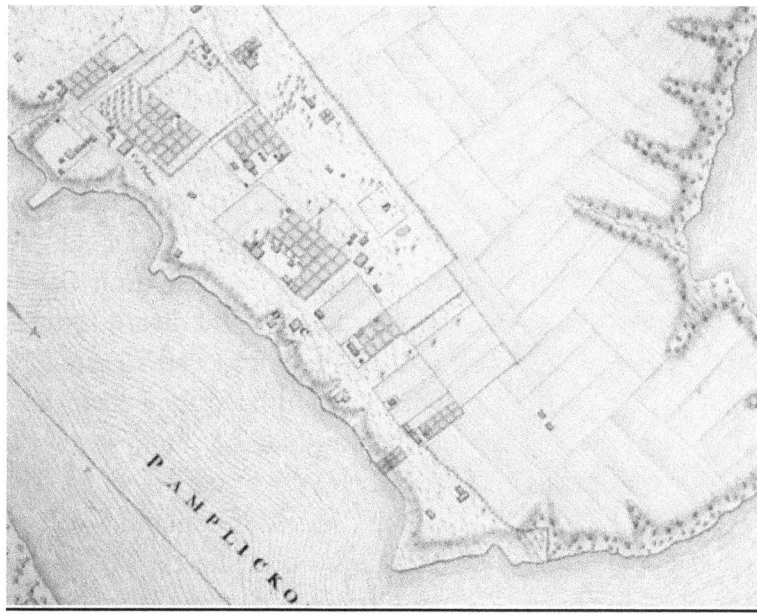

Figure 43: *Close-up of Bath Town from an Eighteenth-Century Map. The image is dedicated to the public domain under CC0.*

To obtain a pardon, Blackbeard sought out Tobias Knight, the Secretary of the Colony, Collector of Customs, Chief Justice, and a member of the Council.³ All those positions made Knight the second most important politician in the colony. Petitioners generally didn't go directly to the governor, they had to go through channels, and Tobias Knight held all the keys that led to the governor's door. His appointment letter is housed in the archives of North Carolina and specifies all his duties and responsibilities:

> Wee do hereby further Impower You to Receive all Prequisites, Fees, and advanteges whatsoever to the said Office of Secret Belonging, or in any wa̍e appertaining.⁴

This would include processing any requests from pirates who were seeking to accept the king's pardon. After visiting Knight, a Vice-Admiralty Court was set up in Bath Town and Blackbeard was tried. This was the usual procedure. That court determined that he acted as a privateersman, not a pirate, and Blackbeard was granted the king's pardon and received a certificate signed by Eden.⁵

In the months to come, Tobias Knight would play a crucial role in the life of Blackbeard, and face serious charges for complicity with pirates. Three others were vital to shaping the political intrigue and legal entanglements surrounding Blackbeard's death. These men were Edward Moseley, Thomas Pollock, and Governor Charles Eden. To properly be able to interpret the events that are about to transpire concerning the attack on Blackbeard at Ocracoke and the possible location of any treasure he may

1711–1712

The Cary Rebellion

Chapter Eighteen

have acquired, it is important to pause for a moment for a brief discussion on the early eighteenth-century politics of North Carolina and the large rift that existed between two political factions as a result of the Cary Rebellion.

The so-called Cary Rebellion started in Bath in 1711. It centered around Thomas Cary, who had held the office of Deputy Governor of North Carolina since 1708, until the Proprietors in South Carolina forcibly replaced him with Edward Hyde in 1711.[6] Cary, along with his supporters, refused to go, and a small armed conflict broke out in Bath. The problems with Cary all stemmed from a disagreement between the older settlers from Virginia and the newer settlers who represented the interests of the proprietors in South Carolina. The older settlers were primarily members of the Church of England, while many of these new settlers were Quakers. Laws passed in the early 1700s prevented Quakers from being sworn in and thus barred them from holding public office.[7]

Thomas Cary was appointed Deputy Governor in 1705 under the understanding that he would be sympathetic to the Quaker cause, but he wasn't and aligned himself with the Anglicans led by Thomas Pollock. In 1707, Cary was removed and replaced by William Glover. However, Glover also supported the anti-Quaker movement and ignored his instructions from the proprietors. Seizing on this opportunity, Cary again agreed to support the Quakers and, in 1708, was elected president of the council with the majority support from the legislators, effectively making him governor. Glover and Pollock fled to Virginia, along with many of Bath's most prominent citizens. However, Lt. Governor Spotswood of Virginia didn't want these troublemakers coming into his colony and issued a proclamation for their arrest.[8]

Even though Carry had agreed to follow the instructions of the proprietors, it soon became apparent that he had no intention of doing so. In 1710, fed up with Carry, the proprietors appointed Edward Hyde as Deputy Governor. Hyde had strong political connections, as he was the cousin of Queen Anne. However, it took him until January 1711 to assume the office, since he was living in England at the time of his appointment.[9] When he finally arrived at Bath, Hyde found the province in total chaos. The courts had been closed and all orders within the precinct had collapsed. Additionally, widespread plundering and destruction had caused planters to lose their crops.[10]

While in office, Cary's chief supporter was Edward Moseley, the Surveyor General for the province. However, when Governor Hyde arrived, he discovered that Moseley had embezzled an enormous amount of money by falsifying surveys. Hyde issued warrants for the arrest of both Cary and Moseley for "High Crimes and Misdemeanors."[11] Hyde also passed a law that made it a criminal offense to "speak seditious words, write or dispense scurrilous Libels against the present Government."[12]

In Bath, Cary formed an armed resistance. With forty-five men, he gathered five cannons, small arms, and gunpowder intending to overthrow Hyde's government. His followers included John Porter, Robert Daniell, and Edward Moseley. Hyde's forces comprised sixty men and two cannons. This included Thomas Pollock, Hyde's political ally and chief opponent to Carry and Moseley.[13] Wanting to enlarge his force, Hyde wrote a letter to Spotswood in Virginia requesting help. In response, a detachment of royal marines was sent from Virginia to assist. A few small engagements took place near Bath, and Cary's forces were eventually defeated. Cary fled to Virginia where Spotswood had him arrested and sent to England for trial.[14] John Porte and Edward Moseley were arrested in North Carolina and also sent to England to stand trial.[15]

The proprietors knew that Hyde needed a first-class lawyer to help him clean up the legal mess that existed in the wake of Cary's Rebellion. They chose Tobias Knight, and on January 24, 1712, appointed him Secretary of "our Province of Carolina that lies North and East of Cape Fear."[16] Shortly after Knight's arrival, Hyde pardoned all the conspirators except Cary and the leaders.[17] Unfortunately, Hyde died of yellow fever in September 1712. Thomas Pollock, the president of the council, was temporarily assigned the governor's duties until a new one could be appointed.[18] Meanwhile, in England, the trial of Cary and his conspirators had lasted over a year. It was such a legal quagmire that the Privy Council concluded the situation had been mishandled on both sides. The charges were dismissed and Moseley returned to North Carolina.[19]

North Carolina was in turmoil. With no official governor and the recent rebellion that just about destroyed its economy, the colony needed a governor who would pull the province together. Charles Eden was appointed governor on May 18, 1713, and was selected to clean things up.[20] He was living in England at the time and had never been to the colonies before.[21] It took him a year to get his affairs in order and travel to the colonies. He took the oath of office on May 28, 1714, at a meeting of the Provincial Council at the house of Capt. John Hecklefield in Little River.[22] When he arrived at Bath, he found a loyal supporter in the person of Tobias Knight. By 1714, Knight and his wife had taken up residence on a 300-acre plantation with a large house at Archbell Point, on the west side of Bath Creek, where it flows into the Pamlico River.[23] When Governor Eden arrived in Bath, he bought a 400-acre plantation with a large brick home on the west

Charles Eden, Governor of North Carolina from 1714 to 1722.

side of Bath Creek, directly north of Tobias Knight's property.[24] Eden and Knight weren't only political allies, they were next-door neighbors.

Eden had walked into a political firestorm. Supporting Eden were his loyal secretary, Tobias Knight, and Thomas Pollock, the acting governor in his absence. His chief rivals were Colonel Edward Moseley, Colonel Maurice Moore, and Jeremiah Vail.[25] The reverend John Urmstone, an influential member of the community, was another foe. He wrote that Knight "grew very impertinent, and hath often opposed me in matters relating to Church discipline." Regarding Eden, he wrote that the governor was, "a complete ruffian, a boatswain's mate, who are commonly the greatest reprobates on a man-o-war, fit only to command the forecastle Gang." He sarcastically added, "Seeing the Genius and tempo of the People are so like the said Gentry, there cannot be a fitter man to govern them."[26] Clearly, Knight and Eden had challenges in administering the province.

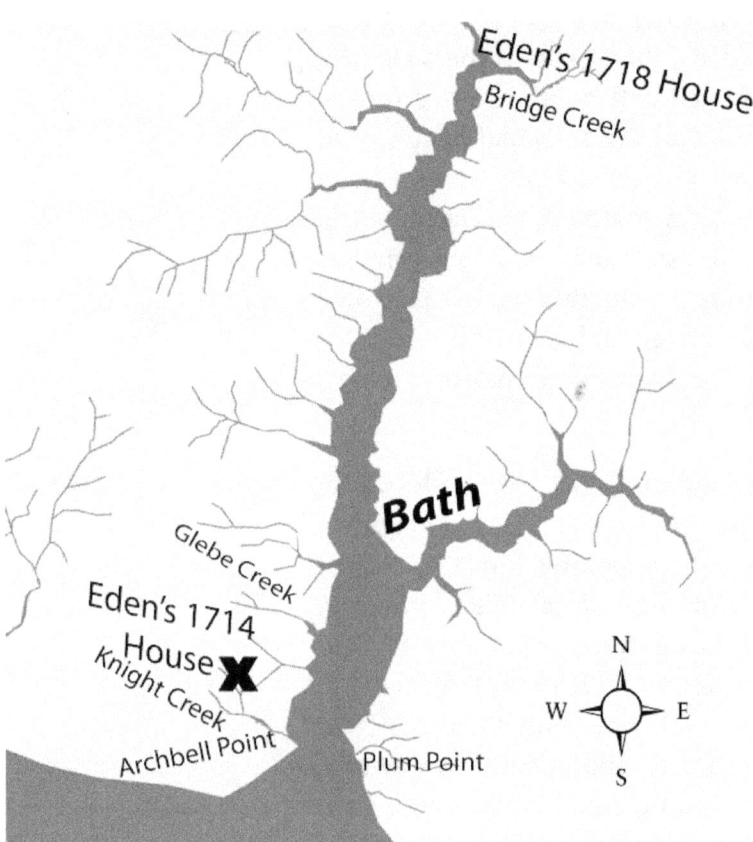

Figure 44: *Map of Bath Showing Homes of Knight and Eden.*

Three years after Eden arrived in Bath, Knight became bedridden by a serious and prolonged illness that lasted until his death in 1719.[27] His last recorded attendance of a Council meeting was August 1, 1717, where previously, he had attended almost all of them.[28] The nature of this illness is unknown. Based on his writings and legal opinions, his mind was as sharp as ever. He remained in his house along with his loving wife Katherine, who cared for him and stayed by his side throughout his illness.[29]

Blackbeard arrives at Bath

By the time Blackbeard arrived in Bath, Governor Eden owned seven lots in the town. Five of them were prime real estate on Bay Street, with water access to Bath Creek. The other two were on King Street at the south end of town.[30] He no longer owned the 400-acre plantation next to Tobias Knight's property. Deeds from Bath show that Governor Eden sold that plantation on April 10, 1718, to two men, Stephen Elsey and James Robins. He bought a 320-acre plantation for £60 at Bridge Creek north of the town of Bath.[31]

Eden's 400-acre property on the west side of Bath Creek is often referred to as "Eden's Tunnel Land." Apparently, the remnants of an underground tunnel leading from the house to the creek could be seen well into the nineteenth century. This tunnel was described in a letter dated March 31, 1857, written by Joseph Bonner, who lived in Bath as a child in 1820. At that time, parts of the old brick mansion that had once belonged to Governor Eden still remained. Bonner wrote:

> I was accustomed to visit this building in my early childhood. Its massive walls, capacious hall s, rich workmanship of the interior, and palace like appearance, indicated that it had been a abode of wealth. A subterranean passage some 60 or more yards in length, communicated from a brick wall near the margin of the creek, with the cellar of the dwelling.[32]

To support this, a newspaper article appeared in 1894 that recounted the statements of two men, Thomas Latham and Jno Burgess, who visited the site of Governor Eden's ruined mansion and stated that the old cellar and subterranean passage were still visible.[33]

These descriptions of a mysterious tunnel running from the edge of the creek directly to Eden's house sparked a flurry of speculation by historian authors and pirate enthusiasts alike. They contend that this tunnel was used by Blackbeard himself to smuggle stolen goods secretly from his sloop on the creek to Eden's cellar without being noticed by the local inhabitants. Some go further and suggest that part of his treasure may still be buried somewhere within the tunnel walls.[34] They seem to have completely overlooked the fact that when Blackbeard was in Bath between mid-June and mid-November 1718, Eden didn't own the property.

News of Blackbeard's acceptance of the pardon from Governor Eden quickly reached the office of the Lt. Governor of Virginia, Alexander Spotswood. His reaction was to issue a proclamation for the arrest of any former pirates who wandered from North Carolina into Virginia. That proclamation was mentioned in a letter dated August 14, 1718, from Lt. Governor Spotswood to the Council of Trade and Plantations. An excerpt from that letter reads:

> The Proclamation prohibiting the unlawfull concourse of persons who have been guilty of piracy was occasioned by the great resort to this Colony, of certain pyrates who being cast away in North Carolina, surrendered there upon H.M. Proclamation; but as there's no great faith to be given to the fore'd submission of men of those principles.[35]

Chapter Eighteen

Very little reliable information exists on precisely what Blackbeard and his crew did while they were in Bath that June and July 1717. Of course, Charles Johnson has plenty to say on the subject. The most popular and retold episode of Johnson's account of Blackbeard in Bath is the one about his marriage to a sixteen-year-old, with Governor Eden performing the ceremony. In this account, this unfortunate young lady was Blackbeard's fourteenth wife, and shortly after the wedding ceremony, he passed her around to some of his crew. Johnson's sensationalized and sexually charged account must have thrilled the readers in eighteenth-century England. It reads:

> Before he sailed upon his Adventures, he marry'd a young Creature of about sixteen Years of Age, the Governor performing the Ceremony. As it is a Custom to marry here by a Priest, so it is there by a Magistrate; and this, I have been informed, made *Teach*'s fourteenth Wife, whereof, about a dozen might be still living. His Behaviour in this State, was something extraordinary; for, while his Sloop lay in *Okerecock* Inlet, and he ashore at a Plantation, where his Wife lived, with whom after he had lain all Night, it was his Custom to invite five or six of his brutal Companions to come ashore, and he would force her to prostitute her self to them all, one after another, before his Face.[36]

Johnson also included an account of Blackbeard's piratical acts toward local mariners, which reads:

> Captain *Teach*, alias *Black-beard*, passed three or four Months in the River, sometimes lying at Anchor in the Coves, at other Times sailing from one Inlet to another, trading with such Sloops as he met, for the Plunder he had taken . . . at other Times he made bold with them, and took what he liked.[37]

To the residents of North Carolina, Johnson's most insulting account insinuates that the wealthy planters would peacefully stand by and allow Blackbeard and his crew to use their wives and daughters sexually, or even worse, that they might have willingly taken money in exchange for their services. Additionally, he asserts that Governor Eden was powerless to stop Blackbeard. That account reads:

> He often diverted himself with going ashore among the Planters, where he revelled Night and Day . . . but, as for Liberties (which 'tis said) he and his Companions often took with the Wives and

Daughters of the Planters, I cannot take upon me to say, whether he paid them *ad Valorem*, or no . . . . nay, he often proceeded to bully the Governor, not, that I can discover the least Cause of Quarrel betwixt them, but it seemed only to be done, to shew he dared do it.[38]

There is one account, however, that was written by a reliable source. Captain Brand, the senior naval officer in Virginia, wrote a report on the entire Blackbeard incident on February 6, 1719, two and a half months after his death. After receiving word of the wreck of the *Queen Anne's Revenge* and the pardon that Gover Eden gave Blackbeard, Brand hired an agent to go to Bath and give him a report. Brand's hearsay information about Blackbeard's life in Bath during June and July 1717 came from that man. Contained within this agent's report was that Blackbeard married while in Bath. An excerpt from Brand's letter reads:

> This serves to acquaint you for their Lordships information of my proceedings here, the notorious Pyrate Thach alias Blackbeard that I advis'd you of being last in No. Carolina in June last has continued in that place with one sloop and about Twenty men and did there condemn her at a court of Admiralty and gave out he design'd to be an inhabitant & leave of his Piraticall Life and the more to put a glass to his designs he *marryed* there, so soon as I received this advice I employ'd a man that was going into No. Carolina to inform himself how this fellow lived.[39]

Brand goes further, stating that he was told of other reports from locals concerning Blackbeard's abuses. However, there were no reports of any such disturbances made to the authorities or the Chief Justice, Tobias Knight. Considering the volatile political environment that existed in Bath, it is possible that these reports were false and came from Edward Moseley's allies to discredit the legal authority of Eden and Knight. If that was the case, Brand was being played. In a later chapter, Brand's integrity when dealing with Knight will be brought into serious question. Brand wrote:

> I recev'd several accounts from people that came from thence that he did Continue in that place insulting and abusing the masters of all trading sloops and taking from there what goods or Liquors he pleased and that he might not be called a Pyrate, paid such prices to those for their Effects as he pleased.[40]

A far more realistic view of how Blackbeard and his men were received by locals in Bath comes from Debbie Devine, who was a direct descendant of Isaac Jackson, a resident of Bath during Blackbeard's visit. The oral

Chapter Eighteen

tradition handed down by her ancestors throughout many generations painted Blackbeard as more of a benefactor than a rogue. In the 1960s, she wrote, "The folks in North Carolina lived by the goodness of the pirates, and many claim ancestry to them. Those sea dogs were the bringers of goods to a poor, very isolated part of the world."[41]

There is far more to support this notion than just Debbie's belief. John Martin, Ben Hornigold's quartermaster, was from Bath. More than that, he was the son of Joel Martin, a wealthy plantation owner and one of the town's original settlers. Bath deeds show that John Martin inherited 220 acres from his father in 1715.[42] Blackbeard's quartermaster, William Howard, was the son of Philip Howard, another wealthy plantation owner who had 320 acres of land on the Pamlico River east of Bath.[43] Since both these men served as quartermasters for Ben Hornigold, it seems logical to conclude that they left Bath together sometime after 1715, after the date of the Bath deed giving John Martin possession of his inheritance. They would have been easily recognized and welcomed home, maybe even as heroes. Perhaps they weren't the only ones. It is possible that another member of the pirate crew was also well-known in Bath and left with them in 1715. This pirate was their captain, Blackbeard.

Nineteen
Philadelphia

The *Adventure* was securely moored at the end of a long, narrow dock that jutted out from the town of Bath. On deck, the crew was busily preparing it to sail. It was mid-July 1718, and Blackbeard and his crew had only been in Bath a short while. On the *Adventure*, the crew could hear the "click clack" of footsteps as a tall, imposing figure walked down the dock toward them. It was their captain, Blackbeard. Once he reached his sloop, he casually glanced back to the shore and sighed. This time, Blackbeard would be sailing without his quartermaster, William Howard. Their courses were destined for different paths. Although he didn't know it at the time, the captain and his quartermaster would never meet again. Seeing that the preparations were complete, Blackbeard reached out, firmly grasped the sidestay, and stepped aboard. The *Adventure* was now ready to go. Once the lines were untied, the little sloop slowly drifted away from the end of the dock until the mainsail filled with wind. Rapidly gaining speed, it gracefully slipped into the channel and headed down Bath Creek.

As they approached the Pamlico River, they could see Tobias Knights's house at Archbell Point on the right bank. Plum Point was on the opposite bank, about three hundred and fifty yards across the creek. The *Adventure* easily passed between these two land features and entered the Pamlico River. Once clear of the shoals, the helm was put over to starboard, and the sloop swung about to the left to begin its sixty-mile journey down the river to the sea. After carefully sailing through Ocracoke Inlet and into the Atlantic Ocean, the sloop turned north. These waters were very familiar to Blackbeard. He had followed this same course when he was captain of the *Revenge*. As it passed the Virginia Capes, the *Adventure* finally reached Cape May and entered the Delaware River, the passage that led to Philadelphia. Just ten months earlier, Blackbeard had wreaked havoc on the shipping at this exact location. However, this time he wasn't hunting for prizes. Blackbeard was going to visit a friend.

Chapter Nineteen

Blackbeard is mentioned as a ship's mate

Blackbeard's shadowy connection to Philadelphia was mentioned in three primary source documents, which were briefly mentioned in Chapter Nine. As you may recall, Blackbeard was recognized by some of his captives who had known him several years earlier. They identified him as being a mate who frequently sailed into Philadelphia. The first primary source document comes from the November 4 to November 11, 1717 edition of *The Boston News-Letter*. An article in that paper chronicled a series of Blackbeard's attacks on vessels near Philadelphia. It also includes nine astounding words. Blackbeard was described as "Teach, who formerly Sail'd Mate out of this Port." The first paragraph of that article reads:

> Philadelphia, October 24th. Arrived Linsey from Antigua, Codd from Liverpool and Dublin with 150 Passengers, many whereof are Servants. He was taken about 12 days since off our Capes by a Pirate Sloop called the Revenge, of 12 Guns 150 Men, Commanded by one Teach, who formerly Sail'd Mate out of this Port:[1]

More clarity on this issue is provided in a letter dated October 24, 1717, which is the second prime source document. It was written by James Logan, an influential Philadelphia merchant, who wrote Robert Hunter, the Governor of New York and New Jersey, to warn him of the pirates in Nassau as well as those attacking vessels near Philadelphia. Logan wrote that their commander was named Teach and that two years ago, he sailed from Jamaica to Philadelphia as a mate onboard a vessel. In addition to Blackbeard, most of his men were also recognized as having been sailors who regularly sailed the waters of Philadelphia. Logan's original letter is housed at the Historical Society of Pennsylvania. A transcript of the entire letter follows:

> We have been very much disturbed this last week by Pirates they have taken and plundered Six or Seven Vessels bound out or into this river Some they have destroyed Some they have taken to their own use & Some they have dismissed after Plunder. Is [Isaac] Flower will I believe be more particular.

> Some of our people having been Several day or on Board them they had a great deal of free discourse to them, they say they are about 800 Strong at Providence & I know not how many at Cape near Carolina, where they are also making a Settlemt Capt Jennings they Say is their Gov in chiefe & heads them in their Settlemt The Sloop that came on our Coast had about 130 Men all Stout Fellows all English without any mixture & double armed they waited they Said for their Consort

a Ship of 26 Guns wth whom when joined they designed to Visit Philadia, Some of our Mastr Say they knew almost every man aboard most of them having been lately in this River, their Comandr is one Teach who was here a Mate from Jamca about 2 ys agoe.[2]

My transcription of this letter into modern text reads:

We have been very much disturbed this last week by pirates. They have taken and plundered six or seven vessels bound both in and out of this river. Some of those vessels they have destroyed and some they have kept for their own use. And some they have dismissed after plundering them. I believe that Isaac Flower, one of the released prisoners, will be able to provide more details.

Some of our people were held for several days onboard the pirate sloop. They had a great deal of free discourse with the pirates, who say they are about 800 strong at Providence. I know not how many may be at Cape Fear near Carolina, where they are also making a settlement. Captain Jennings, they say, is their Governor in chief and heads them in their settlement. The sloop that came to our coast had about 130 men, all stout fellows, and all English, without any mixture. They were all double-armed. The pirates said that they were waiting for their consort, a ship of 26 guns with whom they planned to attack Philadelphia. Some of our ship's masters say they knew almost every man aboard, most of them having lately sailed in this river. Their commander is one Teach, who has sailed here as a mate onboard a vessel from Jamaica about 2 years ago.

The third primary source document is a letter written by Jonathan Dickinson, the Mayor of Philadelphia, to Joshua Crosby, a Quaker merchant from Kingston, Jamaica. Dickinson wrote to tell Crosby of the pirate activity at Cape May. In his letter, Dickinson mentions the name "Blackbeard." This is the earliest known reference to that name in the historical record. This letter, dated October 23, 1717, is also housed at the Historical Society of Pennsylvania. An exact transcript of Dickinson's letter reads:

Thou mentons pyrating trade wth you, from the begining of this Mo until wthin this Week one Capn Tatch alls Blabeard in a Sloop whch they call Revengers Revenge About 130 Men, 12 or 14 Guns having layne of or Capes & Taken six or seven Vessels Inwd & outwd bound

James Logan (1674–1751) was an Ulster Scots Quaker from Ireland who was a statesman, administrator, scholar, and the fourteenth mayor of Philadelphia.

Chapter Nineteen

> My Son Joseph wnt wth Jn Annis out of or Capes 29th bri and Day pyratt took a Sloop to Southward of our Capes.[3]

When transcribed into modern text, it reads:

You mention the pirating trade with you, from the beginning of this month until within this week. One Captain Tatch, alias Blackbeard, in a sloop which they call the Revengers Revenge *with about 130 men, with 12 or 14 guns, having waited off our capes, took six or seven vessels both inward and outward bound.*

My son Joseph went with John Annis out of our Capes on the 29th of September and on that day, a pirate took a sloop just south of our Capes.

So far, these three prime source documents are the only ones known that specifically link Blackbeard directly to Philadelphia. However, they create far more questions than answers. To make things even more confusing, each question has the potential to erupt into a cascade of possibilities. Was Philadelphia his home port, or did he just sail there occasionally? Was Blackbeard known as a captain or just a member of the crew? Was Blackbeard known as an honest seaman, or was he known as a pirate? Without further information, these questions can't be answered with any certainty.

Working backward, the Dickinson letter in which he wrote, "One Captain Tatch, alias Blackbeard" proves to be a vital clue in establishing the fact that at least a few of the people in the Philadelphia area knew our favorite pirate by that name on or before October 1717. In their 1717 publications, *The Boston News-Letter* only referred to him as *Teach*, so Dickinson couldn't have read the name of Blackbeard in the paper. Additionally, his remarkable letter is dated October 23, 1717. One or more of the captives held by the pirates must have told Dickinson that the pirate captain was called Tatch and was also known as Blackbeard. Of course, this doesn't clarify whether those captives already knew him by the name of Blackbeard from years ago or whether they just learned his name as they were being held onboard.

Regarding the first two prime source documents, the word *mate* can have one of two nautical meanings. The general public often refers to every common sailor as a mate. Even today, they are frequently called *shipmates*. However, when used professionally, it refers to one of the ship's officers. The rank of First Mate was often just below the captain within the command structure. When taken literally, the use of the word m*ate* in this context suggests that Teach was more than just a sailor, but not yet a captain.

A ship's officer would likely be known to others ashore, as he would be the one selling cargo and buying supplies.

Logan's letter offers several more fascinating clues. He mentions that not only was Teach recognized, but some of their ship's masters said they knew almost every man aboard, most of them having lately sailed in this river. Were they part of Blackbeard's crew when they sailed there before, or were these sailors scattered among a dozen vessels in the Philadelphia area and just happened to have recently joined Blackbeard's crew? Since the *Revenge* was originally crewed by men hired by Bonnet on Barbados, it is possible that these sailors were all members of crews who were part of the regular Philadelphia/Barbados nautical trade and part of Bonnet's crew. However, it is just as possible that they were all with Blackbeard in Nassau and came aboard the *Revenge* when he took command.

Logan also mentioned that Teach had sailed there from Jamaica about two years earlier. Was Teach simply a mate on board a merchant vessel that sailed from Jamaica to Philadelphia one time in 1715, or was he a frequent visitor? However, there is one more captivating possibility.

The likelihood that Blackbeard came to Philadelphia as a pirate in 1715 or even earlier isn't as remote as one might think. The port town of Marcus Hook held the keys to the door of the Philadelphia docks. Located along the Delaware River about fifteen miles before reaching the city, every large vessel was required to stop there and arrange for a professional pilot to take them into port. Otherwise, the vessel could run aground on the tricky shoals. This is just as true today as it was in the early eighteenth century. In modern times, a ship's captain simply radios the port authority and they send a boat to guide the ship in.

> A *pilot* is an individual who is not a member of the vessel's crew, but who comes aboard to help navigate the vessel in or out of port.

Figure 45: *Close-up of Marcus Hook from a 1771 Map of Pennsylvania, Library of Congress. The image is dedicated to the public domain under CC0.*

Chapter Nineteen

In Blackbeard's day, the town of Marcus Hook had dozens of professional pilots living in town who would board the large vessels and guide them the rest of the way to Philadelphia. Authorities living in the city and towns lining the shore along the route kept a watchful eye on the crowded river. In contrast, downriver from Marcus Hook and all the way to Cape May, the river was more obscure. With scattered shoreline towns and shipping spread out along a wide river, pirates could freely come and go without much notice.

The land surrounding Marcus Hook was originally settled by the Swedes in the 1640s.[4] They named their colony *New Sweden*. In 1655, the Dutch took over the colony.[5] Under the Dutch administration, the town grew and acquired the name of Marretties Hooke. Even though the Dutch were administering the area, the majority population remained Swedish. William Penn began purchasing land from the Dutch government in 1679, and the next year, an early English settler opened a tavern in the village, serving liquor. With a mixture of Swedes and English Quakers peacefully living together, the name was officially changed to Marcus Hook in 1682. By 1699, the entire area had been purchased by Penn and became the Quaker-friendly colony of Pennsylvania.[6] By 1700, Marcus Hook was an eclectic mix of Swedes, early English settlers, Quakers, and a significant Irish population who had recently fled Ireland as a result of the first Jacobite revolution in 1689.

Marcus Hook was a perfect place for pirates to sell their wares. It was populated by people who wanted to make money and weren't too particular about how they did it. The Quakers, Swedes, and Irish had no love for the English government and didn't much care for the taxes that were imposed upon them. Additionally, thousands of settlers, mostly Quakers and Irish Catholics, were flocking to this fledgling colony of Pennsylvania that offered religious freedom for everyone. Those citizens desperately needed merchandise. All of this was the perfect recipe for a port town friendly to pirates.

In 1696, Lt. Governor Markham called the Council of Pennsylvania together to consider charges, made by the Lords of Trade against Philadelphia, for encouraging trade with pirates.[7] However, the governor's son-in-law, Manes Brown, was refused a seat in the House of Representatives because of accusations that he had a strong connection with pirates.[8] In his 1884 book, *History of Delaware County*, Henry Ashmead wrote:

> If tradition be accepted as authority, at the conclusion of the seventeenth and the first and second decades of the eighteenth century the pirates which then infested the Atlantic coast from New England to

Georgia would frequently stop at Marcus Hook, where they would revel, and when deep in their cups would indulge in noisy disputation and broils, until one of the streets in that ancient borough from that fact was known as Discord Lane, which name the same thoroughfare has retained for nearly two centuries.[9]

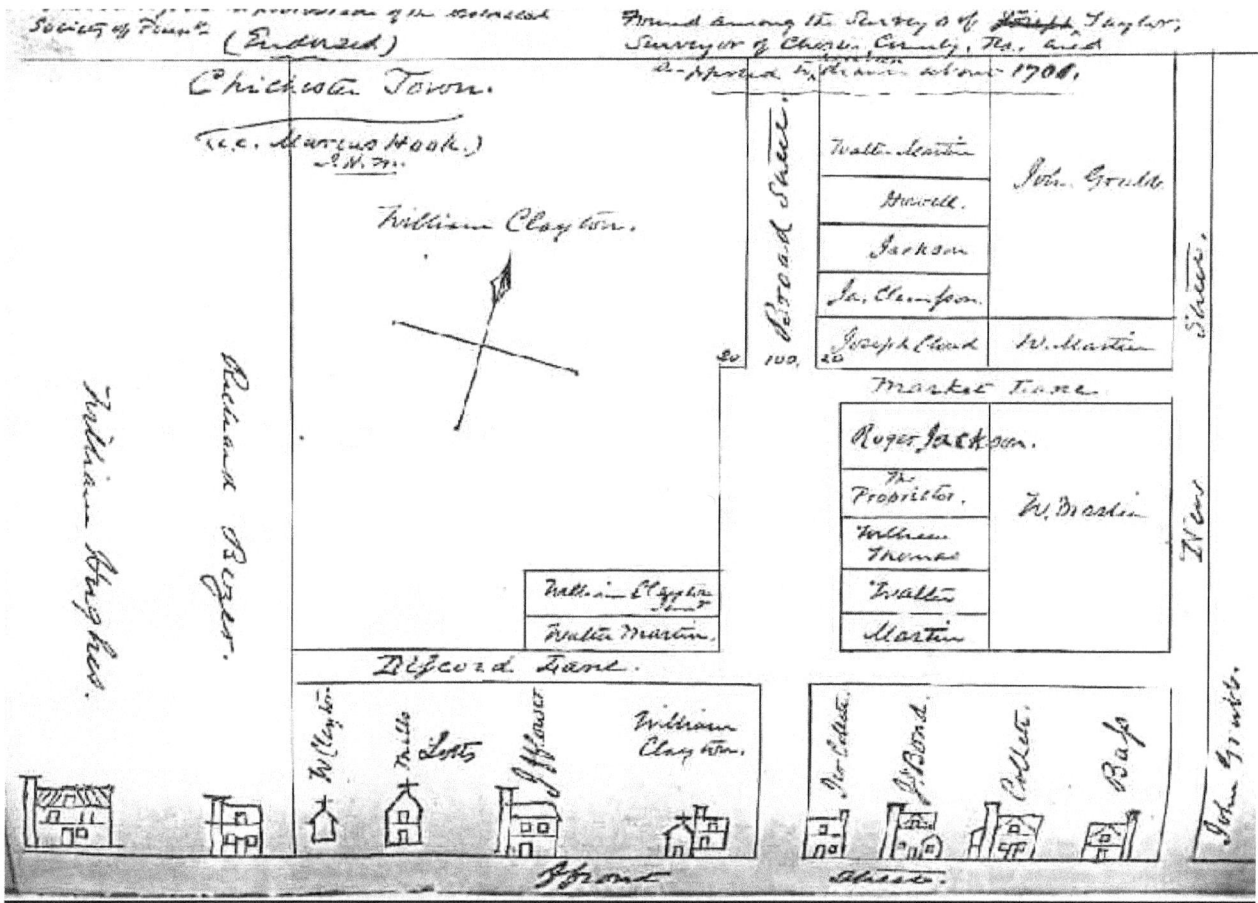

Figure 46: *1708 Map of Marcus Hook, Courtesy of the Marcus Hook Preservation Society.*

As mentioned by Ashmead, the street where the pirates would revel was called Discord Lane. That's where the taverns were located. Today it is called 2nd Street. Discord Lane clearly appears on a map that dates between 1701 and 1708, and it is still called Discord Lane on a map that is dated 1870. According to the minutes of the Council of Pennsylvania, Marcus Hook was officially designated as a market town on Feb. 14, 1700.[10]

That same year, the minutes of the council show that people in that area were accused of illegally trading with pirates. In 1703, the government produced a pamphlet that stated, "these Quakers have a neat way of getting money by encouraging the pirates, when they bring in good store of gold."[11] In John Watson's book, which was first published in 1830, *Annals of Philadelphia and Pennsylvania*, Watson wrote:

> The bay and river doubtless furnished them many a secure place in which they could refit or provide their necessary supplies. Perhaps as jolly sailors, full of money and revelry, they sometimes found places even of welcome, from those who might choose to connive at their real character.[12]

There is no doubt that Marcus Hook was a pirate-friendly town. Pirates could freely walk the streets without question or fear of arrest. The crew could easily sell their goods for large profits and stop off at the tavern for a quick one while they waited to be paid. Under these circumstances, it is easy to imagine Blackbeard frequenting Marcus Hook as a pirate in 1715, or even earlier. That last statement is highly speculative. Is there any evidence that Blackbeard ever visited Marcus Hook? The answer to that question is yes!

On August 11, 1718, Governor Keith addressed the Council of Pennsylvania concerning Blackbeard's presence in the Philadelphia area.[13] Additionally, he issued a warrant for his arrest.

> Upon an Information that one Teach a Noted Pirate, who has Done the Greatest Mischeif of any to this Place, has been Lurking for some Days in & about this Town I have Granted a Provincial warrant for his being apprehended.[14]

But what about before 1718? What evidence is there that Blackbeard came to Marcus Hook or even Philadelphia regularly earlier in his career as a pirate? The only source for such evidence comes from John F. Watson's book, *Annals of Philadelphia and Pennsylvania*. Watson was a long-time member of the Historical Society of Pennsylvania.[15] His book is well-researched and includes an intriguing mixture of oral tradition and prime source documents, such as letters and council meeting notes. What sets his work apart from others is the rich collection of oral histories that provides details not found in the documentation. He apparently interviewed hundreds of people in the Philadelphia area and accurately recorded their stories and recollections. First published in 1830, subsequent editions were published in 1844, 1850, and 1884. These later editions include additional content on Blackbeard.

As mentioned in Chapter Seven, while discussing Thomas T. Upshur's 1901 article, *Eastern-Shore History*, histories based upon oral tradition are often overlooked or even ignored by some historian authors. However, oral tradition, the stories passed down from generation to generation, can sometimes provide valuable and accurate accounts of the events of the past, especially in the centuries before books and journals became prevalent.

Even so, Watson's *Annals of Philadelphia* is significantly different from Upshur's *Eastern-Shore History*. In addition to historical traditions and events, Upshur was looking for ghost stories and local legends. For example, one of his accounts tells of a headless man who demands payment of a toll in order to cross a bridge.[16] Watson was entirely focused on facts.

The initial publication date of 1830 is also important. That was just one hundred and twelve years after Blackbeard's last visit. It may seem like a long time comprising many generations, but this isn't necessarily so. The older people he interviewed may have been only two generations removed from the tellers of those accounts. As I write this book, I am sixty-eight years old. Throughout my formative years, my grandfather, who was born in 1893, told me many stories of his youth, which I would be able to recount word-for-word today. It is conceivable that as Watson was conducting his interviews, the citizens of Pennsylvania were repeating tales of Blackbeard that had actually been experienced by their grandparents.

The most interesting of all accounts in Watson's book concerning Blackbeard's visit to Philadelphia comes from Mrs. Bulah Coates.

> Mrs. Bulah Coates, (once Jacquet—this was the name of the Dutch governor in Delaware, in 1658,) the grandmother of Samuel Coates, Esq., late an aged citizen, told him that she had seen and sold goods to the celebrated Blackbeard, she then keeping a store in High street, No. 77, where Beninghove owned and dwelt—a little west of Second street. He bought freely and paid well. She then knew it was he, and so did some others. But they were afraid to arrest him, lest his crew, when they should hear of it, should avenge his cause by some midnight assault. He was too politic to bring his vessel or crew within immediate reach; and at the same time was careful to give no direct offence to any of the settlements where they wished to be regarded as visiters and purchasers, &c.[17]

Another account in Watson's book also places Blackbeard near High Street.

> The present aged Benjamin Kite has told me, that he had seen in his youth an old black man, nearly 100 years of age, who had been one of Blackbeard's pirates, by impressment. He lived many years with George Grey's family, the brewer in Chestnut street, near to Third street. The same Mr. Kite's grandfather told him he well knew one Crane, a Swede, at the Upper ferry, on Schuylkill, who used to go regularly in his boat to supply Blackbeard's vessel at State Island. He

also said it was known that that freebooter used to visit an inn in High street, near to Second street, with his sword by his side.[18]

There is another account in Watson's book that speaks of a black man who was supposedly a member of Blackbeard's crew. Both accounts may refer to the same person.

> Robert Venables, the old black man who died in 1834, aged 98, told me that he knew personally an old black man, and Carr, a drayman, in Gray's alley, both of whom had been with Blackbeard.[19]

However, the most persuasive sentence in Watson's book doesn't come from an individual, it comes from an entire community. At the time Watson was compiling his memoirs, the people of Marcus Hook strongly believed that Blackbeard had a Swedish girlfriend named Margaret who lived in town and that Blackbeard visited on occasion.

> There is a traditional story, that Blackbeard and his crew used to visit and revel at Marcus Hook at the house of a Swedish woman, whom he was accustomed to call Marcus, as an abbreviation of Margaret.[20]

Figure 47: *Photograph of the Plank House taken by Robert Jacob.*

According to Michael Manerchia, the president of the Marcus Hook Preservation Society, that belief is still prevalent among many of the residents of Marcus Hook today. Additionally, he believes that the very house where Blackbeard visited his girlfriend, Margaret, still exists. Called the Plank House, its construction dates to the time period and is located just one block from Discord Lane at the corner of Market and Broad. It is currently under renovation, with plans to become a museum someday.

1715 is a very significant date for pirates. It was the year the Spanish treasure fleet sank and the year that Nassau developed as a pirate haven. It marks the boundary between the small-time pirates who sailed from Jamaica and hoped they wouldn't be caught and the big-name pirates out of Nassau. This date is also significant to Blackbeard because it was the date mentioned in James Logan's 1717 letter as the last time Blackbeard was in Philadelphia.

Theoretically, Blackbeard could have been one of those small-time pirates. Prior to that, he may have sailed as a privateer in the recent war against Spain. Privateers were legal, but still looked for the best deals when disposing of their captured merchandise. Marcus Hook was clandestinely known as the most profitable port on the coast for privateers to sell questionable goods.

When the war ended in 1714 and no letters of marque were being issued, many privateers turned to piracy. Some even continued the pretense that the war was still going on. Jamaica had been the center of privateer activity and quietly continued as the home port for those who had secretly become pirates. It is widely believed that Blackbeard was one of these privateers-turned-pirates based in Jamaica.

With the end of the war, Blackbeard's actions would have been illegal, so it was imperative for him to maintain a low profile. Taking ships in the Atlantic, he continued to sail to Marcus Hook to sell his goods. His girlfriend, Margaret, was another reason for his visits. This scenario perfectly matches Logan's letter. A mate from Jamaica visited in 1715.

After the Spanish treasure fleet sank, Blackbeard's situation changed. While fishing the wrecks, he partnered with the pirate Ben Hornigold and remained with him until September 1717. His first opportunity to return peacefully to Marcus Hook would have been after he accepted the king's pardon in July 1718. It was finally the ideal time for Blackbeard to reunite with his girlfriend, and I believe that is exactly what he did. The multitude of circumstantial evidence is enough to convince me he also visited Philadelphia near High Street, just as Mrs. Bulah Coates and Benjamin Kite recollected. There are no other reports of Blackbeard's whereabouts during that time.

Unfortunately, this presumptive evidence concerning Blackbeard's Marcus Hook and Philadelphia connections does not lead to any indisputable conclusions. Until further evidence surfaces, Blackbeard's whereabouts between mid-July and mid-August 1718 remain unverified. In any case, when Blackbeard reemerged and once again entered the historical record, he did it as a pirate.

Twenty
I'll Take Two

1786

Benjamin Franklin became the first person to chart the Gulf Stream.

It was hot and the captain and crew aboard the small sloop *Adventure* were becoming restless. They had crossed seven hundred miles of open sea without sighting a single sail since they left the Delaware River. Blackbeard assured his men that there would either be French or Spanish prizes to take within the main shipping lanes. Most vessels leaving the Caribbean followed the Gulf Stream along the Florida coast and turned east for Europe sometime after passing the Virginia Capes. The *Adventure* had already crossed the Gulf Stream and had reached Bermuda without any success. Blackbeard ordered a course change, and the sloop turned south by southwest, hoping to pick up some vessel that left the Gulf Stream for Europe a little early. It was Saturday, August 23, 1718, which was September 3, 1718, in the French Gregorian calendar.

The late afternoon sun was sinking in the sky, which made it difficult to scan for sails. Their course was taking them directly into the sunset and the bright light hurt their eyes. One of the crew enthusiastically shouted, "Sail Ho!" Squinting at the orange horizon, Blackbeard could just barely make out the shapes of two ships, one sailing about three miles behind the other. His crew became energized with excitement. The quiet deck suddenly became engulfed in a flurry of activity as the crew checked their guns and prepared for battle. When the blinding sun finally dropped below the dark blue water, the image of the two ships became clear. They were near the 32-degree north latitude, about thirty miles southwest of Bermuda.

Lucky for Blackbeard, the two ships were French—*La Rose Emelye*, under the command of Captain Jan Geropil with first officer Pierre Boyer, and *La Toison d'Or*, under the command of Captain Eslye Wansbabel. The crew of *La Rose Emelye* consisted of seventeen men, and the ship was armed with 4 guns. They had left Martinique on August 5, 1718, which was August 16, 1718, in the French Gregorian calendar. Their huge cargo consisted of one

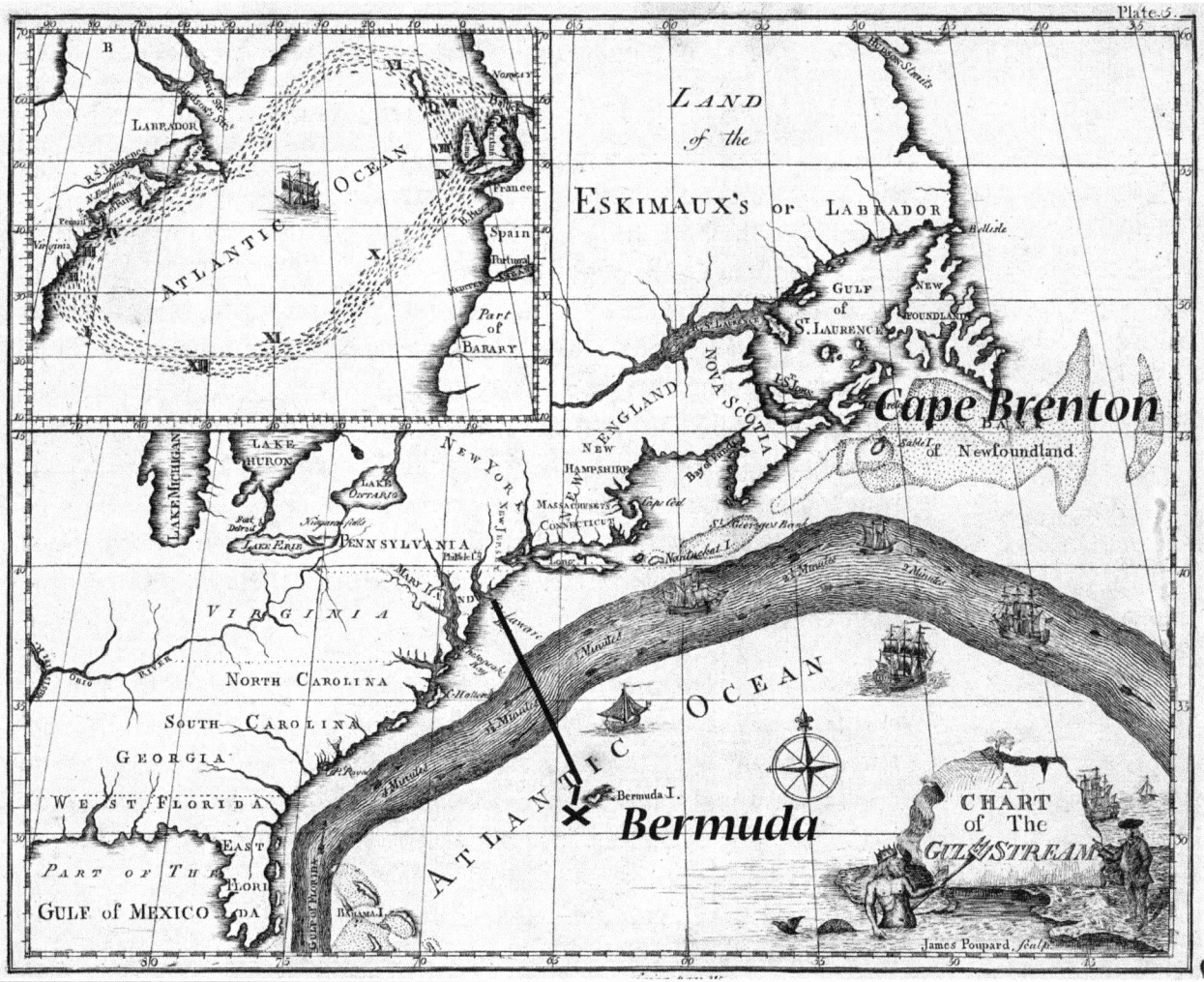

Figure 48: *Benjamin Franklin Map of the Gulf Stream, Library of Congress. The image is dedicated to the public domain under CC0.*

hundred eighty hogsheads of white sugar, around three hundred jute sacks of milled cocoa, and three bolts of cotton.

With all sails set for the maximum speed, the *Adventure* was on a course to intercept the lead ship. It was obvious to Blackbeard that his sloop had been sighted by the French ships, but they didn't seem to be concerned. They didn't alter their courses to evade him or even take any action that would cause Blackbeard to expect a fight was coming. As daylight dwindled into darkness, the distance between the three vessels shortened. At about ten o'clock in the evening, the *Adventure* came within hailing distance of *La Rose Emelye*. Blackbeard looked up at the ship's gunports and side rail to see if they were going to fire. To his surprise, one port opened, and the gun was run out. With no other sign of hostility, this single gun was clearly meant as a warning. A shout came from a man standing at the rail. It was in French, but Blackbeard understood the meaning. The man asked who they were and where they had come from. Blackbeard paused

Chapter Twenty

> Vessels were often identified by the number of guns they carried. This wasn't always consistent. Many vessels had 2 guns mounted at the bow and 2 guns mounted at the stern in addition to the guns they had that were mounted along the sides. Occasionally, a vessel with three guns on one side would be identified as having only 6 guns, not considering the bow and stern guns. Another person may describe the same vessel as having 8 guns or even 10 guns.

for a moment to contemplate the possibilities as he now realized that this ship was prepared to fight, and its size gave the French the advantage.

Blackbeard's sloop had only 6 guns mounted on his sides, which meant that he could only use 3 guns at once. A broadside of 3 guns wouldn't be enough to force a ship that size to surrender. He needed something else to frighten the French into submission. Blackbeard surmised that taking the other French ship might solve his problem. If he took the second ship in clear view of the crew on the first ship, those men might be frightened enough to reconsider their position. Blackbeard didn't answer the French officer's verbal challenge. Instead, he ordered the *Adventure* to change course and sailed straight toward the second ship. On *La Rose Emelye*, Captain Jan Geropil, first officer Pierre Boyer, and the entire French crew watched in horror as Blackbeard's sloop came alongside *La Toison d'Or* and his pirates swarmed aboard. They were amazed at the speed and efficiency of Blackbeard's pirates and at how easily *La Toison d'Or* fell under their control.

A few pirates remained on board *La Toison d'Or* to guard the prisoners and the rest climbed down the ship's side and returned to the *Adventure*. After pushing away from the side of the ship, the *Adventure* came about and sailed straight at *La Rose Emelye*. As Blackbeard approached, he fired two naval guns mounted at his sloop's bow. The cannonballs splashed in the water halfway between the two vessels. As the *Adventure* came closer, Blackbeard fired again, and the cannonballs splashed very close to *La Rose Emelye's* bow. This was followed by several volleys of small-arms fire. Blackbeard's plan had worked. The confidence that Captain Jan Geropil once had was replaced with sheer terror. With no return fire from the French, Blackbeard ceased fire and came alongside of the big ship.

As he had done many times before, Blackbeard ordered the captain and five of his men to lower a boat and row over to his sloop. It was standard practice for Blackbeard to take a prize ship's captain hostage before his pirates boarded. Reluctantly, Captain Geropil complied. Once the French captives were safely held onboard the *Adventure*, the pirate boarding party rowed over to *La Rose Emelye* and climbed up the side of the ship. Each man was exceptionally well armed with a fierce assortment of pistols, cutlasses, muskets, and blunderbusses. Their appearance was overwhelmingly intimidating to the French.

Once aboard, the leader of the pirate boarding party, probably quartermaster William Howard, began inspecting the ship's compartments. He stepped down the ladder into the hold and gawked at the huge cargo of sugar and cocoa. He couldn't believe his luck. The cargo's enormous value was immediately clear to Howard. Howard quickly returned to the *Adventure* to confer with his captain, Blackbeard. It was apparent that the cargo was too large to transfer to their sloop, they would have to keep the ship. Blackbeard and Howard rowed over to *La Rose Emelye*. The French crew were gathered together and Blackbeard told them he was keeping their ship and they would all have to transfer over to *La Toison d'Or*. In his typical flare for the dramatic and to maintain a high level of intimidation, Blackbeard added that they were lucky that there was another ship nearby. He told them that if there hadn't been, he would have set fire to their ship and everyone would have been killed.

As soon as Captain Geropil, First Officer Pierre Boyer, and the rest of *La Rose Emelye*'s crew were safely aboard *La Toison d'Or*, Blackbeard set sail. The French sorrowfully watched as the *Adventure* and *La Rose Emelye* disappeared over the horizon. Captain Geropil realized that their provisions wouldn't last throughout the long voyage back to France, so they sailed northeast toward the nearest French port, Cape Brenton, nine hundred and sixty miles away. With fresh provisions, *La Toison d'Or* sailed back to the port of Nantes in France, where First Officer Pierre Boyer gave a deposition of the incident. His original deposition remains housed in the French Archives.[1] A translation of this deposition was used to produce a list of details taken directly from Pierre Boyer's own words. Although the previous narrative was written from Blackbeard's perspective instead of that of the French, I used Boyer's details to provide the facts and the sequence of events. The literal translation of Boyer's testimony is difficult to read due to the eighteenth-century syntax of the French language.

Pierre Boyer identified the names of the ships and the names of the captains, and mentioned that *La Rose Emelye* was "armed with four cannons and staffed with 17 men." He gave the date of "last April 28th" when they left Martinique and that their cargo consisted of "one hundred eighty hogsheads of white sugar . . . three hundred jutes of milled cocoa and three bales of cotton." Boyer described first seeing the *Adventure,* saying that "the night of the 3rd heading to the 4th being 32 degrees in latitude north . . . at eight to ten leagues in the south-southeast at around six to seven in the evening they noted a ship of around 35 tons armed with six canons." The sloop was sailing on a course of "south-southeast" and when they were "two and half to three leagues which seemed to them to come from Bermuda and around ten in the evening being approached."

Boyer described the battle, starting with his shout where he:

> demanded where the boat originated . . . he had drawn canon on Him. Not responding, he turned the ship on *La Toison d'Or* which

Cape Brenton is in Nova Scotia. After the war in 1714, it was the last remaining French-controlled port along the east Atlantic coast of New France and was the site of a French garrison at Fort Sainte Anne.

was a league behind . . . Which was taken by the said pirate ship. He charged . . . on which he shot three or four rounds of canons and several musket rounds . . . Their commandant told the said captain to board with five of his crew . . . after which he sent a rowboat full of men armed with rifles, pistols and sabers who climbed aboard the said declaring which they searched and robbed all and having examined the cargo of the said boat and found it richly loaded They chased them from their ship with an order to go aboard *La Toison d'Or* . . . they would have set fire to the said boat if they could have . . . The supplies running short they had been obliged to set down at Cape Breton where they made and took supplies.

It is certain that William Howard was with Blackbeard during this attack. The following month, he was arrested in Virginia and Lt. Governor Spotswood mentioned Howard's participation in a letter to the Board of Trade writing:

> One Howard, Tach's Quarter Master came into this Colony with two Negros which he own'd to have been Piratically taken, the one from a French ship and the other from an English Brigantine,[2]

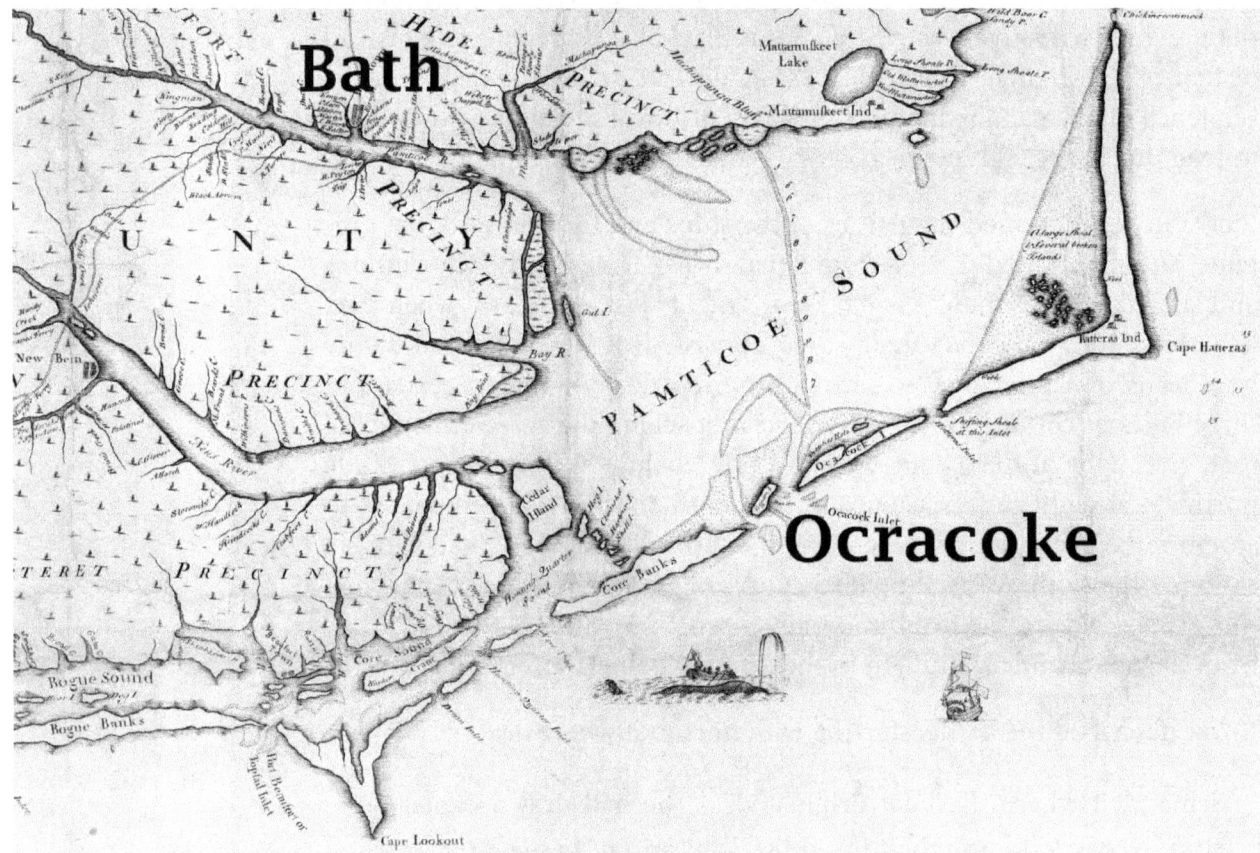

Figure 49: *Ocracoke to Bath from 1733 Moseley Map Courtesy of East Carolina University.*

I'll Take Two

White waves curled and splashed over the bows of Blackbeard's two vessels as they cut through the dark blue sea. One was Blackbeard's *Adventure*, and the other was his captured French merchant ship. In the distance, a familiar shoreline grew closer. They were headed for a break of blue/white water within a solid green line of trees and seagrass. As the two vessels came within a mile or two, the crew could hear the surf as it crashed onto the beach. The inlet was tricky to navigate. There appeared to be two channels, but only the one on the right was safe to sail through. Beacon Island lay to the left of this channel, which had a wooden tower with lanterns. At night, the resident pilots would light these lanterns to mark the channel in the darkness. On the right side of the channel was a ten-mile-long island strip of land that was generally less than a mile in width. Blackbeard knew these waters well. This was Ocracoke Inlet and Ocracoke Island.

Ocracoke was an ideal spot for a base. As shown on Moseley's 1733 map, a freshwater well had been built there many years before Blackbeard arrived. The inshore side of the island was protected from the strong waves of the Atlantic. Blackbeard had sailed there a few months earlier after he marooned twenty-five men on Harbor Island. The *Adventure* carefully worked its way through the inlet followed by the *Rose Emelye*. They turned to the right and slowly sailed to a narrow cove that was adjacent to the well. The sails were taken in, and the vessels glided to a gentle stop as the anchors splashed into the water. Months later, some of Blackbeard's men were put on trial in Williamsburg. Part of the testimony at that trial reads:

> Hesikia Hands late Master of the Sloop Adventure Commanded by Edward Thache being sworne and examined deposed that he was on Board the said Sloop Adventure at the takeing of two French ships in the Month of August last past . . . that Thache plundered one of the ships of some Cocoa and the other brought in with him to North Carolina having first put her Crew on Board the ship first mentioned that soone after Thaches Arrival at Ocacock Inlet.[3]

Telling Blackbeard's story is exceptionally challenging. Surprising and enigmatic events constantly arise that defy rational explanations within the tangled details of his life. As soon as the two vessels came to rest at Ocracoke, Blackbeard sent for a surgeon to treat two of his crewmembers. The first report of this unexpected development came from the November 10 to November 17, 1718 edition of *The Boston News-Letter* which reads, "That Teach the Pirate has brought in a Ship to Oackrycook Inlet and Unrig'd her, and suffer no man to go on board except a Doctor to cure his wounded Men."[4] Captain Brand also mentions this surgeon in his letter dated February 6, 1719. Brand wrote:

Blackbeard calls for a surgeon at Ocracoke

Chapter Twenty

In eighteenth-century terminology, "bursting a piece" means shooting a handgun.

I had certain Information that Thach had been at sea and was come into that place again and had brought in with him a ship loaded with sugar ect. and gave out they found her a Wreck in the sea, that his sloop was lying there before Bath Town in No. Carolina and the ship carried into Oerecrock inlet in Pamplico sound, and no persons suffered to go on Board her or the sloop Except a Surgeon that went to dress two wounded men, which they said was done by bursting of a piece.[5]

To unravel this mystery, we must examine the facts one at a time. The nearest doctor or surgeon to Ocracoke would have been Bath. In Brand's letter, the *Adventure* had sailed to Bath to fetch this doctor. Both sources seem to indicate that the two wounded men were aboard the *Rose Emelye*. The article in *The Boston News-Letter* states "Ship to Oackrycook . . . and suffer no man to go on board except a Doctor." Brand wrote, "and no persons suffered to go on Board her or the sloop Except a Surgeon." Brand's statement implies that both vessels were anchored next to each other when the surgeon was ready to provide treatment. This could only be at Ocracoke. If the two men needing treatment were onboard the *Adventure*, the doctor would have treated them in Bath. There would be no need to take him to Ocracoke. As for Blackbeard, he probably remained at Ocracoke when some of his men traveled to Bath to fetch the doctor. He was well known and other documents state that he wasn't seen in Bath until several weeks later.

Taking Brand's statement literally, the *Adventure* returned from Bath with the surgeon onboard. As the *Adventure* approached Ocracoke, the surgeon, standing on the deck, could see the *Rose Emelye* at anchor near the shore with Blackbeard and the rest of his crew waiting nearby on the beach. This doctor was escorted to the *Rose Emelye* and boarded, while Blackbeard and his men watched anxiously from the shore.

Next, we shall examine the nature of their wounds. *The Boston News-Letter* states that the doctor was needed to "cure his wounded Men." Brand adds the cause of their wounds when he wrote that the doctor "went to dress two wounded men, which they said was done by bursting of a piece." Both sources agree that these men were wounded, but that explanation is highly problematic. First of all, where did they get their wounds? There was no battle when the two French ships were taken and no handguns were fired. It is unlikely that they received their wounds by accident. One man can certainly be accidentally shot at sea, but not two. However, the real problem with the gunshot account was the need for their isolation. Both sources agree that only the doctor was permitted onboard, or as Brand put it, "no persons suffered to go on Board her or the sloop Except a Surgeon." Why would Blackbeard take such extreme precautions to treat two simple gunshot wounds?

No source documentation provides the answer to this question, but there is an alternative medical need for a doctor that makes logical sense. Perhaps Blackbeard suspected that these two crewmen had Yellow Fever. This dreaded disease was prevalent in North Carolina in the early eighteenth century. The province's first governor, Edward Hyde, died of yellow fever on September 8, 1712, in the Chowan Precinct.[6] If Blackbeard suspected that two of his crew had contracted Yellow Fever, he would have isolated them on the *Rose Emelye* and sent for a doctor as quickly as possible to provide treatment. When the doctor arrived, he would have been the only person allowed onboard. This version is highly speculative, but it fits the facts and answers the isolation question. If this was the case, we must conclude that it was a false alarm, since there were no reports of any of Blackbeard's men dying from illness, wounds, or any other reason.

Edmund Hyde, Governor of North Carolina (January 24, 1712–September 8, 1712)

The next reasonable question concerns *The Boston News-Letter*'s article. Why would the paper report that these men were wounded? Again, the answer is all speculation. Perhaps this came from Blackbeard himself. Everyone in Bath would have known that he needed a doctor. After all, just a day or two before, the *Adventure* arrived in town and left shortly afterward with a doctor onboard. Yellow Fever is highly contagious. If a rumor circulated around town that some of his crew had Yellow Fever, no one would have wanted to do business with him. On the other hand, no one would have had any concerns over two of his crew suffering from gunshot wounds. In this scenario, after examining the two afflicted crewmen on the *Rose Emelye*, the doctor came ashore and gave Blackbeard the good news. It wasn't Yellow Fever. Greatly relieved, Blackbeard asked the doctor to tell everyone in Bath that he had simply treated two wounded men. Blackbeard may have even given the good doctor an additional sum of money to ensure that this was the story the doctor told. Brand's letter wasn't written from first-hand knowledge. His letter was three months after *The Boston News-Letter* article. Since his wording is suspiciously similar, it is likely that Brand's only source for his report on the two wounded men came from that article.

Either way, with the medical crisis resolved, Blackbeard's crew began the tedious process of unloading all the cargo from the captured French ship while setting up camp. Throughout the first three weeks of September 1718, sails were turned into tents and dining flies, and open-air campfires were built. The *Rose Emelye* had many barrels, hogsheads, and casks of sugar on board, as well as dozens of bags of cocoa.[7] For protection from

Chapter Twenty

the weather, surf, and ocean mist, all the cargo was safely secured under sailcloth tarps. Once the *Rose Emelye* was completely unloaded, it was towed out into the sound and set on fire. An excerpt from a letter that Lt. Governor Spotswood of Virginia wrote to the Council of Trade and Plantations reads:

> That Tach with divers of his crew kept together in North Carolina went out at pleasure committing robberys on this coast and had lately brought in a ship laden with sugar and cocoa, which they pretended they found as a wreck at sea without either men or papers, that they had landed the cargo at a remote inlett in that Province and set the ship on fire to prevent discovery to whom she belonged:[8]

September 14, 1718
William Bell Incident

Meanwhile, an incident occurred that would have a profound effect on the lives of four of Blackbeard's crewmen, as well as Tobias Knight and Governor Eden. It was a little before daybreak on September 14, 1718. William Bell of Currituck Precinct was peacefully sailing his periauger down the Pamlico River in the darkness. He was accompanied by his son and an Indian boy. About three miles from town, as they passed John Chester's Landing, Bell noticed another periauger come alongside. He could tell that there were four men onboard. Two appeared to be white and the other two were either black or were white men disguised as blacks. A shout came from the other vessel asking if he had anything to drink. Bell knew that he had half a barrel of brandy onboard, but it was exceptionally dark that morning and Bell couldn't see precisely where the brandy was. He politely answered, saying that it was too dark for him to find it. In response to Bell's reply, he was violently attacked.

With a sword in hand, one of the men jumped over to Bell's periauger and ordered Bell to put his hands behind his back. Bell was horrified. This villain swore damnation and threatened Bell with death if he didn't reveal the location of any money he had onboard. Although frightened, Bell wasn't about to be robbed so easily. He grabbed his attacker's sword hand, and a struggle began. Two of the other men jumped over to Bell's periauger to help subdue him. The brief struggle ended when the armed attacker struck Bell with the flat side of his sword, breaking the blade in half. Bell now sat helplessly on the deck as his attackers stood over him. They demanded that he give them any pistols that he had. Bell told them that he had a few pistols locked in a chest. He was forced to unlock the chest and his weapons were taken. The two periaugers were then sailed to the middle of the river where Bell was robbed of about £62 in cash, 58 yards of crape cloth, a bin of pipes, half a barrel of Brandy, and a uniquely fashioned silver cup that resembled a chalice. The attackers tossed Bell's sails and oars overboard and left him in the river. Bell managed to retrieve his oars and immediately rowed three miles up the river and into Bath Creek, where he landed his periauger at the house of Tobias Knight.

As Chief Justice, Knight was the highest legal authority in the province. As mentioned in Chapter Eighteen, Knight was an invalid at the time of Bell's visit. He was suffering from a prolonged illness and was barely able to leave his bed. Regardless of his physical limitations, Knight's mind was as sharp as ever and he took his civic responsibilities very seriously. Knight's wife, Katherine, had been living with him for years. In August 1718, Knight was joined by a permanent house guest named Edmund Chamberlayne. Little is known about Mr. Chamberlayne, except that he seldom left the house. It is logical to assume that he was there to provide some sort of live-in care and to help Knight get around. When Bell knocked on the door, it was probably Chamberlayne who answered. Realizing the serious nature of Bell's visit, he would have been immediately escorted to where Knight was resting. Bell urgently reported the robbery to Knight and identified two of his attackers as Thomas Unday and Richard Snelling, commonly called Titery Dick. He didn't recognize the other two men but told Knight that he believed them to be "Negroes or white men disguised like Negroes." Knight immediately issued a hue and cry for the arrest of the suspects.

All the particulars of Bell's attack came from the trial record of four of Blackbeard's men who were tried in Williamsburg. This trial will be discussed in far greater detail in Chapters Twenty-Nine and Thirty. That Virginia trial record was sent to North Carolina and read into the minutes of a meeting of the Governor's Council.[9] The meeting ended with a statement given by the only impartial first-hand witness, Edmund Chamberlayne, who said:

> Knights house on the 14th of September or when William Bell came and complained that he was robbed and defined a hue & Cry from the said Tobias Knight and did hear the said Tobias Knight examine the said Bell whether he could describe the persons to him that robbed him to which the said Bell replied he could not but said he did violently suspect one Thomas Unday and one Richard Snelling commonly called Titery Dick so be two of them and the others to be Negroes or white men disguised like Negroes.[10]

In common law, a *hue and cry* is a process by which bystanders are summoned to assist in the apprehension of a criminal who has been witnessed in the act of committing a crime.

Unaware of the Bell incident, Blackbeard sailed the *Adventure* to Bath, arriving on September 24, 1718. His first stop was the house of Tobias Knight. Not only was he the Chief Justice, Knight was also the Customs Inspector. Blackbeard claimed that he found the French ship abandoned

Chapter Twenty

at sea with no crew or papers onboard. As an abandoned ship, commonly referred to as a wreck, Blackbeard claimed salvage rights. At the time, Knight had no reason to suspect that Blackbeard had taken the ship through piracy. As will be discussed in Chapter Thirty, Knight will be accused of complicity with Blackbeard, but once the facts are carefully examined, they lead to a different conclusion. Knight said that from the time Blackbeard left Bath in late July, he hadn't seen him in town "until on or about the 24th of September last past where he came and reported to the governor that he had brought ale & ect into this government."[11] Chamberlayne supported this by saying that Blackbeard wasn't seen in Bath "until on or about the 24th day of the said last September when as this deponant is informed he came up to the governor and reported to him that he had brought a wreck into this government."[12]

A letter written by Lt. Governor Spotswood of Virginia on December 22, 1718, one month after Blackbeard's death, provides further details on the French ship that Blackbeard took:

> That Tache with divers of his crew kept together in North Carolina went out at pleasure committing Robberys on this Coast and had lately brought in a ship laden with sugar and cocoa, which they pretended they found as a Wreck at sea without either Men or Papers. that they had landed the Cargo at a remote Inlet in that Province and set the ship on fire to prevent discovery to whom she belonged.
>
> I also expect from North Carolina a considerable quantity of sugar and cocoa, wch. were in the possession of Tach and his crew, and appear to have been the Lading of that ship wch. they lately brought in there under pretence of a wreck, but in reality was taken piratically near Bermuda from the subjects of the French King, and the men put on board a ship of the same nation taken at the same time, as some of Taches crew now in custody alledge.[13]

Blackbeard probably spent the rest of September in Bath. From the beginning of October to November 21, 1718, Blackbeard remained in eastern North Carolina. He spent part of the time in Bath and part of the time at Ocracoke. He also sailed the rivers connecting to Pamlico Sound. Most of the accounts of Blackbeard's activities near Bath would have occurred during that time.

Plum Point stands out among the legendary places Blackbeard was believed to have visited. It is located just across Bath Creek from Tobias Knight's house. Some sources suggest that Blackbeard owned property

Plum Point is rumored to be one of the hiding places of Blackbeard's treasure.

there, but an examination of the deed records shows that Plum Point was owned by William Reed.[14] However, he was never there, and Blackbeard would have been able to land there without any issues. Local tradition contends that Blackbeard careened the *Adventure* there. This seems very likely, as the physical landscape is ideal for careening. There is a gentle slope to the water's edge, and the current is mild. Interestingly enough, there is circumstantial evidence to support this legend. The remains of a large brick oven known as Teach's Kettle could be found on the property in the early twentieth century. According to legend, it was used to boil tar for careening. This structure is lost within a vast forest along a stream known today as Teachs Gut and is mentioned in Catherine Albertson's 1914 book, *In Ancient Albemarle*. She wrote:

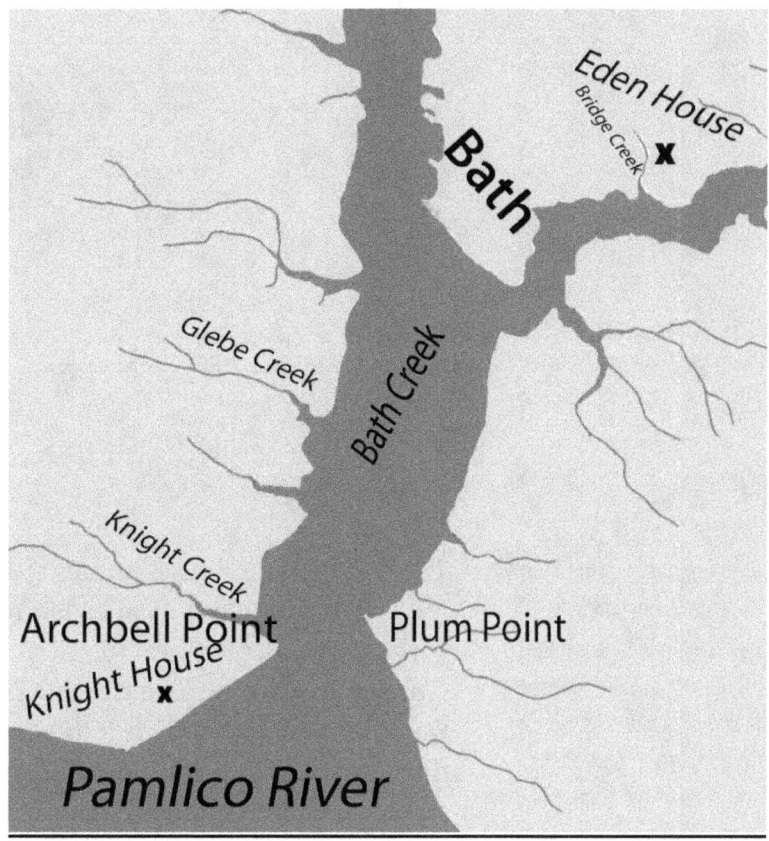

Figure 50: *Map of Bath Showing Plum Point.*

> Near the shores of the creek, just outside the town, there is still to be seen a round brick structure resembling a huge oven, called Teache's Kettle, in which the pirate is said to have boiled the tar with which to calk his vessel.[15]

Most importantly, Plum Point is believed to be the site where Blackbeard buried his treasure. This belief was as strong in 1718 as it is today. Only a few days after Blackbeard's death, locals began digging for his treasure at Plum Point.[16]

As Blackbeard and his crew relaxed in Bath and the surrounding area, they had no idea that the clouds of doom were beginning to form. Forces were at work in both North Carolina and Virginia, which would eventually lead to Blackbeard's demise.

Twenty-One
William Howard's Arrest

Hampton, VA, was first settled in 1610 at the site of an Algonquian village named Kikotan. By 1634, the nearby town was known as Elizabeth City, while the port at the entrance of the Hampton River kept the Algonquian name. It was spelled Kiquotan in the early eighteenth century. The modern spelling is Kecoughtan.

The first of Blackbeard's pirates to be arrested was his former quartermaster, William Howard. While visiting Kecoughtan (present-day Hampton), Virginia, on September 16, 1718, Howard was recognized and Captain Gordon arrested him on the orders of Lt. Governor Spotswood. Although the reason for Howard's trip to Virginia is not mentioned in contemporary documents, it is a logical assumption that he was simply meeting with some old friends at one of the taverns near the docks.[1]

At the time of his arrest, Howard had in his possession £50 in coins. That's about $5,000 in today's money. Additionally, he was traveling with two men of African descent. The authorities assumed these men to be his slaves. They may have been his slaves or just two of his pirate friends who happened to be black. Kecoughtan was the foremost port in the colony of Virginia at that time. Conveniently positioned near the entrance of the Chesapeake Bay as well as the James River, it was also the station of two Royal Navy warships sent to protect the southern half of the North American coast. Those two ships were HMS *Lyme*, a sixth-rated ship of 20 guns commanded by Captain Ellis Brand, and HMS *Pearl*, a fifth-rated ship of 40 guns commanded by Captain George Gordon.[2] After his arrest, Howard was held aboard HMS *Pearl* until October 30, 1718, and then transferred to the gaol in Williamsburg.

Captain George Gordon assumed command of HMS *Pearl* in 1716. His ship was stationed in Virginia in 1717.

Since the sinking of the *Queen Anne's Revenge* and the *Adventure,* many of Blackbeard's former pirates were traveling through Virginia, perhaps on their way home or just in search of a new ship. They had all taken the king's pardon and were given a certificate stating so. They were all technically legal citizens. Spotswood didn't see it that way. He felt that the governor of North Carolina was unjustified in granting them their pardons and that they were still to be considered pirates. On August 14, 1718, Spotswood issued a proclamation for their arrest. In a letter to the Council of Trade and Plantations, Spotswood wrote:

> The Proclamation prohibiting the unlawfull concourse of persons who have been guilty of piracy was occasioned by the great resort to this Colony, of certain pyrates who being cast away in North Carolina, surrendered there upon H.M. Proclamation; but as there's no great faith to be given to the fore'd submission of men of those principles.[3]

Naval authorities in Virginia were on the lookout for Blackbeard's men as they entered the colony. William Howard was perhaps the most well-known of all of Blackbeard's crewmembers. The position of quartermaster is the senior leadership position onboard a vessel second only to the Captain. He had been Ben Hornigold's quartermaster three years earlier and was Blackbeard's quartermaster for the past year. So, when Howard set foot in Virginia, he was easily recognized and the news of his arrival quickly reached the naval authorities, who, in turn, notified Spotswood. At the time of his arrest, Howard must have shown Captain Gordon the certificate he received from Governor Eden, thinking this would be resolved easily.

Williamsburg became the capital of the Colony of Virginia in 1699 when the original capital of Jamestown burned. It was named after King William III.

Details of Howard's arrest and imprisonment come from two primary sources. The first is a letter from Spotswood written to Commissioners for Trade and Plantations dated December 22, 1718.

> Since which one Howard, Tach's Quarter Master came into this Colony with two Negros which he own'd to have been Piratically taken, the one from a French ship and the other from an English Brigantine, I caused them to be seized pertinant to His Majestys Instructions ... There is besides the two Negro Boys about £50 in money.[4]

The second primary source is a report filed by Captain Gordon.

> That Wm. Howard (one of the pyrates named in the last mentioned Certificate) was taken up at Kiquotan in Virginia the 16th. of Septer. 1718, and putt on board the Pearl Man of Warr, in which Ship he remained a Prisoner until the 30th. of Oct. 1718, at which time he was sent up in Irons to Williamsburgh to be tryed for Pyracy.[5]

An *Order to Charge* was prepared for William Howard on October 29, 1718. Howard was formally charged on five counts. First, he joined the pirate crew of Edward Tach. Second, he accepted the position of quartermaster and took the sloop *Betty* off of Cape Charles. Third, on October 22,

Chapter Twenty-One

1717, he took the sloop *Robert* and the Ship *Good Intent* near Delaware Bay. The fourth count concerned his participation in the taking of La Concorde and converting that ship into the *Queen Anne's Revenge*. The final count combined several charges, that he intentionally didn't take the king's pardon, that he took a Spanish sloop near Cuba, and that he took a brigantine off the South Carolina coast in May of 1718.[6]

Fortunately, the original order to charge William Howard is safely housed in the Virginia Museum of History and Culture in Richmond, VA. The following is an abridged version, removing most of the legal language.

Articles exhibited before the Hon'ble his Maj'tys Comm appointed under the great Seal in Pursuance of an Act of Parliament made in the Eleventh and twelfth years of the Reign of King William the third Entitled an Act for the more Effectual Suppression of Piracy

Against

William Howard For Pyracy and Robbery committed by him on the High Seas.

First

That the Said Wm Howard ... did some time in the Year of our Lord 1717 Join and Associate him self with one Edward Tach and other Wicked and desolute Persons & with them did combine to fit out in Hostile manner a Certain Sloop or Vessel Call'd the Revenge to commit Pyracys

And

That ... the said William Howard did Accept the Office of a Quartermaster on Board the said

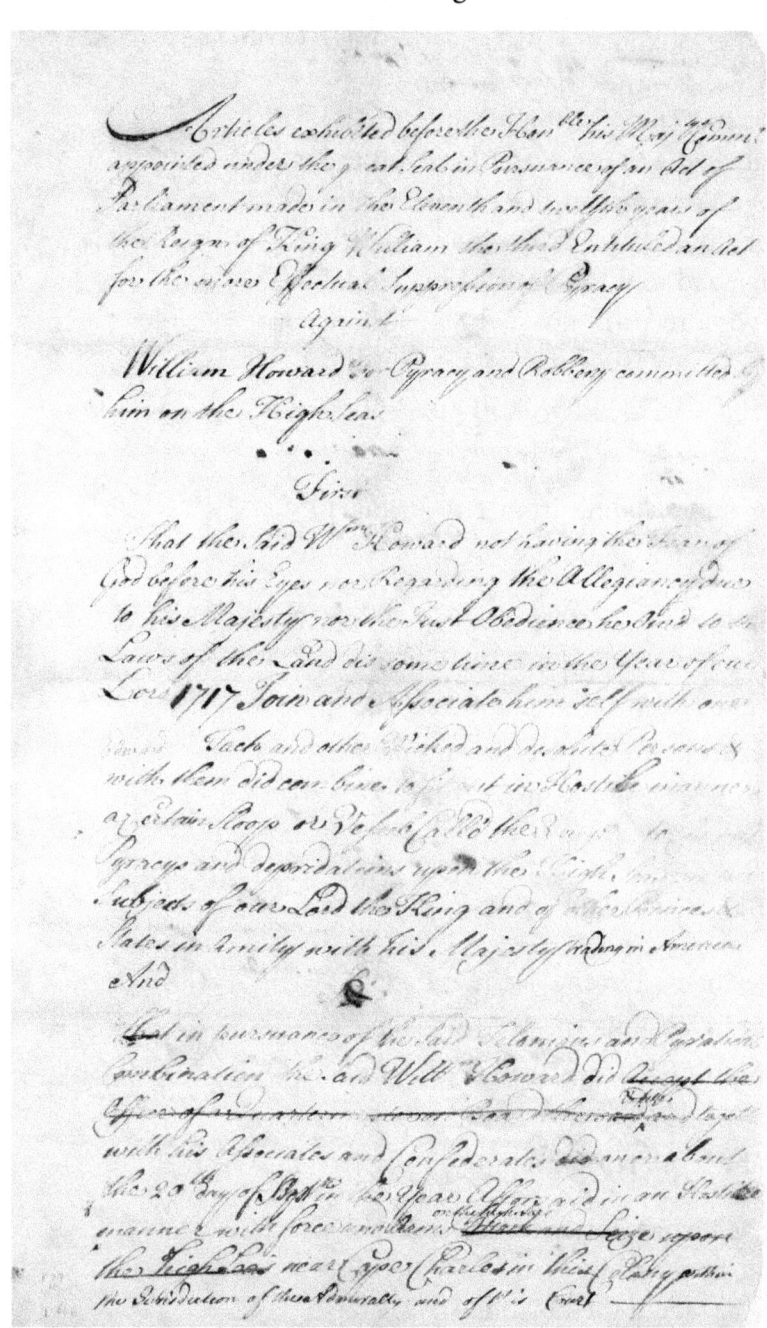

Figure 51: *Photo of William Howard's Charge Sheet, Courtesy of the Virginia Museum of History and Culture.*

Vessel and ... did on or about the 29th day of Septr in the Year Afforsaid ... near Cape Charles in the Colony ... attack and force and Robb a Sloop call's the Betty of Virginia ... and there Rob and plunder of Certain Pipes of Medera Wine and other Goods and Merchandizes and thereafter the said Wm Howard did Sink and destroy the said Sloop with the remaining Part of the Cargo.

That the Said Wm Howard ... on or about the 22d day of Octor in the Bay of Delaware ... Pyratically take Seize and Rob the Sloop Robert of Philadelphia and the Ship Good Intent of Dublin ... did feloniously and Piratically take seize and carry away

That on or about the __ day of December the said Wm Howard ... did Pyratically take and Seize the ship Concord of Saint Malo commanded by Capt D'Ocier ... and having Rob'd and Felonious Spoild the said Subjects of the French King of their Merchandize and Effects consisting of Negros, Gold dust, money, Plate and Jewels did Carry away the said Ship and Convert the Same towards the Carrying on the Prosecuting his ... Pyratical designs

That ... the said Wm Howard not being Ignorant of his Maj'tys Gracious Intentions declared in the said Proclaimation but Dispising his Maj'tys Royal offers of Mercy did after the said fifth of Jany continue to Perpetrate his wicked and Pyratical designs ... with the aforesaid Edwd Tach ... and associates in the aforesaid call'd the ... Queen Anns Revenge on or about the month of April 1718 or about the day of May a Sloop belonging to ye subjects of the King of Spain upon the high seas near the Isld of Cuba did piratically take & seize the same Brginteen bound ... to South Carolina ... near the Coast and within the jurisdiction of the said province

William Howard wasn't alone. Captain Gordon arrested three other pirates at Kecoughtan on November 28, 1717. They were Henry Man, William Stoke, and Adult Van Pelt. Just like Howard, they were temporarily held on the *Pearl* and then taken to Williamsburg to stand trial. Captain Gordon also lists them in his report.[7]

That Henry Man, Wm. Stoke & Adult Van Pelt. (the three other pyrates named in the said Certificate) were taken up near Kiquoatan aforesd. the 28th of Nov. 1718, and brot. on Board his Maty's Sd. Ship the Pearl, as Pyrates, where they remained prisoners until the 15th. Of Dec. following, when they were Sent up to Williamsburgh in Irons

Chapter Twenty-One

Howard's arrest was predictable. What happened next wasn't. John Holloway, a highly respected and influential lawyer in the colony of Virginia, took Howard's case. Holloway had been the Speaker of the Virginia House of Burgesses for fourteen years and served as the Treasure of the Colony for eleven years. He was also the Mayor of Williamsburg. The reason such an important person would take such a case is unknown. Perhaps he just didn't like Spotswood. His fee from Howard was three ounces of gold dust.[8]

> The House of Burgesses was the legislative body of the Colony of Virginia. The term came from the medieval word "burgher," which meant a freeman of a borough.

Holloway immediately went on the offense. He issued warrants for the arrest of the justice of the peace who had arrested Howard and for both Captain Gordon and Lieutenant Maynard of the *Pearl*. Their charge was false imprisonment. He also began a civil lawsuit against them asking for damages in the amount of £500.[9] Later, Captain Brand commented on the charge against his friend, Captain Gordon, in a letter written on July 14, 1719, in which he wrote:

> likewise caus'd Capt. Gordon to be arrested for false improsinment: as they call it in the writ; for takeing up a notorious pyrate which oblig'd me to goe to Williamsburg with him to give in Bail.[10]

Spotswood's council advised against the trial of Howard, but Spotswood initiated proceedings anyway, and in a letter to the Council of Trade and Plantations dated December 22, 1718, he wrote:

> Whereupon he caused not only the Justice who signed the Warrant but the Captain & Lieutenant of the Man of War to be arrested each in an action of £500 dammages & one of the Chief Lawyers here undertook his cause . . . that tho he and the rest of Tache's Crew pretended to surrender and to claim the benefits of his Majestys Proclamation, they had nevertheless been guilty of divers Piracys after the fifth of January for which they were not entitled to his Majestys Pardon. I therefore thought fitt to have him brought to a Tryal, but join'd a strong opposition from some of the council against trying him by virtue of the Commission under the great Seal pursuant to the Act of the 11th & 12th of King William III.[11]

Spotswood also wrote a letter on July 28, 1721, to Secretary Craggs, to complain about Howell's legal proceedings, saying:

> That one Howard, a notorious pirate and Quarter Mast'r to Thach, came into this Country . . . He was taken up and put on board the Pearl Man of War; Mr. Holloway espoused his Cause, and Arrested

Capt. Gordon and Mr. Maynard, his L't, each, in an action of five hundred pounds, for the (pretended) false Imprisonment of this fellow, and I have now by me Holloway's Receipt for three Ounces of Gold dust paid him by Howard.[12]

Spotswood called Holloway "a constant patron and advocate of pirates."[13] Clearly, Spotswood had to do something to prevent Holloway from defending Howard. At the time, all other pirates in the colonies were tried with a grand and petit jury.[14] However, finding a loophole in the law, Spotswood convinced the Council that Howard should be tried without a jury and appointed John Clayton, Attorney General of Virginia, as well as Captain Gordon and Captain Brand to serve on the court.[15]

This became the perfect excuse to exclude Holloway from the proceedings. Spotswood argued that because Holloway had filed suit against Gordon, he shouldn't participate during the trial. In that same letter to Secretary Craggs, he wrote, "to prevent any disturbance on the Bench, w'ch I apprehended would ensue upon their publickly excepting against Mr. Holloway, I sent him a civil Message to desire him not to expose himself by appearing on that Tryal."[16] Holloway was furious, but there was nothing he could do about it.[17]

Of course, *The Boston News-Letter* ran the story, "the Pearl & Lym Men of War were both at Hampton, and the Captains at Williamsburgh upon the tryal of a Pirate."[18] Howard's trial was held in Williamsburg on November 6, 1718. The next day, Spotswood sent a letter to Governor Eden of North Carolina informing him of the trial.

> The evidence given in yesterday upon the Tryall of his Quarter Master William Howard, who now lyes here under the Sentence of death, for being clearly convicted of manifest Pyracys since the fifth of January last, and even of one committed but two days before their running aground at Top-sail-Inlet at which time they Robbed an English Brigantine.[19]

This letter tactfully set the stage for the legal battle between Eden and Spotswood that was yet to come. It also sounded the death knell for Blackbeard.

Twenty-Two
Bonnet on His Own

We last discussed Stede Bonnet as he left Topsail Inlet in mid-June 1718. For clarity, I shall provide a brief recap of Bonnet's actions after the *Queen Anne's Revenge* sank. He took a small boat with a crew of five pirates to Bath, where he received the king's pardon from Governor Eden. While Bonnet was upriver, Blackbeard claimed the Spanish sloop, named it the *Adventure*, and then used it to seize the *Revenge*. Now in control of everyone, Blackbeard left Topsail Inlet through Core Sound, marooned twenty-five pirates on Harbor Island, and sailed on to Ocracoke. Upon Bonnet's return from Bath, he rescued the marooned men and returned to Topsail Inlet, where he reclaimed the *Revenge*.

Bonnet and his men were arrested and tried in October 1718. Piracy was big news at the time and books on pirates sold very well. As mentioned in Chapter Eight, when the trial was over, a copy of the transcript was rushed from Charles Town, South Carolina to London, where it was accurately reproduced in a book titled *The Tryals of Major Stede Bonnet and Other Pirates*, published by Benj. Cowse at the Rose and Crown in St. Paul's Church, in 1719. Easily accessible today, this book is among the best prime source documents on Blackbeard and Bonnet, because it contains invaluable first-hand accounts from the pirates themselves.

For pirates who wish to take vessels legally, sailing with letters of marque as privateers is the only option. Even though war between Britain and Spain hadn't yet been officially declared, everyone knew that a war was inevitable. The governor of St. Thomas wanted to get a head start on the war and was already issuing letters of marque, sometimes

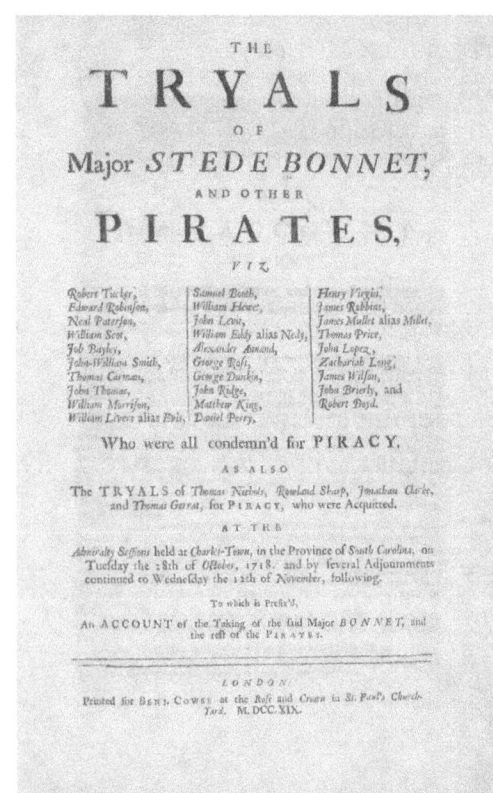

Figure 52: *Cover of the Tryals of Stede Bonnet and Other Pirates, Public Domain Pages from Tryal of Stede Bonnet.*

called commissions, to all former pirates who had accepted the king's pardon, also called his act of grace. When Bonnet returned from Bath with his certificate of pardon, he assured his men that they would not be returning to piracy, but instead, that if they joined him, they would sail to St. Thomas to obtain letters of marque to attack Spanish shipping. This is consistent with the testimony of all of Bonnet's men. Here are a few of their statements.

> Morrison: Capt. Thatch had run the Sloop ashore, and Maj. Bonnet went up to the Governor for the Act of Grace; and when he returned, he told me I might go to St. Thomas's.[1]
>
> Ign. Pell: Maj. Bonnet came with the Boat, and told us, as we were on a Maroon Island, that he was going to St. Thomas's to get a Commission from the Emperor to go against the Spaniards a Privateering, and we might go with him, or continue there: so we having nothing left, was willing to go with him.[2]
>
> Robert Tucker: After Capt. Thatch had taken what we had, and left us, Maj. Bonnet came and told us that he was going to St. Thomas's for the Emperor's Commission, if there was any to be had.[3]
>
> Judge Trott: Bonnet had not above five Hands, and there was of you twenty-five; why would you be all commanded by them? You had no need to yield to them.
>
> Clerk. Daniel Perry, what have you to say?
>
> Perry: When Capt. Thatch left us, it was on a Maroon Island, and Major Bonnet came and told us he had the Act of Grace, and so we might go with him.[4]

The original hand-written trial record is stored at the South Carolina Department of Archives and History in Columbia, SC.

Their testimony concerning St. Thomas is understandable. They all pled "Not Guilty,"[5] and their only defense was that didn't think that they were returning to piracy when they sailed out from Topsail Inlet with Bonnet aboard the *Revenge*. Since they began taking English vessels immediately afterward, leaving Topsail Inlet and sailing to Cape May instead of St. Thomas, the truth of their statements is called into question.

Aboard the *Revenge*, Bonnet was once again in command of the sloop that he had purchased over a year earlier. Robert Tucker was appointed as quartermaster.[6] As they prepared to sail, the name was changed to the *Royal James*, in direct reference to James Edward Stuart, the pretender to

The Royal James *was named after James Edward Stuart.*

Chapter Twenty-Two

the throne.[7] This is clear evidence that Bonnet was a Jacobite. His crew consisted of about fifty men.[8] In addition to the twenty-five he rescued and the five men already with him on his boat, it seems that Bonnet must have also picked up twenty of the men who were still on shore at Topsail Inlet.

Seeking retribution, the *Royal James* sailed through Topsail Inlet to the Atlantic. The channel to Pamlico Sound was too shallow for his large sloop to navigate. Sailing past a string of barrier islands, Bonet and his crew were anxious to kill Blackbeard. After all, they all felt that he had cheated them out of the share of the treasure and he had marooned half of them.[9] Finally, Ocracoke was in sight. A surge of adrenaline swept through them as they approached the island, but quickly turned to disappointment when they realized there would be no fight. Blackbeard had already gone. Following the route that he used ten months earlier while Blackbeard was in command, the *Royal James* sailed north toward the Virginia Capes. About thirty-five miles before reaching Cape Henry, as they were passing Nags Head, they sighted a sail. Blackbeard had taken all of their provisions. The men were starving! Whether they intended to sail to St. Thomas or planned on returning to piracy, it didn't matter. They needed food!

The vessel they sighted was a pink. Easily overtaking the slower vessel, the *Royal James* came alongside and Bonnet ordered their captain to lower a boat and row over to his pirate sloop. Looking at the *Royal James's* open gunports, the captain complied. Bonnet's men took twelve barrels of pork and a good deal of bread. As the pink was being looted, a few of Bonnet's men must have realized that they were actually returning to a life of piracy. Before the two vessels parted, they left Bonnet and remained aboard the prize vessel. David Herriot said:

> That the said Major Bonnet being informed by a Bomb-Boat that brought Apples and Cider, that Thatch lay at Ocracoke-Inlet with only eighteen or twenty Hands, he resolved to pursue him, and cruised after him for four Days: But missing him, made to Virginia; and finding in with the Land, they met a Pink about ten Leagues to the Southward of Cape Henry about July last, whose Name or Master he knows not. And said Bonnet ordered the Pink to send their Boat and come on board them. And the said Bonnet took out of her about ten or twelve Barrels of Pork, and about four hundred Weight of Bread. Says, That several of the said Bonnet's Crew went aboard the said Pink:[10]

This was the start of Bonnet's journey to the hangman's noose. Between July and August 1718, the charges mention that Bonnet took a total of thirteen vessels. This was solely based on David Herriot's deposition. However, when carefully compared to the occasionally vague testimony of the other pirates, the number may be as high as seventeen. A summary of his captured vessels will be provided at the end of the chapter. The prosecuting attorney only charged the pirates with two the piracies of two sloops, the *Fortune* and the *Francis*. There is a very good reason for this. All of Bonnet's other piracies were outside of the jurisdiction of the Carolina court. However, after taking those two sloops near Cape May, he brought them to Cape Fear. Once in Carolina waters, they entered the jurisdiction of the court. At the conclusion of the trial, Judge Trott addressed Stede Bonnet personally, saying:

> Altho' you were indicted but for two has, yet you know that at your Tryal it was fully proved, even by an unwilling Witness, that you piratically took and rifled no less than thirteen Vessels, since you sailed from North Carolina.[11]

While in the waters off Cape Henry in July 1718, the *Royal James* took ten more vessels in less than one week. Bonnet then returned to Cape May, where he and Blackbeard had been so successful the previous October. On July 28, 1718, while sailing north along the eastern shore, he took a schooner off Assateague Island.[12] Bonnet kept the schooner as a consort and continued north. The next day, the *Royal James*, with the captured schooner alongside, entered the Delaware Bay. At thirty-nine degrees latitude, which is about five miles into the bay, they took two snows and a sloop. On July 30, 1718, Bonnet took one more sloop, the *Fortune*, which had just left Philadelphia bound for Barbados. The *Fortune's* captain was Thomas Read, whose son was traveling with him, along with a woman referred to as Dr. Reeve's wife. He offered no resistance. Bonnet kept the sloop and ordered Captain Read, his family, and his crew to come aboard the *Royal James* while five of Bonnet's pirates took control of the *Fortune*. The three vessels sailed to Cape Henlopen, a sharp, hook-shaped point of land a few miles north of Rehoboth Beach in the province of Delaware.

Bonnet off Cape Henry, Virginia, and Cape Henlopen, Delaware

> 30th of the same Month, the said Bonnet and Crew took a Sloop, Burden of fifty or sixty Tons, commanded by Capt. Thomas Read, as they lay off of Delaware Bay, about six or seven Leagues bound from Philadelphia to Barbados, laden with Provisions, and put four or five Hands of the said Bonnet's Crew on board her. That about the 1st Day of August, the said Bonnet and Crew, as they lay at the Hore-Kills in Delaware Bay aforesaid, off Cape Henlopen.[13]

Chapter Twenty-Two

Cape Henlopen was originally settled and named by the Dutch. It was renamed Cape James when it fell under English authority, although most people continued to refer to it by its Dutch name, and still do today.

To the west of Cape Henlopen, a cove provided an ideal place for vessels to rest at anchor, as it blocked both the wind and the ocean waves. The isolated shoreline along the cape was devoid of any sign of human habitation, but a small fishing port called Port Lewis was nestled along the shore about one mile away. In the late afternoon, Bonnet's three vessels quietly slipped into the cove. There were no other vessels in sight, and the pirates didn't expect any action that evening. Bonnet's vessels dropped anchor, and the crew prepared for sleep. An hour after sunset, one of the pirates noticed the sail of a sloop coming into the cove. Whispers among the pirate crew turned into excitement as they realized another prize was coming toward them. It was immediately obvious that this would be an easy prize to take. Hearing the discussion on deck, Captain Bonnet came out of the cabin and joined his crew at the rail. They watched with great anticipation as this sloop dropped anchor among the pirate vessels.

Following his usual tactics, Bonnet lowered a boat with five pirates aboard and rowed over to this unidentified sloop. As Bonnet's boat slowly approached, he cautiously surveyed the sloop in the moonlight for any sign of hostile action. As he grew nearer, Bonnet heard someone on the sloop say, "Here's a canoe a-coming." Once alongside, Bonnet looked up at the sloop's rail in the dark and saw a shadowy figure looking down at him. This man shouted down to Bonnet, asking who he was. Bonnet joyfully answered that he was Captain Thomas from St. Thomas. The man on the sloop remained cautious and asked who the other two vessels belonged to. Bonnet once again happily answered that the other sloop was Captain Read's from Barbados and the schooner was from North Carolina under the command of Captain Yates. Bonnet then asked the man, "Where did your sloop came from?" The man Bonnet was speaking with was not the captain, he was James Killing, one of the sloop's mates. Killing replied, "We are from Antigua." Then, both Bonnet and Killing exchanged friendly greetings. Killing was somewhat relieved.

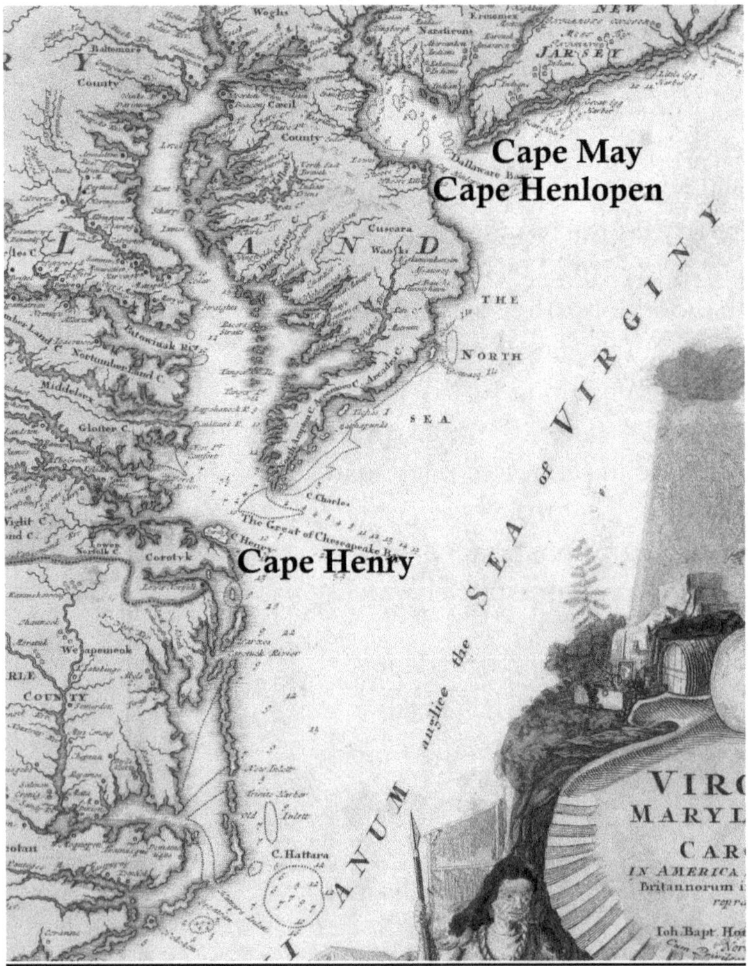

Figure 53: *Close-up of Virginia and Delaware Capes from a 1720 Map, Courtesy of Barry Lawrence Rudman.*

Bonnet saw a rope thrown over the rail, which slapped against the side of the hull. This was followed by an invitation to come aboard. His deception had worked.

Bonnet confidently climbed up the side of the *Francis* and stepped over the rail. With a look of certainty, Bonnet stood firmly on the deck face to face with Killing, who immediately noticed that his visitor had several pistols on his belt and a cutlass at his side. Bonnet clasped his hand on the hilt of his cutlass, smiled, and politely said, "You are all my prisoners." Killing glanced at the rail and saw the other well-armed pirates coming aboard behind Bonnet. He knew that his men were not armed and would have no chance in a fight with these pirates. Killing looked into Bonnet's smiling eyes and with a sigh, calmly said, "Will you give us good quarter?" Bonnet promptly answered, "Yes, providing you remain civil." With that simple exchange of words, the pirate assault was over and the *Francis* was under Bonnet's control. It had taken him less than a minute. As it turned out, their initial impression was correct. This was an easy prize to take.

Bonnet takes the Sloop Francis, *Captain Peter Manwareing.*

James Killing was held prisoner aboard Bonnet's sloops for over eight weeks. He appeared as a witness at the trial of Bonnet and his men. A portion of his testimony follows:

> The thirtieth day of July, between nine and ten o'clock, there running a strong tide of ebb, we came to an anchor about fourteen fathoms of water near Cape James. In about half an hour's time, I perceived something like a canoe. So they came nearer. I said here's a canoe a-coming; I wish they be friends. I hailed them; and asked from whence they came. They said Captain Thomas Richards from St. Thomas's, and Captain Thomas Read from Pennsylvania. They asked me from whence we came. I told them from Antigua. They said we were welcome. I said they were welcome, as far as I knew. So I ordered the men to hand down a rope to them. As soon as they came on board, they clapped their hands to their cutlasses; and I said we are taken.[14]

The *Francis's* captain, Peter Manwareing, was also held as a prisoner for over eight weeks and appeared as a witness at the trial of Bonnet and his men. His testimony supports that of his mate and adds a few other details:

> That being bound from the Island of Antegoa to Philadelphia, on or about the last Day of July last past, he anchored at Cape James about Nine o'clock at Night, seeing three Sail of Vessels at Anchor within the said Cape. In about an Hour after a Canoe with five Hands in her

Chapter Twenty-Two

came near the Sloop; and being hailed by this Informer what Boat that was, said they belonged to Capt. Thomas from St. Thomas's. This Informer asked what the other two Vessels were? They answered, Capt. Read from Barbadoes, and a schooner from North Carolina. Who is Commander? They answered, Capt. Yates. And coming to the said sloop, came on board, and told this Informer they were their Prisoners; being well armed with Guns, Swords, and Pistols. And this Informer having no Arms, desired they would give them good Quarters. Yes, they replied, provided you are civil;[15]

That evening, Captain Manwareing and two of his men were taken aboard the *Royal James,* where they met Bonnet's prisoners from the *Fortune*—Captain Thomas Read and his son, and the woman identified as 'Dr. Reeve's wife.' Over on the *Francis*, a few of the pirates entered the cabin, where they found a good supply of pineapples, drew their cutlasses, and proceeded to slice the pineapples into pieces. They invited James Killing to sit down and eat with them, but he was in no mood to be social. When the pirates asked what alcoholic beverages they had, Killing replied, "Rum." The pirates made some bowls of punch, toasted the Pretender's health, expressed their wish to see him as king of England someday, and sang songs late into the evening. These men were clearly Jacobites. Killing describes these events in his testimony:

> So when they came into the cabin, the first thing they began with was the pineapples, which they cut down with their cutlasses. They asked me if I would not come and eat along with them. I told them I had but little stomach to eat. They asked me, why I looked so melancholy. I told them I looked as well as I could. They asked me what liquor I had on board. I told them some rum and sugar. So they made bowls of punch, and went to drinking of the Pretender's health, and hoped to see him king of the English nation; then sung a song or two.[16]

The next morning, the *Francis* was brought to anchor near the other vessels. Their captured schooner was used to transfer much of the *Francis*'s cargo over to the *Royal James*. As that was happening, Dr. Reeve's wife and Read's son were put aboard a dory along with five pirates and rowed to the nearby village of Port Lewis to be released. Before they left, Bonnet told Read's son to deliver a warning to the people in that town, that if any of the pirates were harmed, Bonnet would kill all his prisoners and destroy the entire town. Manwareing's testimony says:

That on the next Day, being the 1st Day of August last, the said Bonnet sent his Boat with five Men in her armed, to put Dr. Reeve's Wife and Capt. Read's Son ashore at the Hare Kills alias Port-Lewis: And the said Bonnet, at the same time of sending off his Boat, told Capt. Read's Son in this Deponent's hearing, That if any of the Inhabitants offered to hurt the Hair of the Head of any Person belonging to his said Bonnet's Crew, he the said Bonnet would put to death and destroy all the Prisoners he had on board, and would also go ashore and burn the whole Town.[17]

Later that day, the pirates took everything of value off the schooner, including the foremast and the mainsail.[18] Apparently, Bonnet felt that he had no further use for it. However, Bonnet had use for the other two sloops. On the evening of August 1, 1718, the *Royal James*, *Fortune*, and *Francis* all raised anchor and made sail. Captains Manwareing and Read would remain as prisoners for the next eight weeks, along with James Killing and a good part of their crews. Clearing the point, they turned south. Bonnet was returning to the familiar waters of North Carolina. They all arrived at Cape Fear on August 12, 1718. It had been many months since the *Royal James* had been careened and it was in desperate need of repair. The three sloops remained peacefully anchored there until September 26, 1718, when they were discovered by the two pirate hunting sloops of Colonel William Rhett.[19]

At the start of September 1718, a report reached the authorities in Charles Town that the pirate Stede Bonnet and his sloop of 10 guns, along with two captured sloops, were careening at Cape Fear. Colonel William Rhett was commissioned by the governor to go after them. Rhett was given overall command of two sloops, the *Henry* with 8 guns and seventy men, commanded by Captain John Masters, and the *Sea-Nymph* with 8 guns and sixty men, commanded by Captain Fayer Hall. On September 10, 1718, they set sail from Charles Town with Colonel Rhett aboard the *Henry* and sailed to Sulivan's Island to take on supplies. Colonel Rhett was present at the trial of Stede Bonnet and his men. His detailed report was included in the trial record.[20] All the following details come from his report.

A short time after Rhett's sloops reached Sulivan's Island, word reached him that the entrance to Charles Town was once again under attack by pirates. This time, it was Charles Vane in a brigantine of 12 guns, and his partner Charles Yates in a sloop of 8 guns. It took five days for Rhett to get his two sloops underway, and by that time, Vane was already gone. Stede Bonnet's sloops were no longer a top priority, Charles Vane was. Some of the crewmen that Vane had temporarily detained while their vessels were

Colonel William Rhett (1666–1723) was the captain of his merchant ship, a surveyor, comptroller of customs for Carolina and the Bahama Islands, and the Colonel of the local militia. In 1706, he led the naval fleet that defended Charles Town against the combined Spanish and French attack.

Charles Vane attacks vessels at Charles Town

Chapter Twenty-Two

On the 1709 and 1733 maps of North Carolina, Cape Fear is spelled Cape Fair; however, it is spelled Cape Fear in the trial records and other contemporary documents.

being looted had overheard the pirates talking about careening their vessels somewhere to the south. This was reported to Rhett, who immediately sailed south in search of Vane's pirates. After sailing a considerable distance with no sign of any of the men he was searching for, Rhett ordered his two sloops to come about and head north. Perhaps he missed them and he would come upon Vane careening in one of the many secluded inlets along the Carolina coast. Rhett never found Charles Vane, but he did find Stede Bonnet.

Rhett's sloops reached the mouth of Cape Fear River on September 26, 1718. The sun was low in the western sky as light was rapidly fading. Rhett stared over the low-lying beach grass that covered a spit of land near the entrance of the river. In the distance, Rhett could just barely make out the masts of three sloops hidden deep within the recesses of an estuary. The *Henry* and the *Sea-Nymph* put their helms over and headed directly to the three anchored sloops. Rhett wasn't positive, but he strongly suspected that these were the three sloops belonging to Bonnet that he had originally set out to find. However, there were many sand bars hidden below the surface and both of Rhett's sloops ran aground as he came within sight of the sloops. By now, it was totally dark, and he knew there was no chance of getting free until the tide came in. Rhett also knew that the three sloops wouldn't be able to get past him.

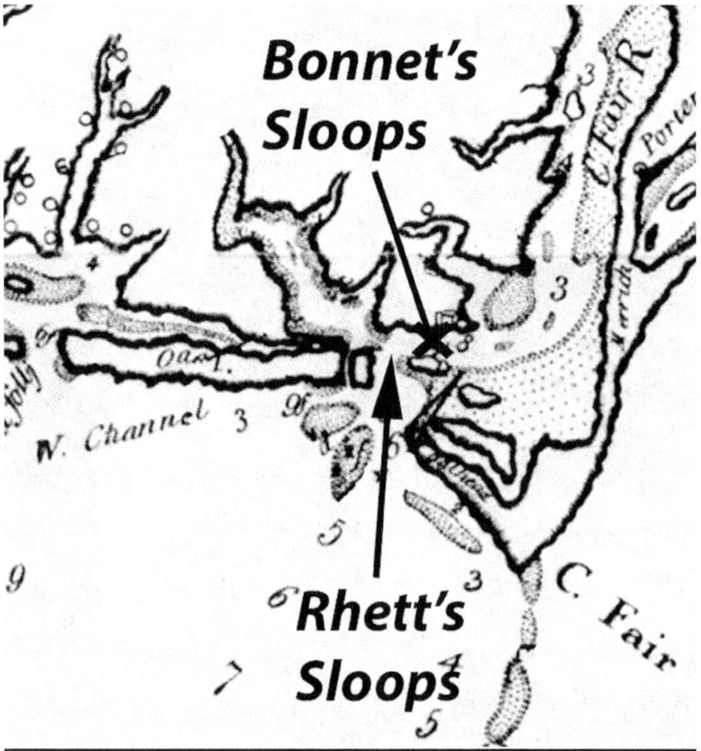

Figure 54: *Close-up of Battle Site at Cape Fear from James Wimble 1738 Map, Courtesy of East Carolina University.*

Colonel Rhett's attack on Stede Bonnet at Cape Fear

Over on the *Royal James*, Bonnet noticed the two sloops approaching just before the last bit of light vanished in the night sky. Wondering who they were, Bonnet sent three boats loaded with his pirates down the river to investigate or possibly even to capture the unidentified sloops. The pirates were able to come right up to Rhett's sloops in the darkness and were shocked to see that both sloops were filled with well-armed men. The pirates knew exactly what they were up against. They rowed back to the *Royal James* as quickly as they could. After reporting the true nature of the two sloops that stood between Bonnet and the sea, the pirates spent the rest of the evening preparing for battle. Early the next morning, before the sun broke the horizon, the rising tide had freed Rhett's sloops. They raised their sails and moved in, maneuvering to take positions on both sides of the *Royal James* so they could board it. Realizing what Rhett was attempting, Bonnet sailed his sloop closer to shore to keep his distance. The *Henry* opened the engagement with a broadside, which forced the *Royal James* to run aground. Colonel Rhett noted they were "in the same shoal Water." The *Henry* suddenly struck a sandbar, as did the *Sea-Nymph*.

Rhett found himself in a terrible position. The *Royal James* lay grounded in a perpendicular position just about thirty feet in front of his bow. That rendered all of his big guns totally useless, as they were mounted on both sides of his sloop, not at the bow. The *Sea-Nymph* lay aground next to him and slightly ahead, but they were just about out of ammunition. Rhett's horrible position was made worse by the fact that the *Royal James* had settled aground, listing away from the *Henry*. This means that their deck was angled in such a position that the pirates could fire their small arms from behind the cover of their tilted sloop. The *Henry* was listing in the same direction, which had the opposite effect, exposing all the men on board to the fire from the pirate sloop. A fierce gun battle between the two sloops, with muskets, pistols, and blunderbusses firing, continued for five hours. At several points during the battle, the pirates taunted Rhett's men and beckoned with their hats for them to come and board them.

As the tide rose, the *Henry* drifted free. Once in deeper water and clear from the battle, Rhett's men worked feverishly to make hasty repairs to their shattered rigging. Once Rhett had control of his sloop's movements, he closed in to give the *Royal James* a broadside. The pirates knew they had no chance and raised a flag of truce. Colonel Rhett accepted their surrender. The casualties were ten killed and fourteen wounded on the *Henry* and two killed and four wounded on the *Sea-Nymph*. The pirates suffered seven killed and five wounded, two of which died soon afterward.

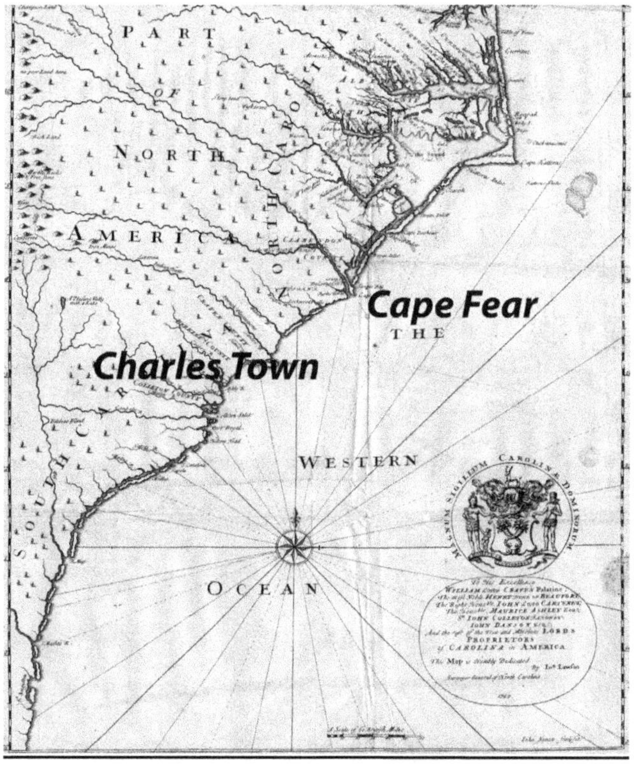

Figure 55: *1709 Map of North Carolina, Courtesy of the University of North Carolina at Chapel Hill.*

Colonel Rhett was pleased to learn that he had actually captured Stede Bonnet. He wasn't sure until the battle was over and his men had taken possession of the *Royal James*. A little further upriver, the two prize sloops, *Fortune* and *Francis*, were still at anchor. Rhett's men took control of them as well. After eight and a half weeks of captivity, Captains Manwareing and Read, along with their crews, were finally rescued. Colonel Rhett sailed from Cape Fear with all five sloops on September 30, 1718, and arrived at Charles Town on October 3, 1718, "to the great Joy of the whole Province."

Nicholas Trott, the Chief Justice of South Carolina, was appointed Judge of the Vice-Admiralty, and trial preparations began.[21] David Herriot, former master of the sloop *Adventure* taken by Blackbeard, and Ignatius Pell, Bonnet's boatswain, agreed to testify against the others in exchange for amnesty. Charles Town didn't have an actual prison, so most of the pirates were kept in the cellar of the half-moon battery in the center of the city's defensive wall. However, Stede Bonnet was still considered a gentleman, and as such, was separated from the others and detained at the marshal's house. Additionally, David Herriot and Ignatius Pell were also separated for their own protection and held at the marshal's house along with Bonnet. The house was under the guard of two sentries. In an odd turn of events, Bonnet and Herriot escaped on October 24, 1718.[22]

The Half-Moon Battery was constructed in 1701. Charles Town's customs house and exchange was built over the original foundations in 1767. Today, the building is a museum called *The Old Exchange & Provost Dungeon*.

Bonnet's escape is understandable. He was clearly guilty and would most likely be convicted and hanged. However, David Herriot turned King's Evidence, and was to be released after he testified. One can only speculate as to his reasoning. Perhaps Bonnet persuaded him that the prosecutors wouldn't keep their word and would hang him even if he did testify against the others. Perhaps Herriot was far more involved as one of Bonnet's leaders, and he knew that his testimony wouldn't stand up against cross-examination. In any event, Bonnet and Herriot managed to obtain a small boat and made it to Sullivan's Island.[23]

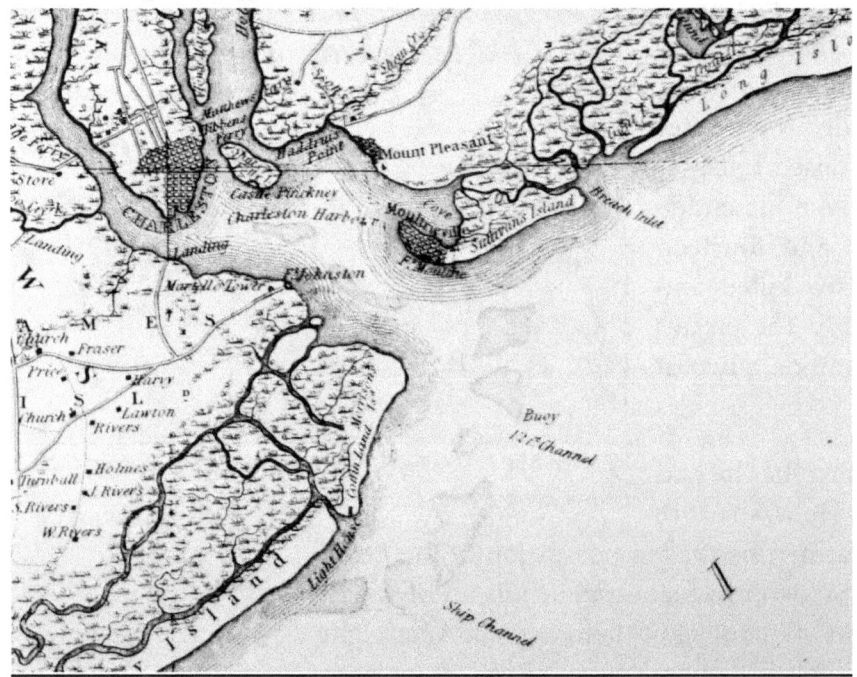

Figure 56: *Close-up of Charles Town and Sullivan's Island from the 1820 Mills Atlas. The image is dedicated to the public domain under CC0.*

The pirates weren't tried all together, they were separated into groups and tried in several different trials. The first of these trials began on October 28, 1718, as Bonnet and Herriot were still on the loose.[24] Before he escaped, Herriot had given a deposition, which was used by the prosecution in his absence. Most of the men pled "Not Guilty."[25] Their explanation was that they thought they were going to St. Thomas to get a privateer commission and that they were forced to take a few vessels in order to get provisions. When Judge Trott asked them why they fired upon Colonel Rhett if they were innocent, they replied that they believed his sloops to be pirates. To which Trott replied, "And so one Pirate might fight with another; but how could you think it was a Pirate when he had King George's Colours?"[26]

Meanwhile, Colonel Rhett and his men tracked Bonnet and Herriot down and came upon them on the night of November 5, 1718. The soldiers opened fire and killed Herriot with their first volley. Bonnet surrendered and was brought back to Charles Town the next day.[27] This explains why there is no testimony from Herriot in the trial record, only his original deposition. Herriot was killed before he could have appeared in court. Bonnet's trial began on November 10, 1718. He also pled "Not Guilty."[28] His defense was simply that even though he was the captain, he never gave his consent to commit any piracies. Bonnet said:

> I am not guilty of what I am charged with. As for what the Boatswain says, relating to several Vessels, I am altogether free; for I never gave my Consent to any such Actions: For I often told them, if they did not leave off committing such Robberies, I would leave the Sloop; and desired them to put me on shore. And as for taking Capt. Manwareing, I assure your Honours it was contrary to my Inclination. And when I cleared my Vessel at North Carolina, it was for St. Thomas's.[29]

Of the thirty-three crewmen on trial for piracy, twenty-nine were found guilty and sentenced to hang. Four of the men managed to prove that they had been forced to join the crew and were discharged.[30] On Tuesday, November 11, 1718, the admiralty court found "Stede Bonnet alias Edwards, alias Thomas, Guilty." The next day, November 12, 1718, "the Judge of the Court of Vice-Admiralty pronounced Sentence of Death upon the Prisoner."[31]

In case you were wondering what Charles Johnson has to say on this subject, you just read it. In the chapter on Stede Bonnet in Johnson's book, *A General History of the Pyrates*, from the time the *Queen Anne's Revenge* sank to the end of Bonnet's life, Johnson's book is practically identical to *The Tryals of Major Stede Bonnet and Other Pirates*. Since that book was published in London in 1719, it is certain that Johnson simply copied it almost word for word.

Bonnet spent his last few days writing letters to everyone he thought might intercede on his behalf to get a pardon. This included a letter to Colonel Rhett[32] and one to Governor Robert Johnson.[33] I have seen a curious letter supposedly written by Stede Bonnet being sold in gift shops and posted on

"King George's Colours" refers to the British flag

The Boatswain is the senior position or rank assigned to a sailor on a vessel, responsible for the maintenance of deck equipment and the ropes used for raising and controlling the sails. It is the oldest rank in the English Navy, dating back to 1040.

websites for decades, as the letter he wrote to Governor Johnson. I often wondered if it was genuine or simply the invention of someone wishing to sell some very unique souvenirs. Now, after some research, it appears that it just might be the real thing. In 1809, Less than one hundred years after Bonnet's imprisonment and execution, David Ramsay wrote a book titled *History of South Carolina from its First Settlement in 1670 to the Year 1808.* Ramsey mentions the letters that Bonnet wrote and said that he had one of the original letters, "which has been preserved, and by the politeness of Judge Bee."[34] That letter was the one he wrote to Colonel Rhett and is reproduced word for word in his book. Although it's not the same letter written to Governor Johnson that is mentioned above, it is very similar in style and composition. Among the many whining and complaining paragraphs within Bonnet's letter, there is one very interesting part which reads:

> God the knower of all secrets, will lay to my charge; and must intreat you to consider that I was a prisoner on board Captain Edward Thatch, who with several of Captain Hornigold's company which he then belonged to, boarded and took my sloop from me at the island of Providence, confining me with him eleven months.[35]

The earliest source I have found for the Bonnet–Johnson letter is in Shirley Carter Hughson's book, *Blackbeard & the Carolina Pirates*, published in 1894. The entire letter appears in Hughson's book exactly as seen in the recent giftshop version. He cites the original source for this letter as Charles Johnson, writing, "This letter was published by Captain Charles Johnson some years after his execution."[36] The letter isn't even mentioned in any edition of *A General History of the Pyrates*, but Johnson often published other articles and broadsheets. Perhaps Hughson was referring to one of those. Preceding the letter, Hughson wrote, "A few days before his execution he addressed a letter to Governor Johnson, in which he begged most piteously for mercy, pleading the ridiculous excuse that his crimes had been committed under compulsion." Bonnet's letter follows:[37]

Honoured Sir;

I Have presumed on the Confidence of your eminent Goodness to throw myself, after this manner at your Feet, to implore you'll be graciously pleased to look upon me with tender Bowels of Pity and Compassion; and believe me to be the most miserable Man this Day breathing; That the Tears proceeding from my most sorrowful Soul may soften your Heart, and incline you to consider my Dismal State, wholly, I must confess, unprepared to receive so soon the dreadful

Execution you have been pleased to appoint me; and therefore beseech you to think me an Object of your Mercy.

For God's Sake, good Sir, let the Oaths of three Christian Men weigh something with you, who are ready to depose, when you please to allow them the Liberty, the Compulsion I lay under in committing those Acts for which I am doom'd to die.

I entreat you not to let me fall a Sacrifice to the Envy and ungodly Rage of some few Men, who, not being yet satisfied with Blood, feign to believe, that I had the Happiness of a longer Life in this World, I should still employ it in a wicked Manner, which to remove that, and all other Doubts with your Honour, I heartily beseech you'll permit me to live, and I'll voluntarily put it ever out of my Power by separating all my Limbs from my Body, only reserving the use of my Tongue to call continually on, and pray to the Lord, my God, and mourn all my Days in Sackcloth and Ashes to work out confident Hopes of my Salvation, at that great and dreadful Day when all righteous Souls shall receive their just rewards: And to render your Honour a further Assurance of my being incapable to prejudice any of my Fellow-Christians, if I was so wickedly bent, I humbly beg you will, (as a Punishment of my Sins for my poor Soul's Sake) indent me as a menial Servant to your Honour and this Government during my Life, and send me up to the farthest inland Garrison or settlement in the Country, or in any other ways you'll be pleased to dispose of me; and likewise that you'll receive the Willingness of my Friends to be bound for my good Behavior and Constant attendance to your Commands.

I once more beg for the Lord's Sake, dear Sir, that as you are a Christian, you will be as Charitable as to have Mercy and Compassion on my miserable Soul, but too newly awaked from an Habit of Sin to entertain so Confident Hopes and Assurances of its being received into the arms of Blessed Jesus, as is necessary to reconcile me to so speedy a Death; wherefore as my Life, Blood, Reputation of my family and future happy State lies entirely at your Disposal, I implore you to consider me with a Christian and Charitable Heart, and determine mercifully of me that I may ever acknowledge and esteem you next to God, my Saviour, and oblige me ever to pray that our heavenly Father will also forgive your Trespasses.

Chapter Twenty-Two

Now the God of Peace, that brought again from the Dead our Lord Jesus, that great Shepherd of the Sheep thru' the Blood of the everlasting Covenant, make you Perfect in every good work to do his Will, working in you that which is well pleasing in his Sight thro' Jesus Christ, to whom be Glory forever and ever, is the hearty Prayer of

*Your Honour's
Most miserable, and,
Afflicted Servant
Stede Bonnet*

A memorial to Stede Bonnet is located at White Point Gardens Park near the spot where he was executed.

The Tryals of Major Stede Bonnet and Other Pirates ends with the following statement, "On Wednesday December the 10th, 1718, the said Major Stede Bonnet was executed at the White-Point near Charles-Town according to the above Sentence."[38]

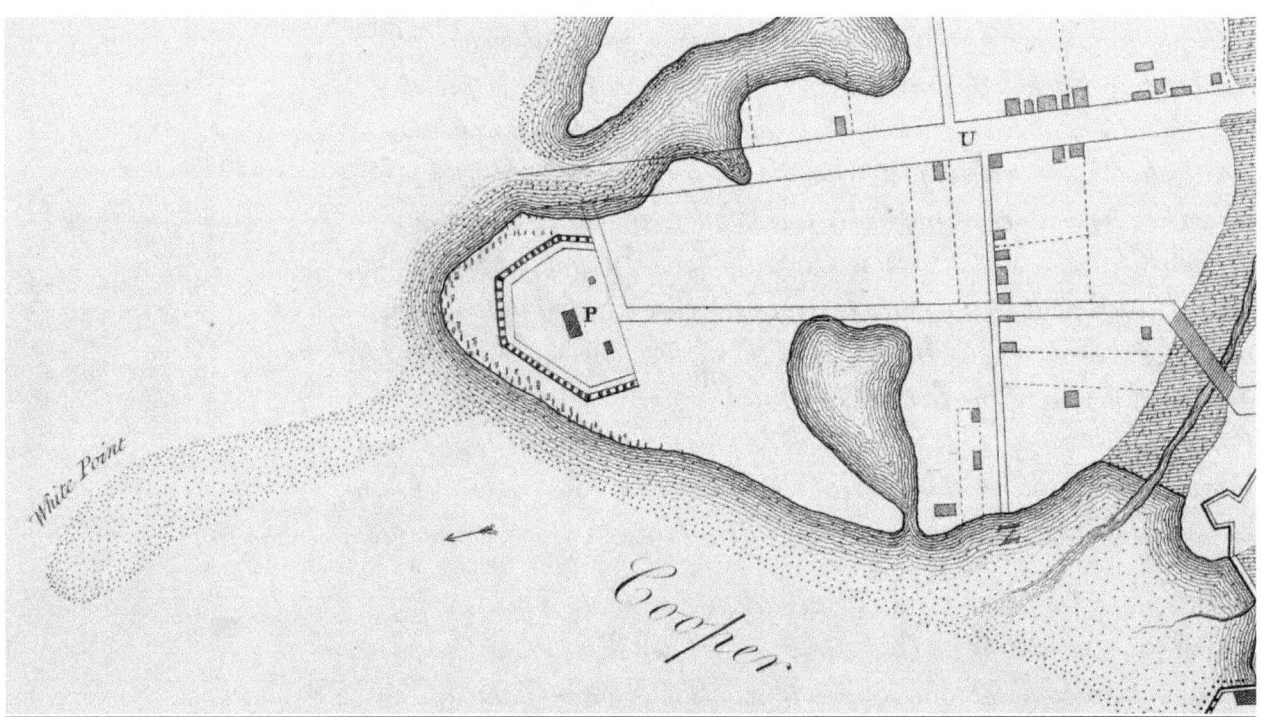

Figure 57: *Close-up of White Point from a 1739 Map of Charles Town, Courtesy of the Library of Charleston.*

Twenty-Three
Charles Vane's Visit

Charles Vane was clearly identified by the British government as the most troublesome pirate in Nassau since February 1718. As you may recall from Chapter Fifteen, Vane and his followers took control of the fort at Nassau, pulled down the British flag, and hoisted a "Black Flag with the Death's Head."[1] Captain Pearse of HMS *Phenix* arrived at Nassau on February 23, 1718. Shortly after his arrival, he had problems with Vain, too. Benjamin Hornigold had warned Captain Pearse and one of his officers, Lt. Symonds, that Charles Vane would be the greatest problem.[2] Captain Pearse had several incidents with Vane while in Nassau. Vane was forced to surrender to Captain Pearse at Providence and sign the list of those who were going to accept the king's pardon. His name appears on line one hundred twenty-two out of two hundred nine.[3] However, Before Pearse left Nassau on April 6, 1718, Vane gave him more trouble, causing Pearse to put a mark next to his name where it appeared on his list, indicating that he returned to piracy. That incident was documented in his letter to the admiralty dated June 3, 1718, as follows:

HMS *Phenix* was a Sixth-Rate warship of 24 guns stationed in New York in 1718.

> The said Pirate Sloop who had been out since the 24th Inst. March came to an anchor at the Easternmost part of the harbour, with 'an other called the Lark, which they had taken three days before off Harbour Island, onboard of which they had Removed themselves, she being a good Sailer & Equipt with 9 Guns & 8 Patraroes, there was a Barr between me & them on which was no more than eight foot water, which made them very Insolent. The said Pirate Sloop was Commanded by one Cha. Veine & Man'd with 45 Men.[4]

I also discussed the Jacobite relationship between Vane and Blackbeard in an earlier chapter. Although there was no contemporary documentation of a meeting between Blackbeard and Vane, they were both in Nassau at

Chapter Twenty-Three

the same time in May 1718. It is highly likely that they met somewhere. After all, Blackbeard and Vane were old friends.[5] More importantly, they were both Jacobites.[6]

Throughout this book, I have briefly mentioned the Jacobite sympathies of several pirates, especially Blackbeard and Bonnet. They both gave their vessels names that directly related to Jacobites. The *Queen Anne's Revenge* was in reference to the last of the Stuart monarchs, and the *Royal James* was none other than James Edward Stuart himself. As you may recall, Bonnet's men were heard drinking toasts to the "Pretender," who was also James Edward Stuart. It's now time for me to give these claims a little more weight. The strongest link between the Jacobite cause and the pirates of Nassau is through George Cammock and Charles Vane.

From July 1717 to November 1718, James Edward Stuart was living in exile in Urbino, Italy.

Captain George Cammock was a distinguished British naval officer with a magnificent career. He entered the navy in 1682 and became a captain in 1702.[7] However, when King George I was crowned, Cammock was accused of being a Jacobite and forced to leave the Royal Navy. Following his resignation, he ran guns in the uprising of 1715.[8] In the early months of 1718, Captain Cammock wrote a letter to Mary of Modena, the former Queen of England, widow of James II, and mother to James Edward Stuart. In his letter, Cammock stated that the pirates at New Providence "did with one heart and voice proclaim James III for their King," and that they had written him inviting him to serve as a Jacobite Captain-General and to organize "resistance" against the Hanoverian "usurper."[9] He described how the pirates had seized control of Nassau at the fort and noted that many of the pirates there had rejected the Act of Grace offered by King George I. Finally, he proposed that if given the funding from the exiled James Edward Stuart, the pirates under his command could seize Bermuda and "destroy the West India and Guinea trade."[10] Notice that he mentioned that the pirates had written him, inviting him to be their Captain-General. This letter either no longer exists or is hidden away in some obscure location. Some historian authors believe the author of that letter to be Charles Vane.[11]

Woodes Rogers was appointed Governor of the Bahamas on January 6, 1718, and held the office until June 1721.

For the three months between Captain Pearse's departure on April 6 and the arrival of Woodes Rogers and his fleet on July 26, 1718, Nassau was in a state of chaos. Ben Hornigold was in support of the king's forces, and Charles Vane was against them. Vane sailed out of Nassau on his sloop, *Ranger*, in May 1718, and took twelve vessels. Unlike Blackbeard and most other pirates of the day, Vane treated his captives harshly, often beating them. What was very troubling to the British authorities was that his victims reported that he was heard drinking damnation to King George, reinforcing the suspicion that he was a Jacobite. Lt. Governor Bennett of Bermuda wrote a letter to the Council of Trade and Plantations that says:

> Deposition of Samuel Cooper, Mariner, of Bermuda, 24th May, 1718. Deponent was on board the Diamond sloop, Capt. Tibby, when she was captured off Rum Key, one of the Bahama Islands, by Charles Vain, Commander of the pirate sloop, Ranger. The pirates robbed and beat Tibby and the rest of the company. They had taken 12 vessels on their cruise, 7 belonging to Bermuda, including Edward North, Daniel Styles and James Basden. They beat the Bermudians and cut away their masts upon account of one Thomas Brown who was (some time) detain'd in these Islands upon suspicion of piracy etc. Brown, they said, had subscriptions of hands to the number of 70 to go out under his command upon the account of piracy and would give no quarters to Bermudians etc. Their expressions at drinking were Damnacon to King George and that they designed to be with us (meaning the inhabitants of these Islands) this summer etc. Signed, Samuel Cooper.[12]

The pirate life at Nassau all came to a crashing halt on July 25, 1718, when Woodes Rogers and his fleet arrived at the entrance to Nassau's harbor. Contrary to the plan, HMS *Rose* sailed in ahead of the rest of the fleet and dropped anchor in the narrow channel leading to the town. Within moments, the captain was confronted with the sight of a burning ship headed his way. Charles Vane had taken a French guard ship and intentionally set it on fire, then headed it straight at the *Rose*. To escape being burned, the Rose had to cut the anchor cable and sail out of the way. The next morning, As Rogers approached the entrance to the channel aboard his ship, HMS *Milford*, Vane sailed past him onboard his sloop, flying his pirate flag, and firing his guns in defiance. This is described in a letter from Governor Woodes Rogers to the Council of Trade and Plantations:

> Pursuant to my Instructions I take leave to acquaint your Lordships, I arriv'd in this port 26th July in company with the men of warr ordered to assist me. I met with little opposition in coming in, but found a French ship (that was taken by the pirates of 22 guns) burning in the harbour, which we were told was set on fire to drive out H.M.S. the Rose who got in too eagerly the evening before me, and cut her cables and run out in the night for fear of being burnt, by one Charles Vane who command'd the pirates and at ours and H.M.S the Milford's near approach the next morning they finding it impossible to escape us,

Chapter Twenty-Three

Charles Vane took vessels at Charles Town in September 1718.

he with about 90 men fled away in a sloop wearing the black flag, and fired guns of defiance when they perceiv'd their sloop out sayl'd the two that I sent to chase them hence.[13]

On September 10, 1718, Charles Vane repeated the actions of his friend, Blackbeard, by boldly attacking the shipping at the entrance of Charles Town. By now, Vane was sailing in a brigantine of 12 guns and ninety men. He was also sailing with his new partner, Charles Yeats, who had a large sloop with 8 guns. His first prize was a small ship from Antigua, commanded by Captain Cook. Next came a vessel from Ipswich, commanded by Captain Coggeshall, and a sloop from Barbados, commanded by Captain Dill. This was followed by another sloop from Curaçao, commanded by Captain Richards. Then, they took a brigantine slave ship from Guinea, carrying ninety slaves, commanded by Captain Thompson. The slaves were put onboard Yates' sloop.

Apparently, Yates could no longer put up with Charles Vane's personality, and as soon as night fell, he set sail and sped away, taking the ninety slaves with him. Seeing that Yates was fleeing, Vane quickly made sail and chased him, but when it became apparent that he couldn't catch Yates' sloop, Vane came about sharply, so all his guns on one side of his brigantine faced the fleeing sloop. Vane fired a broadside, but Yates was out of range and he vanished in the darkness. Successfully evading Vane's pursuit, Yates hid in

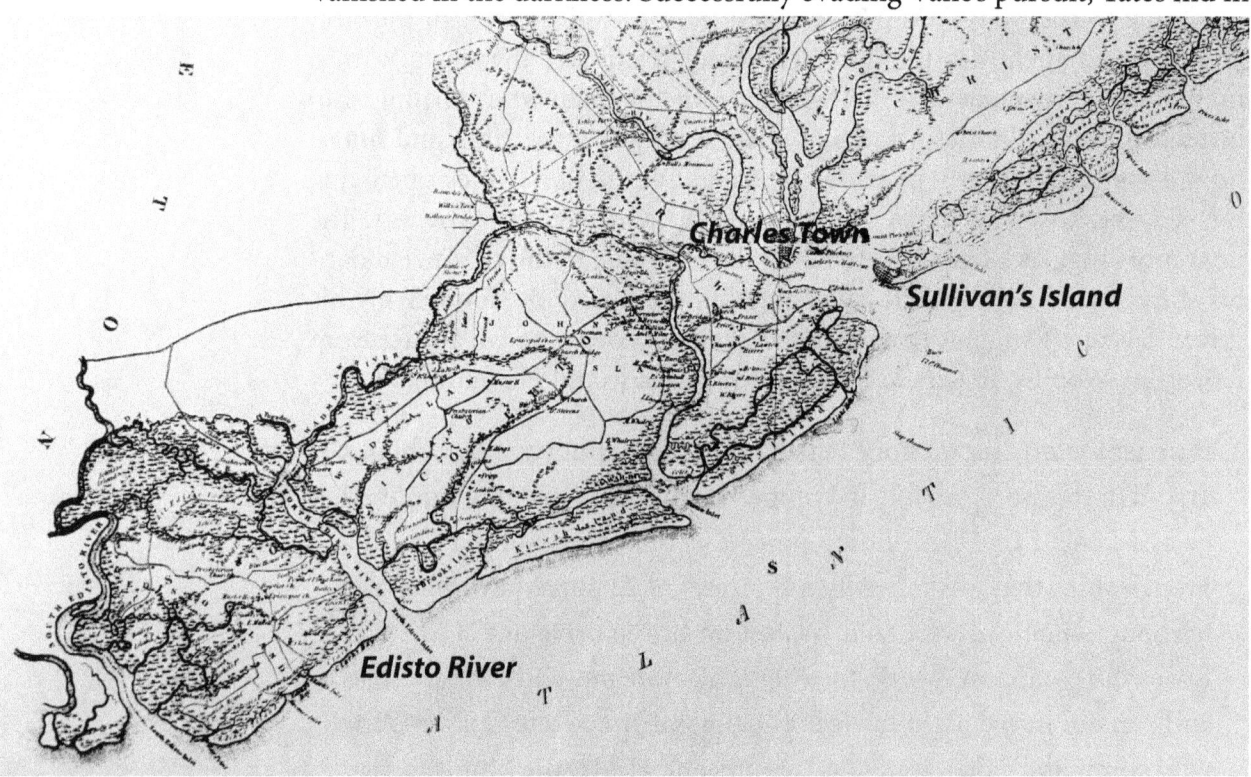

Figure 58: *Close-up of Edisto to Charles Town from the 1820 Mills Atlas of Charles Town. The image is dedicated to the public domain under CC0.*

the Edisto River. After Vane left the area, Yates sailed into Charles Town and turned himself in. The slaves were returned to their owner and Yates was granted the king's pardon.

The Edisto Rover is thirty-five miles from Charles Town Harbor.

After losing Yates in the darkness, Vane returned to Charles Town and took two more vessels, a large pink of 16 guns named the *Neptune,* under the command of Captain King, and a vessel of 10 guns named the *Emperor,* under the command of Captain Power. Details of Vane's piracies come from two source documents—Colonel Rhett's report found in *The Tryals of Major Stede Bonnet* and an article in the October 13 to October 20, 1718 edition of *The Boston News-Letter.* In his report, Colonel Rhett stated,

> And just then arrives a small Ship from *Antegoa,* one *Cook* Master, who gave us an account, That in sight of our Bar he was taken and plunder'd by one *Charles Vane* a Pirate, in a Brigantine of twelve Guns and ninety Men, and who had also taken two other Vessels bound in here; one a small Sloop, Capt. *Dill* Master, from *Barbadoes*; the other a Brigantine, Capt. *Thompson* Master, from *Guinea,* with ninety odd Negroes, which they took out of his Vessel, and put on board another Pirate Sloop they then had under the Command of one *Yeates,* with fifteen Men: which was fortunate to Capt. *Thompson's* Owners. *Yeats* having often attempted to leave this Course of Life, took this Opportunity; for in the Night he got away from the Brigantine, and carried the Sloop and Negroes into *North Edisto* River, to the Southward of this Port. The Owners got their Negroes; and *Yeates* and his Men had Certificates given them from the Government.
>
> *Vane* mean while continued cruising off our Bar, in hopes to catch Yeats: and it unfortunately happen'd that four Ships, bound to *London,* and who had waited some time for a fair Wind, got then over the Bar, and two of them were taken; namely, the *Neptune,* a large Pink with sixteen Guns, Capt. *King* Commander; and the *Emperor,* with ten Guns, Capt. *Power* Commander; but both very deep loaded.[14]

The Boston News-Letter, Monday, November 17 to Monday, November 24 1718 reads:

> Rhode-Island, October 10 Arrived here Thomas Cranston from South Carolina who informs, that two Pirates had taken eight vessels on that Coast about the 10th of September last, the Pirates are Charles Vean Commander of a Briganteen of 12 Guns and 90 Men,

Chapter Twenty-Three

the other Charles Yeatts in a large sloop of 8 Guns and 20 Men his Consort, they took one Coggeshall of Ipswich loaden with Logwood, and threw half of it overboard, designing to make his Vessel fit for Piracy, but finding the would not do, he had her again to go about his business: They took one Dill of Barbadoes, one Richards of Curacoa, a Ship from Antigua, and another from Guinea; he does not know the names of the rest, neither of the Men nor Vessels taken: Vean took 90 Negroes out of the Guinea Man, which he put on Board his Consort Yeatts, and he being weary of that way of living slipt his Cables and departed, Vean being jealous of his design went after his Sloop as far as the Breakers, firing a broad side at her, which did no harm and got clear of Vean, put into Edisto within ten Leagues of Charlestown, and sent an express to the Governour to see if they might have the benefit of the King's Act of Grace, they would surrender themselves, their Vessel, the 90 Negroes, and all they had to his Mercy; which was granted by the Governour, and they came in accordingly and surrendered, Capt. Cranston also informs, that the Governour thereupon immediately Man'd and fitted out two Sloops, one of 8 Guns and 80 Men, another of 8 Guns and 60 Men, Commanded in Chief by Col. Reat, in pursuit of the Vean.[15]

Vane had taken seven vessels within sight of Charles Town harbor. What makes it even worse is that he took those vessels right under the nose of Colonel William Rhett. As mentioned in the previous chapter, Colonel Rhett was given a commission and two sloops to go after Stede Bonnet. Vane was plundering vessels at the entrance to Charles Town harbor, while Rhett was only about one mile away on Sullivan's Island, fitting out his sloops for battle. Rhett set sail as soon as his sloops were ready, but he was too late. Charles Vane was gone.[16]

Charles Vane's deeds made him public enemy number one. First of all, he had blasted his way out of Nassau in front of the new governor and his fleet. Then, he took seven vessels in sight of the entrance of the largest port in the south while the only military authority stood helplessly on Sullivan's Island as his sloops were in the process of being refitted as war sloops. with two sloops. But Vane was far more threatening than that. By now, the British authorities must have been aware of his Jacobite connection with Cammock. Vane wasn't just a pirate. He was a political revolutionary who had the potential to spark a Jacobite movement among the pirates in the Caribbean. A pirate fleet under Vane could disrupt the entire economy of the most profitable colonies in the British Empire. Even worse, a strong

Jacobite movement in the colonies could quickly spread to Scotland, Ireland, and even England. Charles Vane represented a crisis of the highest order. His next actions, if true, sealed the fate of Blackbeard.

Charles Vane with Blackbeard at Ocracoke in early October 1718

According to Charles Johnson, Vane visited his old friend, Blackbeard, at Ocracoke Inlet in October 1718. Upon his arrival, Vane fired a salute and Blackbeard answered. For the next several days, the two crews enjoyed each other's company. Johnson mentions this event only in the chapter on Charles Vane. There is no mention of it in the chapter on Blackbeard. Additionally, the Ocracoke visit appears in the first edition and remains unchanged in subsequent editions. This is what Johnson wrote:

> Captain Vane went into an Inlet to the Northward, where he met with Captain Thatch, or Teach, otherwise call'd Black-beard, whom he saluted (when he found who he was) with his great Guns, loaded with Shot, (as is the Custom among Pyrates when they meet) which are fired wide, or up into the Air: Black-beard answered the Salute in the same Manner and mutual Civilities passed for some Days; when about the Beginning October, Vans took Leave, and sailed further to the Northward.[17]

There are no primary source documents that directly support this Ocracoke visit between Vane and Blackbeard, only the word of Johnson. In the centuries that followed *A General History of the Pyrates*, some historian authors have included this visit in their narrative and others omitted it. And in true Blackbeard fashion, some historian authors embellish by adding details to the story and expanding it far beyond Johnson's initial version. The most common elaborations include visiting women from the local community and the consumption of multiple alcoholic beverages. It appears that everyone had a great time.

Among the first successful books on pirates written after Charles Johnson's books was *The Pirates Own Book*, published by Charles Ellms in 1837. His work is basically a rewrite of Johnson's book, bordering on plagiarism. As with Johnson, there is no mention of this Ocracoke party in the chapter on Blackbeard, but the chapter on Vane recounts Johnson's version almost word for word.[18] However, even though there is absolutely no text in Blackbeard's chapter mentioning the Ocracoke party, it does contain a fascinating full-page woodcut titled, "The Crews of Blackbeard's and Vane's Vessels Carousing on the Coast of Carolina."[19] A picture is worth a thousand words. This illustration shows a real party. There are women, dancing, a fiddler, and lots of bottles and barrels that appear to contain adult beverages.

Chapter Twenty-Three

Figure 59: *1837 Woodcut Illustration of the Celebration at Ocracoke from* The Pirate's Own Book. *The image is dedicated to the public domain under CC0.*

The clothing shown in this 1837 woodcut print is highly inaccurate for the early eighteenth century.

In his 2008 book, *The Last Days of Black Beard the Pirate*, historian author Kevin Duffus calls this party a banyan.[20] According to him, the word "banyan" was originally of Hindu origin and referred to a merchant who conducted business under a large tree. Seventeenth-century sailors working for the Portuguese East India Company picked up the term and began using it in reference to shore leave. By the eighteenth century, the term had spread to English sailors and was commonly used among them. This would include, of course, English pirates. Kevin Duffus also provides a vivid description of this pirate banyan, writing that there were "sailcloth awnings and tarpaulin ground-cloths, open cook fires with large cuts of meat roasting on iron spits, stacks of firewood and driftwood piled here and there, small kegs for stools circled around games of dice or cards, ceramic jugs of wine, rum, and brandy, and stacks of swords, cutlasses, and muskets, ready for action."[21]

Although this is speculation, it is very well-informed speculation. His description of their camp is a fairly accurate depiction of all camps made by sailors or pirates while they were ashore. While careening, no one stayed onboard the vessel. It was lying on its side on the beach. A camp would have to be constructed ashore. Sailcloth would be used to make tents and dining flies. There would be several open-air cooking fires scattered about the camp, each having either a tripod with an iron pot hanging by a chain or a roasting spit with some sort of meat turning over the fire. Of course, a generous supply of firewood would be a necessity. Gaming with dice or cards was exceptionally popular in the eighteenth century, especially among sailors and pirates. Improvised tables would have been made from barrels and planking. Reports of goods stolen by pirates indicate that wine, rum, and brandy were the most common alcoholic beverages and they were always in either ceramic jugs or glass bottles. The "stacks of swords, cutlasses, and muskets, ready for action," certainly be expected of pirates, who were at their most vulnerable when in camp ashore. Today, pirate reenactors throughout the world participate in festivals, and this precise description applies to the reenactor camps.

Many pirate enthusiasts often refer to this pirate banyan as more of a pirate orgy than anything else. This idea is spurred on by the famous illustration in Charles Ellms' 1837 book, *The Pirates Own Book*, shown above. Although the notion is romantic and exciting, it is very unlikely. There just weren't very many women living in the area. The closest town of Bath was fifty miles away. The pirates would have to spend several days sailing up and down the rivers of eastern North Carolina, advertising their party and inviting local women to come aboard for a few days. This would also require the pirates to take them home afterward. Since this was a clandestine gathering, the pirates wouldn't have wanted everyone to know about it. Duffus points out that if any women attended, they would have been the wives of the pilots living nearby.[22]

I mentioned earlier that direct prime source documentation of this pirate banyan doesn't exist. However, circumstantial evidence suggests that it did take place. There is no record of Blackbeard's or Vane's activities during the first week of October 1718. This indicates that they would have had the opportunity to attend this banyan. Stronger circumstantial evidence can be found in a letter from Lt. Governor Spotswood to the Council of Trade and Plantations in which he says:

> I hope, prevented a design of the most pernicious consequence to the trade of these Plantations, wch. was that of the pyrats fortifying an Island at Ouacock Inlett and making that a general rendevouze of such robbers.[23]

It is apparent from this letter that news of the pirates at Ocracoke Inlet reached Lt. Governor Spotswood in Virginia. He doesn't specifically name Charles Vane, but he does mention a general rendezvous at Ocracoke Inlet. The reports of the pirate banyan likely came to Virginia from sailors passing through the inlet on their way to Virginia ports. Ocracoke Inlet was one of the only two practical passages from Pamlico Sound to the Atlantic Ocean and most of the major waterways in North Carolina stem

North Carolina Rivers connected to *Pamlico Sound* include the *Neuse* and the *Pamlico*.

from that sound. Vessels were constantly sailing through the inlet, which was located only about three miles away from where the pirate banyan would have been held. The pirate vessels at anchor would have been easily seen by all the sailors on any passing vessel, and they may have even seen the pirates themselves.

I believe that this pirate banyan happened. I also believe that word of this banyan reached Charles Johnson through the stories that sailors told in the taverns of London. This account doesn't seem like something that Johnson would invent. There was no sensationalism in his simple account. Johnson didn't add any violence or sex. There were no accounts of pirates fighting among themselves or any mention of an orgy. That came many years later. Johnson's simple, dry, and emotionless account seems too believable not to be real. When coupled with Spotswood's comment about a pirate rendezvous, the pirate banyan between the crews of Charles Vane and Blackbeard becomes extremely believable.

Twenty-Four
Brand and Maynard's Invasion of a Province

Lt. Governor Alexander Spotswood of Virginia sat in his opulent palace in Williamsburg, wondering what he should do about the pirate crisis in North Carolina. Reports about Blackbeard had been flowing into Virginia ever since the *Queen Anne's Revenge* sank. Captain Brand, commanding officer of HMS *Lyme*, had already written the first report on Blackbeard at Topsail Inlet, and a week later, he hired a spy to go to Bath to gather any information he could on Blackbeard's actions. Brand wrote a letter dated February 6, 1719, describing all the information he received from his spies, as well as a highly detailed account of the battle at Ocracoke. This letter still exists in the Admiralty, and a copy can be found in the North Carolina Archives. Throughout this chapter, all accounts attributed to Brand came from that letter.[1]

As mentioned in Chapter Eighteen, Brand's spy reported that Blackbeard continually insulted and abused the masters of all the sloops in the area. He added that Blackbeard would take whatever goods he pleased. Brand wrote:

> So soon as I received this advice I employ'd a man that was going into No. Carolina to inform himself how this fellow lived, and if, after the manner he gave out, I recev'd several accounts from people that came from thence that he did Continue in that place insulting and abusing the masters of all trading sloops and taking from there what goods or Liquors he pleased.

When Blackbeard arrived at Bath in June 1718, he entered a political firestorm between Governor Charles Eden and his chief rivals, Colonel Edward Moseley, Colonel Maurice Moore, and Jeremiah Vail.[2] The

Chapter Twenty-Four

Colonel Edward Moseley and Colonel Maurice Moore were brothers-in-law.

information provided by Brand's spy is questionable. There were no local reports recorded of any incidents involving Blackbeard or any of the pirates. It appears that they were very well-behaved. The only reports of indiscretions made by pirates are those that came to Brand by way of his spy. Considering the political turmoil in Bath, it is highly likely that those reports of Blackbeard insulting and abusing people in Bath were told to Brand's spy by Moseley, Moore, or Vail. As will be described later in this chapter, all three will play a vital role in the planning and execution of the attack on Blackbeard at Ocracoke.

North Carolina wasn't the only province where Blackbeard's pirates dwelt. Some of his former pirates were crossing the border into Virginia. Upon hearing this, Spotswood was afraid that they might begin to organize pirate groups in the Virginia waters. Spotswood wrote, "Some of the same Gang having passed throughout this Countrey in their way to Pensilvania, and contrary to my Proclamation assembling in great numbers with their arms."[3] As mentioned in the chapter on William Howard's arrest, Spotswood issued a proclamation on August 14, 1718, for the arrest of any pirate coming into Virginia. In a letter to the Council of Trade and Plantations, Spotswood wrote:

> The Proclamation prohibiting the unlawfull concourse of persons who have been guilty of piracy was occasioned by the great resort to this Colony, of certain pyrates who being cast away in North Carolina, surrendered there upon H.M. Proclamation; but as there's no great faith to be given to the fore'd submission of men of those principles.[4]

September 1718 was a turning point in the events that led to Blackbeard's death. On September 16, 1718, Wiliam Howard, former quartermaster to Blackbeard, was arrested by Captain Gordon's men. The details of his arrest and trial are explained in Chapter Twenty-One.

On September 24, 1718, while Howard was sitting in chains onboard the *Pearl*, Blackbeard sailed his sloop into Bath with news that he had found and salvaged a large French sugar merchantman. This was the date mentioned by Tobias Knight at his trial several months afterward. Knight said, "24th of September last past where he came and reported to the governor that he had brought ale & etc into this government."[5] His statement was supported by Edmund Chamberlayne, who testified on behalf of Tobias Knight. Chamberlayne said that he didn't see Blackbeard "until on or about the 24th day of the said last September when as this deponant is informed he came up to the governor and reported to him that he had brought a wreck into this government."[6]

A second spy hired by Brand sent word that Blackbeard had arrived at Ocracoke with a ship loaded with sugar and that he had sailed his sloop to Bath to file a salvage claim. When approved, Blackbeard would have clear ownership of the valuable sugar. However, Blackbeard's story was suspicious, and some believed that was more likely that Blackbeard took the sugar merchantman through an act of piracy rather than just finding it abandoned. Brand wrote:

> In August I Employed a second person going into No. Carolina, to make particular Enquiry after the Pyrate... I had certain Information that Thach had been at sea and was come into that place again and had brought in with him a ship loaded with sugar etc. and gave out they found her a Wreck in the sea, that his sloop was lying there before Bath Town in No. Carolina and the ship carried into Oerecrock inlet in Pamplico sound.

Around the same time, Moseley, Moore, and Vail became more involved with the intrigue that always surrounded Blackbeard. In Chapter Twenty, the assault on a man named William Bell was discussed. The day after his attack, Bell identified "Smith, Undey, Titery Dick and others" as his attackers.[7] Moseley saw this as an opportunity to blame Blackbeard for this assault and implicate his political enemy, Tobias Knight, in the process. Someone convinced Bell to change his story and to name Blackbeard and four of his crew as his assailants. This assault took place on September 14, 1718. Moseley alleged that Blackbeard had secretly returned to Bath to meet with Knight at his home and plan the disposal of the sugar and the sharing of the profits. All of those charges came out six months later in a trial in Williamsburg and a trial in Bath.[8] At his trial, Tobias Knight specifically mentions that "Mr. Maurice Moore Jeremiah Veal and others of that family," as those behind Bell's change of mind.[9] These trials will be discussed in great detail in Chapter Thirty.

William Bell incident was used to implicate Knight in collusion with Blackbeard.

Moseley's motivations were political and personal. Governor Eden and Tobias Knight were Moseley's greatest political adversaries, and bringing charges against Knight for collusion with Blackbeard was the first step in implicating Governor Eden. After all, he had granted the king's pardon to Blackbeard, who apparently was continuing his acts of unlawfulness. Before, Moseley and his companions had simply fed Brand's spy information on petty charges. Now they had something more substantial to report to Spotswood. Sometime at the end of September or the beginning of October 1718, Moseley shared all his contrived evidence with Spotswood. Although Brand didn't mention any names, he did comment that the governor had been receiving letters of complaints from some of the citizens in North Carolina. Brand wrote:

> When I was to wait on the Governour he lett me know he had received some Letters of Complaints, and affidavids from some of the Inhabitants of No. Carolina setting forth the great Difficulty and hardship they Labour under from the insults of this notorious Pyrate.

These letters must have contained all the minute details of Bell's assault. That is to say, the details that Moseley wanted Spotswood to read. The details came out at the trial in Williamsburg of the four pirates who allegedly took part in this assault. Whereas the sending of letters to Spotswood isn't in doubt, a personal visit by Moseley or one of his friends remains unproven. It is far more intriguing to imagine either Col. Moseley, Col. Moore, or Jeremiah Vail traveling to Williamsburg to personally discuss the details in a clandestine meeting with Spotswood. As expected, no contemporary documentation has been found that directly supports this. Such a journey would have been conducted with the utmost secrecy. If discovered, Governor Eden would have looked upon it as treason. Many modern historian authors, such as Alan Watson, believe that this undisclosed meeting with Spotswood actually took place.[10] Based on the circumstantial evidence, I agree with Watson. In a letter dated September 14, 1721, Captain Gordon mentioned that when Captain Brand traveled from Virginia to Bath in November 1718, he was accompanied by a "Gentlemen of that country."[11] It is likely that this gentleman was one of the three conspirators mentioned above.

All this data alleging Blackbeard's continued crimes hit Spotswood within a few weeks. Reports from Brand and Moseley arrived from North Carolina. Among the most significant was the report about his assault on William Bell. Then, the news that Blackbeard may have piratically taken a sugar merchant ship, the *Rose Emelye*, reached Spotswood. Moseley quickly linked this crime to Tobias Knight as the receiver of stolen goods. Governor Eden's lack of response contributed to Spotswood's concerns. As if that wasn't enough, William Howard's arrest added to the evidence piling up against Blackbeard. Back in June, Blackbeard had surrendered and accepted the king's pardon from Governor Eden of North Carolina. However, the King's Proclamation clearly stated that for a pirate to qualify for amnesty, the individual in question must not have committed any acts of piracy after January 5, 1718.[12] William Howard's charges included a brigantine that he took while sailing with Blackbeard in May 1718.[13] Then, just as Howard's trial was concluding, news of the arrest of Stede Bonnet reached Spotswood.[14] Bonnet had received the king's pardon from Governor Eden, just as Blackbeard had. If Bonnet had accepted the pardon and immediately returned to piracy, Spotswood must have wondered what would keep Blackbeard from returning to piracy as well.

In Spotswood's opinion, Blackbeard and his crew weren't entitled to receive the proclamation in the first place. The only logical conclusion Spotswood could reach was that the government of North Carolina was either incompetent or complicit with the pirates. Convinced that he needed to intervene personally, Spotswood went to the council for their approval and mentioned the weakness of the North Carolina government. However, Spotswood was met with tremendous opposition from his council. They knew that it would be illegal for any force from Virginia to take any action in North Carolina without the approval of that province's governor. Since Spotswood had no trust in Governor Eden, he wasn't even going to ask him for permission. In a letter to the Council of Trade and Plantations, Spotswood wrote:

Lt. Governor Spotswood becomes involved in the plan to eliminate Blackbeard.

> Tache's Crew pretended to surrender and to claim the benefits of his Majestys Proclamation, they had nevertheless been guilty of divers Piracys after the fifth of January for which they were not entitled to his Majestys Pardon. I therefore thought fitt to have him brought to a Tryal, but joind a strong opposition from some of the council against trying him ... That Tache with divers of his crew kept together in North Carolina went out at pleasure committing Robberys on this Coast and had lately brought in a ship laden with sugar and cocoa, which they pretended they found as a Wreck at sea without either Men or Papers. that they had landed the Cargo at a remote Inlet in that Province and set the ship on fire to prevent discovery to whom she belonged: and having at the same time received Complaints from divers of the trading people of that province of the insolence of that Gang of Pyrats, and the weakness of that Government to restrain them, I judged it high time to destroy that Crew of Villains, & not to suffer them to gather strength in the neighbourhood of so valuable as Trade as that of this Colony.[15]

Spotswood turned to his two naval officers, Captain Brand and Captain Gordon. Brand commanded HMS *Lyme*, a sixth rated ship of 20 guns, and Gordon commanded HMS *Pearl*, a fifth rated ship of 40 guns.[16] They had both recently served as judges at Howard's trial and had in-depth knowledge of the size and armament of Blackbeard's sloop as well as the strength of his crew.[17] Additionally, Brand had been gathering intelligence for months. However, that intelligence was out of date. Brand sent another agent to North Carolina to make a full appraisal of the situation. This agent also recommended two pilots that could guide Brand through the waters of Pamlico Sound. Brand wrote:

> Upon which the Governour and my self began to project some ways for the taking this fellow and his crewe, in the mean time it was thought proper that an Express should be sent into No. Carolina to one of the injured Gentlemen, to send us a full account of that fellows Condition as to the force of his sloop and number of men, and where his usual place of Rendevouze was an likewise to send us two able Pilots in five or six days the messenger returned and with Letters of advice that his usual Rendevouse was at Bath Town and refer'd us to one of the Pilots for other Particulars. The Pilot being Come we Learned the Condition of the Pyrate and what Else might be of use for our Carrying on this service.

While Spotswood waited for Brand to complete his intelligence profile, the governor asked his assembly to pass an act offering rewards for the apprehension and conviction of any pirates, particularly Blackbeard.[18] However, Spotswood intentionally concealed his plans to go after Blackbeard in North Carolina from his council. He knew that he had no jurisdiction in another province and that his council would forbid him from acting, especially since his council was already in the process of petitioning King George for Spotswood's removal on multiple charges of corruption.[19] His explanation for proceeding with a military expedition into a neighboring province without the council's knowledge or approval was the need for security and that there might be some pirate-friendly factions within the Virginia government who could have informed Blackbeard of the impending attack. Spotswood wrote:

> I did not communicate to the Assembly nor Council the Project then forming against Tach's Crew for fear of his having Intelligence there being in this Country & more especially among the present Faction an unaccountable Inclination to favour Pyrates of which I begg leave to mention some Instances.[20]

Captains Gordon and Brand, dressed in their finest uniforms, confidently approached the governor's palace in Williamsburg. They were greeted by one of the governor's servants and escorted into an antechamber that was connected to the main hallway. After a few moments, a door opened, and Governor Spotswood entered. The two naval officers came to attention. Spotswood took his seat at the head of the table as the two officers patiently waited for permission to be seated. Spotswood's servant closed the door

Captains Gordon and Brand plan the invasion with Spotswood.

as he left the room. After a friendly nod and a few pleasant words from the governor, Gordon and Brand took their seats. There was no scribe in the room, just the three men. This meeting was highly confidential. They would be discussing the plot to seize Blackbeard and all his men.

Captain Brand was the senior officer and began briefing his plan to Spotswood. Gordon sat quietly and listened as Brand methodically explained the intelligence he had received and proposed his plan of attack. The first point that Brand covered was the depth of the water at Ocracoke. It was far too shallow for either the *Lyme* or the *Pearl* to navigate. They would have to use shallow-draft sloops. Local pilots who knew the intricate channels of Pamlico sound would have to be hired. Spotswood asked Brand if the navy would supply such sloops. Brand respectfully answered that the navy didn't have any small sloops, so they would have to be rented from civilian mariners in Virginia. Spotswood calmly asked Brand if he and Captain Gordon would pay for the sloops. Brand's expression changed to one of apprehension. In a moment of hesitation, Brand and Gordon exchanged glances. Neither one of them was willing to pay for the sloops, but neither did they wish to anger the governor. Brand cautiously explained to Spotswood that the cost would have to come out of their pockets and that they were unwilling to pay such an expense. Spotswood forcibly asked what they were willing to provide. Sensing a way out of the tense situation, Brand enthusiastically answered that they would, of course, supply the crew and the officers needed to sail the sloops. Spotswood agreed and consented to pay for the sloops and the pilots himself. Spotswood wrote:

> Having gained sufficient Intelligence of the Strength of Tache's Crew, and sent for Pylots from Carolina, I communicated to the Captain of His Majestys Ships of war on this station the Project I had formed to extirpate this Nest of Pyrates. It was found impracticable for the Men of war to go into the shallow and difficult Channells of that Country, and the Captains were unwilling to be at the Charge of hyring sloops which they do not want to do, and must therefore have paid out of their own Pocketts but as they readily consented to furnish more, I undertook the other part of supplying at my own charge Sloops and Pilots Accordingly I have hired two sloops and put Pilots on board, and the Captains of his Majestys Ships having put 55 Men on board under the command of the first Lieutenant of the Pearle & an officer from the Lyme.[21]

Captain Brand also mentioned Spotswood's purchase of the sloops in his letter. He added that Captain Gordon provided thirty-five men and his

Chapter Twenty-Four

Lieutenant, who would be the operational commander of both sloops. Brand provided twenty-five men and a midshipman. The sloops had no naval guns, but the men had a full complement of small arms. They were fitted out for combat at Kecoughtan, the *Lyme* and the *Pearl's* home port. The initial plan was to split their forces into three elements. After clearing Cape Henry, one sloop would enter Roanoke Inlet and proceed to Ocracoke from the north, while the other sloop would sail directly to Ocracoke in the Atlantic and enter the inlet from the south. Brand would travel to Bath over land. Brand wrote:

> The Governour hired two sloops Capt. Gordon and I undertook to fit Each of us one his to be Commanded by his Lieut. and 35 men, and mine Commanded by a midshipmen and twenty five men; at the Coming away of the Pilot from No. Carolina Thach had been gone down to Ocrecock Inlett Ten days, for a second Tripp of Goods the first then being in Bath Town and it was his opinion we should find him there by the time we were ready, and to Carry on this service that we might not be disappointed in the taking of this Pyrate, either at sea or on shore we did agree so soon as the sloops were ready to proceed, one to enter at the Barr of Roanock and the other at Ocrecock Inlett where their prize Lay, and I undertook to go into No. Carolina by Land in order for the securing him if I found him in Town.

Command of both sloops was given to Lieutenant Robert Maynard, one of Gordon's officers.[22] Maynard would personally take command of the larger of the two sloops and Midshipman Hyde, one of Brand's officers, took command of the other.[23] The armament onboard the sloops and the size of the crews were mentioned by Lt. Maynard in a letter he wrote to his friend, Lt. Symonds. Maynard was Gordon's lieutenant who was given operational command of the sloops. In his letter, Maynard mentioned that the sloops had no naval guns, only small arms. He also mentioned that there was a total of fifty-four men in his command. That is different from the sixty men mentioned by Brand. Perhaps a few of his sailors came down with illness before they sailed. Maynard also mentions that he departed Virginia on November 17, 1718. His letter, dated December 17, 1718, was printed in the April 25 1719 edition of *Weekly Journal and Saturday Evening's Post*. A portion of his letter reads:

> This is to acquaint you, that I sail'd from Virginia the 17th past with two Sloops, and 54 Men under my Command, having no Guns, but only small Arms and Pistols. Mr. Hyde commanded the little Sloop with 22 Men and I had 32 in my Sloop.[24]

In the seventeenth century, a midshipman was a rating for an experienced seaman. The word was derived from the area aboard a ship where he generally worked, amidships, or at the middle of the ship. Beginning in the eighteenth century, the term was used as the rank of a commissioned officer candidate.

The names of these two sloops were the *Jane* and the *Ranger*. Although, for about two hundred and sixty years, only the *Ranger* was identified. There was no mention of their names in any of Brand's or Gordon's correspondences. In the media, the first reference was in 1724 in Charles Johnson's book. He wrote, "Mr. *Maynard's* other Sloop, which was called the *Ranger*."[25] After Johnson, most publications followed his lead. In 1960, Hugh Rankin wrote, "The Pearl was given command of the larger sloop, which was nameless in the official dispatches, while the first officer of the Lyme was in charge of the smaller vessel, the Ranger."[26] Lee's 1974 book, *Blackbeard the Pirate*, reads, "The smaller of the two sloops was as the *Ranger*, the other never identified by name."[27] However, the truth was out there all the time. In researching his 1985 book, *Pirates on the Chesapeake*, Donald G. Shomette found the answer in the logbook of HMS *Lyme*.[28] A copy of the original November 1718 page from the *Lyme's* logbook is in a collection of documents housed in the North Carolina State Archives. It clearly reads: "Modt gales yos'd saild the Ranger & Jane sloops with 22 of our men & 32 of Pearle in quest of ye Pirate Teech in N Caroline."[29]

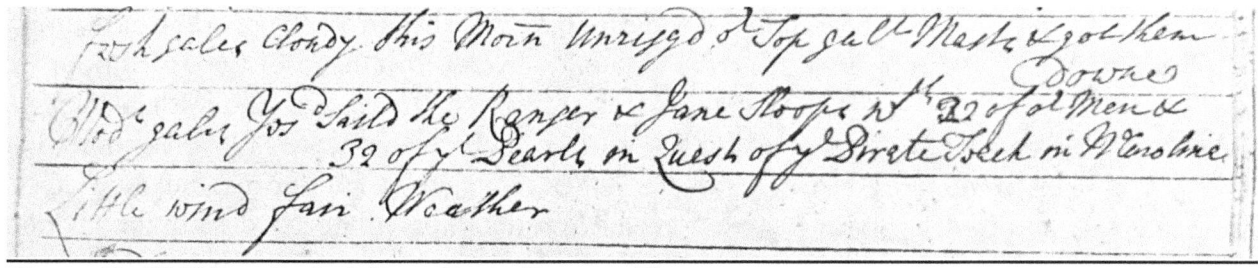

Figure 60: *HMS Lyme Logbook Entry Showing Ranger and Jane.*

Before the sloops departed, Captain Brand gave Lieutenant Maynard discretionary command. That is a military term that means Maynard had the authority to change the plan if he became aware of any new intelligence. The term Brand used was that the officers could govern themselves accordingly. That's precisely what Maynard did. The sloops sailed out of Kecoughtan at about 3:00 pm on November 17, 1718. Once they were past Cape Henry, Maynard chose not to split up his two sloops. Both sloops entered Roanoke Inlet together on November 20, 1718. Brand wrote:

> All things being ready on the 17th November about 3 a Clock the sloops sail'd and that night I set out for No. Carolina ... The Commanders of the Sloops had directions to advise with the Pilots and if they Could Learn any other Intelligence to govern them selves accordingly ... On the 20th. November our two sloops were entring the narrows of Roanock.

Chapter Twenty-Four

As mentioned above in Brand's letter, he left Kecoughtan on the evening of November 17, 1718, writing, "and that night I set out for No. Carolina." The size of Brand's force and the route they took have been the subjects of enormous exaggeration over the years. Charles Johnson doesn't mention Brand at all in his 1724 book, *A General History of the Pyrates*. The earliest reference to Brand in a published book was in Shirley Hughson's 1894 book, *Blackbeard & the Carolina Pirates*. In Hughson's account, Brand accompanied Maynard and was present at the battle of Ocracoke. Brand only came to Bath by sloop afterward. Hughson wrote, "From Ocracoke Inlet Brand sailed into Bath with nine prisoners."[30]

Patrick Pringle was the first historian author to provide an accurate account. In his 1953 book, *Jolly Roger: The Story of the Great Age of Piracy*, Pringle wrote, "Captain Brand went overland to Bath, where he met Edward Moseley."[31] This was soon followed by Hugh Rankin's 1960 book, *The Pirates of North Carolina* where he wrote, "Brand hurried to Bath."[32] Although accurate, both of these accounts are devoid of any details. The first historian author to add weight to Brand's expedition was Robert E. Lee. In his 1974 book, *Blackbeard the Pirate*, Lee wrote, "Spotswood gave Brand overall command of the expedition. He was to travel overland with a force to Bath."[33] This word "force" is somewhat ambiguous and opens the door for a flood of enhanced descriptions. Colin Woodard expanded this force as a contingent of marines. In his 2007 book, *The Republic of Pirates*, Woodard wrote, "Brand, as the senior-most officer, would lead a contingent of marines overland to Bath. A second force would travel by sea to Ocracoke, ensuring no pirates escaped into the Atlantic."[34]

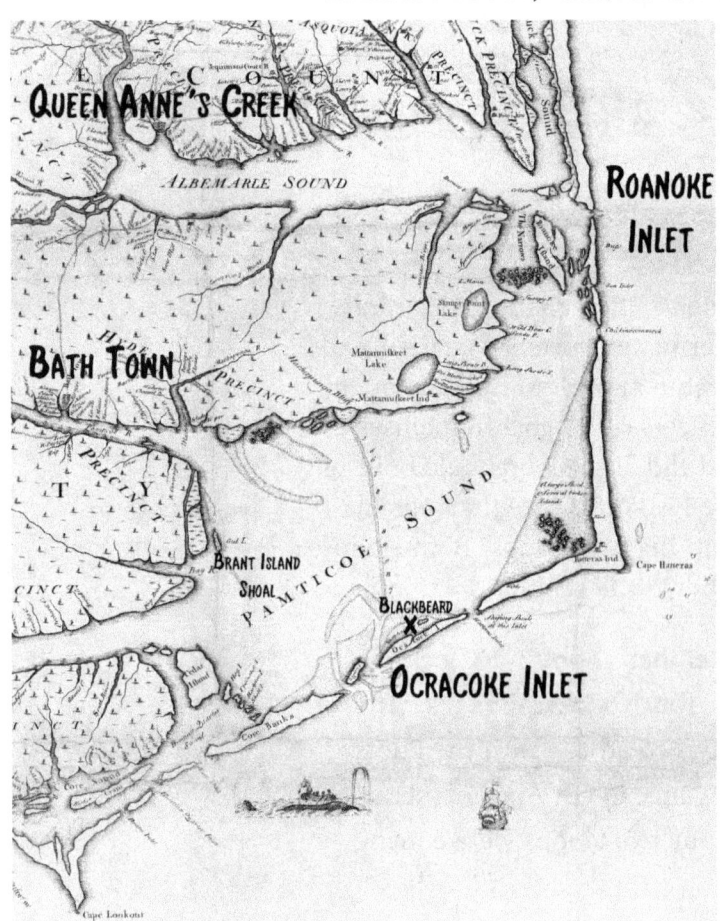

Figure 61: *Close-up of Roanoke Inlet to Ocracoke Inlet from 1733 Moseley Map, Courtesy of East Carolina University.*

The historian author who took this "force" to the extreme was Angus Konstam. In his 2006 book, *Black Beard: America's Most Notorious Pirate*. Konstam wrote:

Captain Brand would command the main force of the expedition, which would march across country from the James River to Bath Town on the Pamlico River. It would cross the boundary line between the two colonies of Virginia and North Carolina near Windsor . . . Brand would be relying on speed and surprise to counter any move by the North Carolina authorities or the pirates to prevent him from carrying out his job.

To avoid any trouble with Govern Eden, Brand's column would avoid the settlement of Albemarle where the North Carolina governor had his residence. The captain's force consisted of around two hundred men, half sailors and the rest a company of Virginia militia.

Maynard could blockade the town from the sea while his commanding officer swept into the town from the landward side.[35]

This account reads like a well-planned military invasion, focused on seizing control of an enemy territory as if a state of war existed between the two colonies. It would be impossible for a force that large to hack its way through the dense woods of the North Carolina countryside and cross the wide rivers just to avoid Albemarle. Additionally, Eden lived in Bath in 1718. He didn't relocate to the Albemarle area until 1720.

According to fellow officer Captain Gordon, Captain Brand traveled by horse and was accompanied by a servant and a gentleman, presumably either Moseley, Moore, or Vail. There was no force to speak of. There was no contingent of marines, and there certainly wasn't any two-hundred-man army. In a letter dated September 14, 1721, Gordon wrote:

> Capt. Brand who went by land with a single gentleman, and a servant to apprehend Thatch with the assistance of the Gentlemen of that country who were weary of that rogues insolence: being informed of his being on shoare often then on board.[36]

Brand most likely took the easiest route. He traveled by boat from Kecoughtan to the south side of the James River. Mounting his horses, the three of them followed an old and well-used road that led from Suffolk to Queen Anne's Creek. Today, that town is named Edenton. This road approximately traced the modern road, NC Route 32. Queen Anne's Creek was located at the end of Albemarle Sound. At the time, it consisted of a cluster of small buildings, including "publick houses" which fronted a row of docks that jutted out into a natural harbor. To the west, just beyond the harbor, the Albemarle Sound twisted north to become the Chowan River. Many large creeks flowed into the sound and the river. The shoreline was lined with plantations and large country homes. It had been the

Chapter Twenty-Four

Established in 1712 as the Towne on Queen Anne's Creek, it was later known as Ye Towne on Mattercommack Creek and as the Port of Roanoke. In 1722, its name was changed to Edenton in honor of Governor Charles Eden, who moved there in 1720 and died there in 1722.

Lt. Robert Maynard's route to Ocracoke

home of the former governor, Edward Hyde, and was the home of Colonel Edward Moseley.

Late in the afternoon of November 21, 1718, Captain Brand and the two men who accompanied him rode quietly to the edge of town. He was greeted by several of Moseley's comrades. For the rest of that evening, Brand, Moseley, Moore, and a few others discussed their upcoming trip to Bath and how they should best deal with Governor Eden. Brand had no idea that Lt. Maynard and his sloops had already reached Ocracoke and were preparing for battle.

That same day, November 21, 1718, while sailing down Pamlico Sound, Maynard hailed a passing vessel and asked if they had seen Blackbeard. The captain told Maynard that he had seen him stranded on Brant Island Shoal and that another sloop was there attempting to help him get free. Maynard enthusiastically ordered all the sails to be trimmed to increase speed. With the pilots onboard to guide them through the treacherous shoals, Maynard's sloops raced toward Brant Island Shoal. At around 4:00 pm, disappointment set in among the crew as they came within sight of the shoal. Everyone strained their eyes trying to locate a sail or even a mast, but not a single vessel was visible. Maynard's report on the pirates mentioned Ocracoke as their location. Since they weren't at Brant Island Shoal, Maynard's pilots set the course for Ocracoke. Brand wrote:

> On the 20th. November our two sloops were entring the narrows of Roanock they spoke with a Vessell that acquainted them they had seen Thach's sloop the Monday befor on Brant Island Shoals a ground and a sloop with him helping him of on the 21st about three or four a Clock they being the Length of those Shoals and not seeing him there they look'd into Ocrecock where they saw two Sail at anchor, then they haul'd in for them, and night coming on both our Sloops anchor'd.

The ground swell, created by the subsurface shoals, gently rocked the sloops from side to side. It was twilight. The sun had just set over their starboard stern quarter and the sky was a velvet purple. The cold November air penetrated their clothing. The pilot pointed in the direction of their destination. Maynard's intense gaze remained fixed on that spot until they were close enough to see the shore clearly. As the navy's sloops got closer, the outline of two masts gradually came into focus. They didn't expect to see two masts. Maynard was overcome with eagerness and trepidation. Questions swirled about in his thoughts as he contemplated the upcoming battle. Were these indeed the pirates he was searching for? Who did that second sloop belong to? Was that another pirate? Would he have to do battle with both of them? Did they notice his approach? If so, would they fight tonight or just attempt to escape?

Maynard quickly devised a plan of action to counter each of these possibilities. As his sloops came within three hundred yards of the pirates, there was no movement on the other sloops. No one was on deck raising sails. No one was loading guns. On the contrary, they appeared to be bedding down for the evening. It became apparent to Maynard that the pirates weren't alarmed by his presence. They would still be there in the morning. At Maynard's command, the *Jane* and the *Ranger* turned into the wind and glided to a stop. Their crews took in sails and dropped their anchors. The next twelve hours would be incredibly tense for all of Maynard's men. The night before a battle always is.

The night before the battle of Ocracoke, November 21, 1718

Meanwhile, aboard the *Adventure*, some of the pirates may have noticed the two sloops silently coming to anchor across the main channel. There was no cause for concern. Ocracoke Inlet was a difficult channel to navigate, even in the daylight. Very few vessels risked sailing through the inlet at night. Seeing two sloops at anchor near the channel in the late evening was fairly routine. Next to the *Adventure* was the sloop that belonged to Samuel Odell. It is highly likely that this was the sloop that helped the *Adventure* get off Brant Island Shoal a day or two earlier. Blackbeard simply wanted to repay Odell's kindness. As the evening wore on, the twenty-one men aboard the *Adventure* drank toasts, played games, and shared sea stories, just as every crew aboard an anchored vessel has done for thousands of years. The only light on deck came from the signal tower on Beacon Island, which marked the channel's entrance. All was quiet from the two sloops anchored across the channel. Blackbeard and his men peacefully went to sleep, not expecting to have any difficulties in the morning.

Charles Johnson has a different ending for Blackbeard's last night on Earth.[37] He wrote:

> The Night before he was killed, he set up and drank till the Morning, with some of his own Men, and the Master of a Merchant-Man, and having had Intelligence of the two Sloops coming to attack him, as has been before observed; one of his Men asked him, in Case any thing should happen to him in the Engagement with the Sloops, whether his Wife knew where he had buried his Money? He answered, That no Body but himself and the Devil, knew where it was, and the longest Liver should take all.

Twenty-Five
Battle of Ocracoke

Bitterly cold, the morning air was still and calm and the glassy water had a silvery shimmer from the pre-dawn light. Aboard the *Jane* and *Ranger*, Maynard's men were busy checking their weapons and preparing for battle. All the flints were replaced with fresh ones. A misfire caused by a dull flint could cost a man his life. Everyone was trying to make as little noise as possible. Their adversaries were only about three hundred yards away and the slightest sound could carry a great distance, especially since the tranquil water was silent. Maynard came to the rail and peered at the two sloops that lay at anchor near the shore. It was dark when his sloops had arrived the night before and he wasn't able to see any details. Now, as the sun began to rise, he could see a fairly large camp on the shore just behind the sloops. As the light grew brighter, the particulars of the camp became visible. Maynard could see several tents and dining flies made of sailcloth, as well as barrels and cargo boxes scattered about the shore. The open-air firepits with cooking pots and spits were strategically placed to keep the tents from catching fire. From the size and nature of the camp, Maynard could tell that this wasn't temporary. It had been there for weeks. The night before, Maynard wasn't certain that he had located the pirate base he was sent to destroy. Once he saw the camp, Maynard was convinced that he had found Blackbeard.

Three prime source documents enable historian authors to describe the battle at Ocracoke accurately. Captain Brand, the overall commander, wrote an official report of the engagement to the admiralty on February 6, 1718.[1] This report contains many details of the pursuit of Blackbeard, from the early accounts of Blackbeard in Bath, all the way through the battle at Ocracoke and Brand's actions at Bath afterward. The second document is a letter written by Maynard himself to a friend and fellow officer, Lt. Symonds, and is dated December 17, 1718, just twenty-five days after the battle. This letter was published in the April 25, 1719 edition of *Weekly*

Journal and Saturday Evening's Post.[2] The third document is a letter written by Captain Gordon, Maynard's commanding officer. Although Gordon wasn't present, Lt. Maynard and his men gave their commanding officer a full briefing when they returned to Virginia. Gordon's letter was addressed to the Admiralty and dated September 14, 1721.[3] Each of these is cited above. To reduce the number of citations in this chapter, I will identify each quote as coming from Brand, Maynard, or Gordon. Quotes from other sources will be cited as usual.

These other sources include articles that appeared in *The Boston News-Letter*. These articles add a few details, although they were all written based on hearsay. Lt. Governor Spotswood added a few comments on the battle, which he must have read in reports sent to him from Brand. Of course, Charles Johnson's *A General History of the Pyrates* offers an account, which is unreliable and somewhat inaccurate.

Johnson incorrectly gave the route of Maynard's sloops as sailing to Ocracoke in the Atlantic and coming to anchor at the mouth of the inlet on the ocean side of the island. He wrote that they got sight of the pirate sloop from their anchorage. The ocean side of Ocracoke consists of a long beach with rough surf and dangerous riptides, like all of North Carolina's outer banks. The nearest safe anchorage would be well offshore, about two or three miles away from Blackbeard's camp. The island itself is covered with lush vegetation, including live oaks. In addition to being completely contrary to Brand's report, it would have been impossible for Maynard to see the *Adventure* at anchor on the other side of the island. Johnson wrote:

> The 17th of November, 1718, the Lieutenant sail'd from Kicquetan, in James River in Virginia, and, the 21st in the Evening, came to the Mouth of Okerecock Inlet, where he got Sight of the Pyrate.[4]

By 9:00 am, Maynard's men were ready in their two sloops, the *Jane*, called the *Pearl's sloop* in Brand's report, and the *Ranger*, which Brand called the *Lyme sloop*. The sun was higher in the sky and in line with their target. This made it more difficult to see the pirate sloops and their camp. As they prepared to get underway, the wind proved to be most uncooperative. It was blowing directly toward them, which rendered the use of their sails totally impractical. Fortunately, the *Jane* and the *Ranger were* equipped with oars, so they ran them out and began slowly rowing their way across the channel. The constantly shifting sandbars inside Ocracoke Inlet made navigation exceptionally complicated. If the pilots gave Maynard and recommendation as to the best course to take, he ignored them. His two sloops were rowing straight for the pirate sloop *Adventure*. Without warning, Maynard's sloop, *Jane,* struck a sandbar and came to a sudden stop. It had run aground. The *Ranger* continued on for a little while, then ran aground too. Maynard's men desperately threw part of their ballast overboard to lighten the vessel and float free. Meanwhile, the men of the *Ranger* were doing the same thing except they were throwing their water kegs overboard. Brand wrote:

> The next morning Weigh'd about nine theirs going at the Wind Lient. Maynard ordered the Lyme, sloop to make the best of their way for Thach's Sloop and Board him, and he should Endeavoure to keep near him the small sloop made the best of their way for the Pirate the Pearls sloop in following them run a ground and was forced to heave part their Ballast over Board to get of again before the Lyme Sloop got near the Pyrate She ran a ground and was obliged to Stave her water to get off.

From the *Jane*, Maynard could see that the *Adventure* had 9 guns mounted. After the battle, Maynard learned that there were twenty-one men aboard. Maynard wrote, "I attack'd him at Oherhock in North-Carolina, when he had on Board 21 Men, and nine Guns mounted." In a letter to the Council of Trade and Plantations, Spotswood wrote, "they came up with Tach at Ouacock Inlett on the 22nd of last month, he was on board a sloop weh. carried 8 guns and very well fitted for fight."[5] Most likely, the *Adventure* had four on each side and one at the bow. This would explain the difference between Maynard's report of 9 guns and Spotswood's report of 8 guns. Quite often, swivel guns mounted at the bow weren't included in the total count of guns on a vessel.

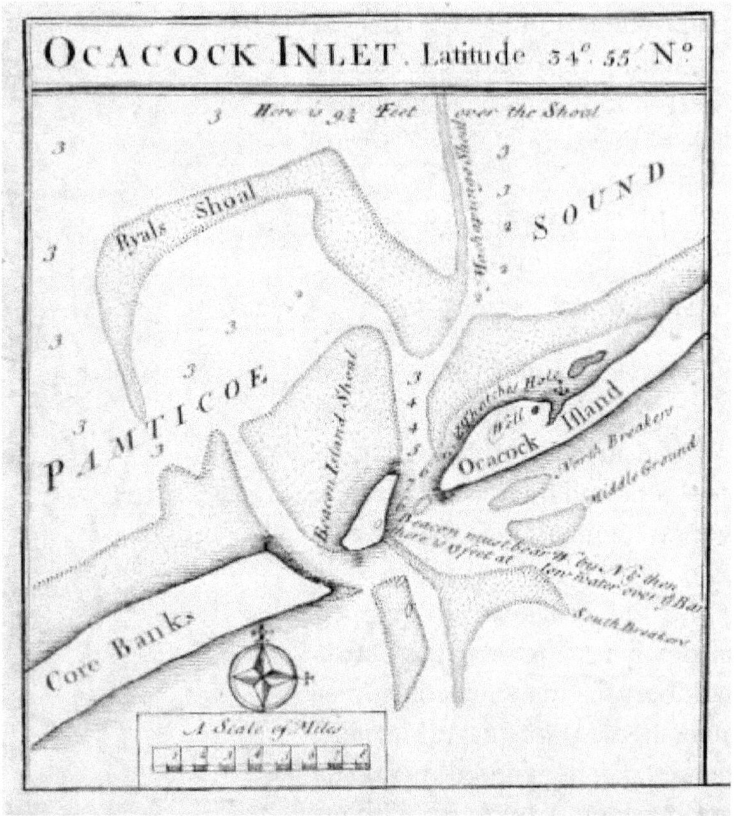

Over on the *Adventure*, the crew began to wake and some of them sleepily wandered up on deck. One of the crew happened to notice the two sloops grounded less than 100 yards away. He called for the captain to take a look. With an air of confidence, Blackbeard stepped up from the cabin and walked to the rail. Peering at the sloops, he immediately became suspicious. Innocent merchant sloops would be in the channel, not aground near his

Figure 62: *Inset of Ocracoke from 1733 Moseley Map, Courtesy of East Carolina University.*

sloop. To make matters worse, based on their position, it appeared that they were heading in his direction before they hit the sandbar. Reality set in. Blackbeard realized that these sloops had been positioning for an attack before they ran aground. Without a moment of hesitation, Blackbeard spun about to address his crew, saying, "Prepare for battle." His serious tone of voice and expression reflected a sense of urgency, which sparked the crew into immediate action. As some of the men on deck began to get the sails ready, Blackbeard went back down into the cabin and flung a baldric over his shoulder. It had a cutlass and several loaded pistols. By now, the rest of his crew were busy loading their small arms and getting their cutlasses ready, too. The naval guns were already loaded. Blackbeard always kept them that way as a precaution. Among this flurry of activity, Samuel Odell was in a daze. He didn't know exactly what he should do. Odell wasn't a pirate; he was simply a guest who had been invited onboard the day before. Odell had never experienced a shipboard battle. He sat helplessly as he contemplated his fate.

Maynard's sloop was within hailing distance of Blackbeard, less than 100 yards away. This entire battle took place within the space of a football field. The two commanders exchanged words. Maynard wrote, "At our first Salutation, he drank Damnation to me and my Men, whom he stil'd Cowardly Puppies, saying; He would neither give nor take Quarter." Not giving or taking quarter means that he wouldn't take any prisoners or surrender. Spotswood echoes Maynard's statement writing, "As soon as he perceived the King's men intended to board him, he took up a bowl of liquor and calling out to the Officers of the other sloops, drank Damnation to anyone that should give or ask quarter."[6] Of course, there are several versions published in the media of precisely what they said, predominantly in Charles Johnson's *A General History of the Pyrates* and in *The Boston News-Letter*. An article that appeared in the February 23 to March 2, 1719 edition of *The Boston News-Letter* reads:

> Teach called to Lieutenant Maynard and told him he was for King GEORGE, desiring him to hoist out his Boat and come aboard, Maynard replyed that he designed to come aboard with his Sloop as soon as he could, and Teach understanding his desires, told him that if he would let him alone, he would not meddle with him; Maynard answered that it was him he wanted, and that he would have him dead or alive, else it should cost him his life; whereupon Teach called for a glass of Wine, and swore Damnation to himself; if he either took or gave Quarters: Then Lieut Maynard told his Men, that now they

Verbal salutation between Maynard and Blackbeard

knew that they had to trust to, and could not escape the Pirates hands if they had a mind; but must either fight and kill or be killed:[7]

Every edition of Johnson's *A General History of the Pyrates* has this version of the story:

> Upon which *Black-beard* hail'd him in this rude Manner: *Damn you for Villains, who are you? And, from whence came you?* The Lieutenant made him Answer, *You may see by our Colours we are no Pyrates.* *Black-beard* bid him send his Boat on Board, that he might see who he was; but Mr. *Maynard* reply'd thus; *I cannot spare my Boat, but I will come aboard of you as soon as I can, with my Sloop.* Upon this, *Black-beard* took a Glass of Liquor, and drank to him with these Words: *Damnation seize my Soul if I give you Quarters, or take any from you.* In Answer to which, Mr. *Maynard* told him, *That he expected no Quarters from him, nor should he give him any.*[8]

If this dialogue was presented as a script, it would look like this:

Blackbeard: Damn you for Villains, who are you? And, from whence came you?

Maynard: You may see by our colors we are no Pyrates.

Blackbeard: Send your Boat over to me, that he might see who you are.

Maynard: I cannot spare my boat, but I will come aboard of you as soon as I can, with my sloop.

Upon hearing this, Blackbeard took a glass of liquor, and drank to him with these Words:

Blackbeard: Damnation seize my soul if I give you quarters or take any from you.

Maynard: I expect no quarters from you, nor should he give me any.

The only real discrepancy between Johnson's version and Maynard's is that Johnson has Blackbeard calling Maynard's men villains, whereas Maynard said that he called them cowardly puppies.

While Blackbeard and Maynard were exchanging words, the *Adventure*'s crew raised the sails and cut the anchor cable. The wind was still very light, but it was in Blackbeard's favor, and the pirate sloop began to move along the channel. Both the *Jane* and the *Ranger* came free of the sandbar and continued rowing toward Blackbeard's sloop. The *Ranger* (referred to as the small sloop or *Lyme's* sloop) entered the channel and closed to within fifteen feet (half a pistol shot) of the *Adventure*. The *Ranger's* coxswain put the helm a lee, and the sloop swung around, placing it in a position that would allow the *Ranger's* men to board Blackbeard's sloop. Opening fire with small arms, the *Ranger's* crew shot away the *Adventure's* jib sail and fore halyards. As the jib sail came crashing down to the deck, the *Adventure* answered with a full broadside, supported with small arms fire. The *Ranger* shuttered and rocked back as the rail exploded in a hail of splinters. Musket balls pelted the deck or whizzed overhead. A few rounds found their targets. Midshipman Hyde, the sloop's commander, and the coxswain both fell dead on the deck. Several other members of the crew were wounded. On the *Adventure,* the mainsail was still undamaged, and the sloop shot past the *Ranger* and down the channel toward the *Jane*. However, the wind was exceptionally light, and with no jib sail, Blackbeard lost control of the *Adventure* in the narrow channel and was forced to sail near the shore (flew to).

Brand wrote:

> The Pyrate saw they design'd for him, cut his Cable and got under sail, the Pearls Sloop was off and rowing up after the other sloop, the Pyrate had fired severall shot at the small Sloop as she was standing for him, the Pyrate endeavouring to get out at the same channell as our sloop came in put for it the Lymes Sloop put her helm a Lee to Board Thach they then being within half Pistol Shott. The Pyrate at the same time fired his Broad side kill'd the Commander of my Sloop, wounded Wm. Baker, Brother of Capt. Flerculer Baker whom I had appointed Lieutenant for this service and kill'd my Coxin whom had the third command and wounded several of my fellows, they had their Gibb and foresheets shot away so the Pyrate shot by them Lieut. Maynard was then pulling with his Oars for the two sloops and Thachs fore Halliards and Gibb being shott by the fire they receiv'd from the Lyme Sloop he flew to, which gave Wm. Maynard an opportunity to board Thach.

Maynard's account is similar, except that in his version, it was his men on the *Jane* (the *Pearl* sloop) who shot away Blackbeard's Jib and fore halliards. Maynard wrote:

Chapter Twenty-Five

> Immediately we engag'd, and Mr. Hyde was unfortunately kill'd, and five of his Men wounded in the little Sloop, which, having no-body to command her, fell a-stern, and did not come up to assist me till the Action was almost over. In the mean time, continuing the Fight, it being a perfect Calm, I shot away Teach's Gib, and his Fore-halliards, forcing him ashoar,

Maynard's letter was published in The Boston News-Letter.

The Boston News-Letter published an article that echoes Maynard's account. Part of this article, from the February 23 to March 2, 1719 edition, reads:

> We have this further account of it by a Letter from North Carolina of December 17th to New-York, viz. That on the 17th of November last, Lieut. Maynard of the Pearl Man of War Sail'd from Virginia with two Sloops and 54 Men under his Command, no Guns, only small Arms, Sword and Pistols, Mr. Hyde Commanding the Little Sloop with 22 Men and Maynard had 32 in his Sloop, and on the 22d Maynard Engaged Teach at Okercock in North Carolina, he had 21 Men, Nine Guns Mounted, Mr. Hyde was killed, and one more and Five wounded in the Little Sloop, and having no body aboard to Command them, they fell a Stern and did not come up to Assist Lieut. Maynard till the Action was almost over, Maynard shot away Teach's Gibb and Fore-halliards, and put him ashore.[9]

At this point, Captain Brand claims that his ship's sloop, *Ranger*, saved the day. Brand contends in his letter that his sloop sailed directly between Maynard's and Blackbeard's sloops and conducted boarding action that ended the battle. This is completely contrary to Maynard's and Gordon's reports. Maynard wrote that "the little Sloop, which, having no-body to command her, fell a-stern, and did not come up to assist me till the Action was almost over." Captain Gordon agreed. **The lack of details in Brand's account suggests that this was merely a self-serving attempt to make himself look good. He wanted the credit for the victory and possibly a larger share of the reward money as well. This account foreshadows Brand's questionable conduct and integrity, which will be discussed in later chapters.** Brand wrote:

> The little sloop not being above her length from their shott in between The Pyrate and Wm. Maynard Enter'd their men, did good service and suffered much,

With the *Ranger* temporarily out of action, the *Jane* rowed in a straight line for the *Adventure*, which was cornered near the shore. Maynard's sloop came alongside in a position that would allow his men to board. Blackbeard had other ideas. His guns were loaded with swan shot and partridge shot. At a range that could only be described as dangerously close, the *Adventure* fired another broadside. This time, it was Maynard's sloop that felt the full brunt of Blackbeard's devastating blast. In an instant, twenty-one of his men lay on the blood-soaked deck, either killed or wounded. Dense white/gray smoke from Blackbeard's guns hung in the air, obscuring visibility between the two sloops. Maynard took advantage of the concealment provided by the smoke to rush his survivors down into the hold. Maynard himself took cover in the aft cabin. Only two men remained on deck, his midshipman at the helm, and Mr. Butler, his pilot. Each had instructions to signal Maynard if the pirates boarded.

> Swan shot is slightly smaller than buckshot and is used for hunting large fowl. Partridge shot is even smaller and used for hunting partridges. Both were loaded in artillery and used for anti-personnel at close range.

The Boston News-Letter has a slightly different version. Their article contends that Maynard heroically remained on deck and that his pilot was named Abraham Demelt. Other than these two discrepancies, the rest aligns with Gordon's account. The article in the February 23 to March 2, 1719 edition reads:

> Lieutenant Maynard, ordered all the rest of his Men to go down in the Hould himself, Abraham Demelt of New-York, and a third at the Helm stayed above Deck;[10]

Charles Johnson adds that Blackbeard's men threw grenades and case bottles filled with gunpowder and small shot over to the *Jane*. This is very possible, as grenades were common among warships of the day. Johnson wrote:

> When the Lieutenant's Sloop boarded the other, Captain *Teach*'s Men threw in several new fashioned sort of Grenadoes, *viz.* Case Bottles fill'd with Powder, and small Shot, Slugs, and Pieces of Lead or Iron, with a quick Match in the Mouth of it, which being lighted without Side, presently runs into the Bottle to the Powder, and as it is instantly thrown on Board, generally does great Execution, besides putting all the Crew into a Confusion;[11]

The pirates on the *Adventure* braced to receive the return fire from Maynard's sloop but all was quiet. Blackbeard anxiously strained to see if there was any movement through the smoke. As the smoke gradually dissipated, the only sailors that Blackbeard could see on the *Jane*'s deck appeared to be dead. Believing that most of his opponents were killed and that the sloop was effectively his, Blackbeard ordered his helmsman to sail the *Adventure* into the *Jane*. Holding a rope, one of the pirates jumped over to the *Jane* and securely tied the two sloops together, then Blackbeard and eight more of his pirates stepped aboard. Everything remained still and quiet on the *Jane* as Blackbeard cautiously surveyed the casualties lying on the deck. Meanwhile, Mr. Butler was anxiously watching the pirates from his topside hiding place. As the last of the pirates stepped aboard, Butler shouted, "Now!" Maynard and his twelve remaining men erupted from the hold and the cabin, and a fierce and desperate hand-to-hand battle began.

Gordon wrote:

> Thatch observing all his men of which gave them a broadside; his guns being suffitiently charged with Swan Shot, partridge shot, and others; with this broad side he killed and wounded (most by the Swan Shot) one & twenty of his men, Mastr: Maynard in finding his men thus exposed, and no shelter order his men down into the hold, goeing himself into the cabin at aft; ordering the midshipman, that was at the helm, or Mr: Butler, his pilate to aquaint him with anything that should happen. Thatch observing his decks clear of men presently concluded the vessel his own, and then sheers on board Lieut: Maynards sloop enters himself the first man, with a rope in his hand to lash or make fast the two Sloops: Mr. Butler aquainting Lieut. Maynard with this, turned his men upon deck, and was himself presently among them:

Maynard wrote, "I boarded his Sloop, and had 20 Men kill'd and wounded. Immediately thereupon, he enter'd me with 10 Men; but 12 stout Men I left there." The word "boarded" has caused some confusion among many who read his letter. In naval terminology of the eighteenth century, boarding can have two meanings. One is the commonly understood meaning that his men physically boarded the other vessel. Boarding can also mean that his vessel had taken a position which would allow his men to physically board the other vessel. This is what Maynard meant when he said, "I boarded his Sloop." As seen above, Brand mentions this too when he wrote, "Lymes Sloop put her helm a Lee to Board Thach."

The battle was on! Hand-to-hand combat is always frantic and chaotic. Sounds of men shouting in aggressive anger or screaming from deadly wounds were everywhere. Other sounds included cutlass blades clanging and pistols firing. Maynard wrote that his men "Fought like Heros, Sword in Hand." As the battle raged on the *Jane*, the *Ranger* drifted into the *Adventure* and one of the *Ranger's* crewmen jumped onto the deck of the pirate sloop. This battle was fast-paced and confusing. One of Maynard's men saw a Ranger crewman standing on the deck of the *Adventure* and mistook him for a pirate who was about to join the fight. He took careful aim with his musket and fired. This unfortunate sailor was instantly killed, becoming the only casualty of friendly fire in the engagement. During the course of the fight, some of the pirates sought escape and jumped overboard, but Maynard's men shot them with their muskets A highly descriptive illustration of the fight between Blackbeard and Maynard was drawn in 1920 by Jean Leon Gerome Ferris. It provides an accurate visual sense of this deadly struggle.

Figure 63: *The capture of the Pirate, Blackbeard, 1718. Painting by J. L. G. Ferris. The image is dedicated to the public domain under CC0.*

Gordon wrote:

> In less than six minutes tyme Thatch, and five or six of his men were killed; the rest of these rogues jumped in the water where they were demolished, one of them being discovered some days after in the reeds by the fouls hovering over him: the sloop in which the Lymes people were in, had the missfortune to have the thre officers that commanded them killed a head of his Sloop & another shot through the body in Thatchs sloop by one of our men, takeing him by Mistake, for one of the pirates. This far is the true and real steps of that action given in upon oath at his Majesties Court of Admin: in Virginia

Meanwhile, over on the *Adventure*, those who hadn't boarded Maynard's sloop began to realize that they were about to lose the battle. Earlier, Blackbeard had instructed a black pirate to blow up the sloop rather than be taken prisoner. As he approached a barrel of gunpowder with a lighted match, a "planter" who had been forced onboard stopped him. The following account was written by Spotswood in a letter to the Council of Plantations and Trade:

> His [Blackbeard's] orders were to blow up his own vessel if he should happen to be overcome, and a Negro was ready to set fire to the Powder had he not been luckily prevented by a Planter forced on

Chapter Twenty-Five

board the night before & who lay in the Hold of the sloop during the actions of the Pyrats Tach.[12]

As expected, Charles Johnson added this story to his account of the battle, only he mentions that this person was stopped by two prisoners. The identity of these mysterious prisoners, one being the planter mentioned above, will be discussed in a later chapter. For now, Johnson wrote:

> For before that, Teach had little or no Hopes of Escaping, and therefore had posted a resolute Fellow, a Negroe, whom he had bred up, with a lighted Match, in the Powder-Room, with Commands to blow up, when he should give him Orders, which was as soon as the Lieutenant and his Men could have entered, that so he might have destroy'd his Conquerors: and when the Negro found how it went with Black-beard, he could hardly be persuaded from the rash Action, by two Prisoners that were then in the Hold of the Sloop.[13]

Back on the *Jane*, the fighting continued. Maynard and Gordon didn't provide very many details of the battle. Their only details are those already mentioned above. Neither of them mentions the name of the man who actually killed Blackbeard. More sensational accounts can be found in *The Boston News-Letter* and in Johnson's *A General History of the Pyrates*. The former source contends that Blackbeard was killed by a highlander, while the latter source has him dying from a combination of many wounds. *The Boston News-Letter* in the February 23 to March 2, 1719 edition reads:

> Maynard making a thrust, the point of his Sword went against Teach's Cartridge Box and bended it to the Hilt, Teach broke the Guard of it, and wounded Maynard's Fingers but did not disable him where upon he Jumped back, threw away his Sword and fired his Pistol, which wounded Teach. Demelt struck in between them with his Sword, and cut Teach's Face pretty much; in the interim both Companies ingaged in Maynard's Sloop, one of Maynard's Men being a highlander, ingaged Teach with his broadsword, who gave Teach a cut on the Neck, Teach, well done Lad, the Highlander repli'd, if it be not well done, I'll do it better, which that he gave him a second stroke, which cut off his Head, laying it flat on his Shoulder.[14]

Johnson's *A General History of the Pyrates* reads:

> *Black-beard* and the Lieutenant fired the first Pistol at each other, by which the Pyrate received a Wound, and then engaged with Swords,

Blackbeard's death

till the Lieutenant's unluckily broke, and stepping back to cock a Pistol, *Black-beard*, with his Cutlash, was striking at that Instant, that one of *Maynard*'s Men gave him a terrible Wound in the Neck and Throat, by which the Lieutenant came off with a small Cut over his Fingers. They were now closely and warmly engaged, the Lieutenant and twelve Men, against *Black-beard* and fourteen, till the Sea was tinctur'd with Blood round the Vessel; *Black-beard* received a Shot into his Body from the Pistol that Lieutenant *Maynard* discharg'd, yet he stood his Ground, and fought with great Fury, till he received five and twenty Wounds, and five of them by Shot. At length, as he was cocking another Pistol, having fired several before, he fell down dead;[15]

Maynard summed up the battle beautifully. He ended his letter by writing:

I should never have taken him, if I had not got him in such a Hole, whence he could not get out, for we had no Guns on Board; so that the Engagement on our Side was the more Bloody and Desperate.

Blackbeard's sloop was trapped along the shore and forced into a small dead-end inlet. He had no room to maneuver. Blackbeard was cornered in a spot that Maynard described as "a Hole." This term is still used in modern times when someone says, "He dug himself into a hole." With no naval guns on the *Jane*, Maynard had no chance in a running battle. Hand-to-hand combat was his only option. Maynard realized this and used it to his advantage. He was able to close with Blackbeard's sloop to such a close range that the *Adventure* couldn't effectively fire another broadside. When the two sloops were less than ten feet away, Maynard tricked him into boarding the *Jane* by hiding his men below deck. It was a brilliant tactical maneuver. As a result of Maynard's comment, the spot where Blackbeard was defeated was identified as "Thatches Hole" on maps throughout the eighteenth century. Regardless of the specifics, it is certain that after a six-minute hand-to-hand struggle, the world's most famous pirate lay dead on the deck of Maynard's sloop.

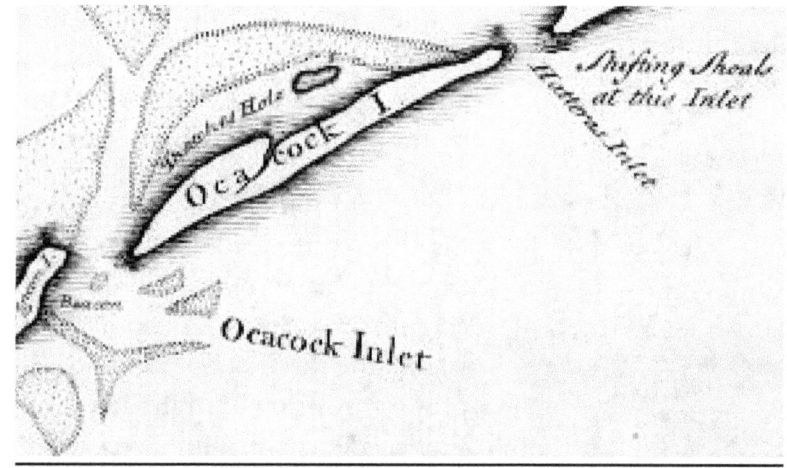

Figure 64: *Close-up of Thatches Hole from 1733 Moseley Map, Courtesy of East Carolina University.*

Twenty-Six
Aftermath in Ocracoke and Bath

Suddenly, all was quiet. Just a moment before, a fierce battle had been raging with all the sounds of battle, the shouts of angry men, pistols firing, swords clanking, and dying men screaming. But with the demise of their leader, the remaining pirates dropped their weapons and ceased their resistance.

The seas were still calm, and the breeze was light. As the morning sun rose toward noon, the piercingly cold late-fall air began to warm. The fragrant aroma of the myrtle bushes that grew all along the shoreline was replaced by the putrid smell of black powder gun smoke and death.

Lieutenant Robert Maynard had done it! Blackbeard along with most of his crew were dead. One can imagine how he felt. As with every victor in battle, his emotions were a mix of pride in his accomplishments and sorrow for losing a large portion of his men. Near the western tip of the island, the three sloops were still clustered together in the same positions they were during the fight. In the distance, Maynard could see the captured French merchant ship laying quietly at anchor. Adjacent to that ship, the beach was cluttered with tents, scattered cargo boxes, and barrels taken from their prize ship, cooking pots and tripods, and common items and accouterments found in every long-term camp.[1]

Now that the fighting was over, it was time to shift from his intense focus on winning the fight to compassion for the wounded, the rounding up of the prisoners, and the burial of the dead. Some accounts mention that the dead pirates were thrown overboard, but this is extremely unlikely. Lt. Maynard's men were there for quite a while. They would not want to see or smell bodies floating by their sloops. All the casualties would have been taken ashore and eventually buried. During this process, the gruesome task of cutting off Blackbeard's head was performed before his headless body was buried. Over a month later, when Maynard finally left North Carolina, he would hang Blackbeard's head from the bowsprit of the sloop. Maynard also described Blackbeard, the second description in the historical record that came from a person who actually saw the pirate in person, as a man with a long beard, and that he tied it up in black Ribbons. Maynard mentions both of these in his letter to Lt. Symonds, writing:

> He went by the Name of Blackbeard, because he let his Beard grow, and tied it up in black Ribbons . . . I have cut Blackbeard's Head off, which I have put on my Bowspright, in order to carry it to Virginia.[2]

The casualty list comes from two primary sources, Lieutenant Maynard's letter dated December 17, 1718, and Captain Brand's letter dated February 6, 1719. Maynard lists twelve pirates killed and nine taken prisoners writing, "We kill'd 12 besides Blackbeard, who fell with five Shot in him and 20 dismal Cuts in several Parts of his Body. I took nine Prisoners, mostly Negroes, all wounded."[3] Checking the numbers, it is certain that he meant to write "12 including Blackbeard." The pirates started out with a total of twenty-one men. Adding twelve dead plus nine prisoners makes twenty-one. Twelve besides Blackbeard would add up to twenty-two.

Explaining the casualties

Brand reported ten pirates killed and nine captured. Brand wrote, "Pyrate had nineteen men, Thirteen White and Six neogroes, ten white men kill'd and the rest of the Prisoners were all wounded."[4] Notice the discrepancy between the two reports. They each have nine captured, but Maynard lists twenty-one men at the start of the battle with twelve men killed, while Brand lists nineteen men at the start of the battle with ten men killed. The difference is two men. This can easily be explained if two of the pirates were unidentified. Brand wasn't present and was going off a list given to him by Maynard. Most likely, Maynard left the two unknown pirates off his list.

The surviving pirates and some of Maynard's men spent the next several days burying the dead. During that time, another pirate was discovered dead in the reeds. They became aware of his presence by the vultures circling over him.[5] This fact was mentioned by Captain Gordon in his account of the battle.

As for the naval personnel, Maynard lists eight men killed and eighteen men wounded from his sloop *Jane* with one killed and five wounded from the *Ranger*. "In the whole, I had eight Men kill'd, and 18 wounded. Mr. Hyde was unfortunately kill'd, and five of his Men wounded in the little Sloop."[6] Brand's report lists nine killed from the *Jane* and two killed onboard the *Ranger*. "The Pearl sloop had kill'd and died of there wounds

Chapter Twenty-Six

nine. My sloop had two kill'd in both sloops there was upwards of twentie wounded."[7] The discrepancy can easily be explained by the death of two of the wounded men between December when Maynard wrote his letter, and February, when Brand wrote his letter.

Over the next several weeks, Maynard searched the pirate camp, the *Adventure*, and the prize ship. He didn't find the treasure he hoped to find. Inside a tent on the beach, he found one hundred and forty bags of cocoa and ten casks of sugar.[8] Aboard the *Adventure*, he discovered some goods and some gold dust, worth about £2,500.[9] He also found the silver cup that Blackbeard had stolen from Captain Taylor on the *Great Allen* when he captured her near Saint Vincent.[10] Then, Maynard made the find of his career. It was a letter.

When Maynard first came across the letter, it must have seemed totally insignificant—a single sheet of paper that could have been anything from a cargo manifest taken from one of his prize vessels to a page ripped out of a book. Anyone could have easily overlooked this paper, but for some reason, Maynard chose to read it. At that time, he couldn't have begun to comprehend the enormous significance of that letter or the role that it would play in the attempt to discredit a government official and shift political power within the province. The letter was written by Tobias Knight and addressed to Edward Thache. It read:[11]

Tobias Knight's letter to Blackbeard

Nov. 17th 1718

My Friend

If this finds you yet in harbour I would have you make the best of your way up as soon as possible your affair will let you, I have something more to say to you than at present I can write, the Bearer will tell you the end of our Indian Warr and Garret can tell you in part what I have to say to you so refer you in some measure to him.

I realy think these three men are heartily sorry at their difference—with you and will be very willing to ask your pardon if I may advise be friends again its better so then falling out among your selves. I expect the governor this night or tomorrow who I believe would be likewise glad to see you before you goe, I have not time to add save my hearty respects to you and am your real friend and servant

T Knight

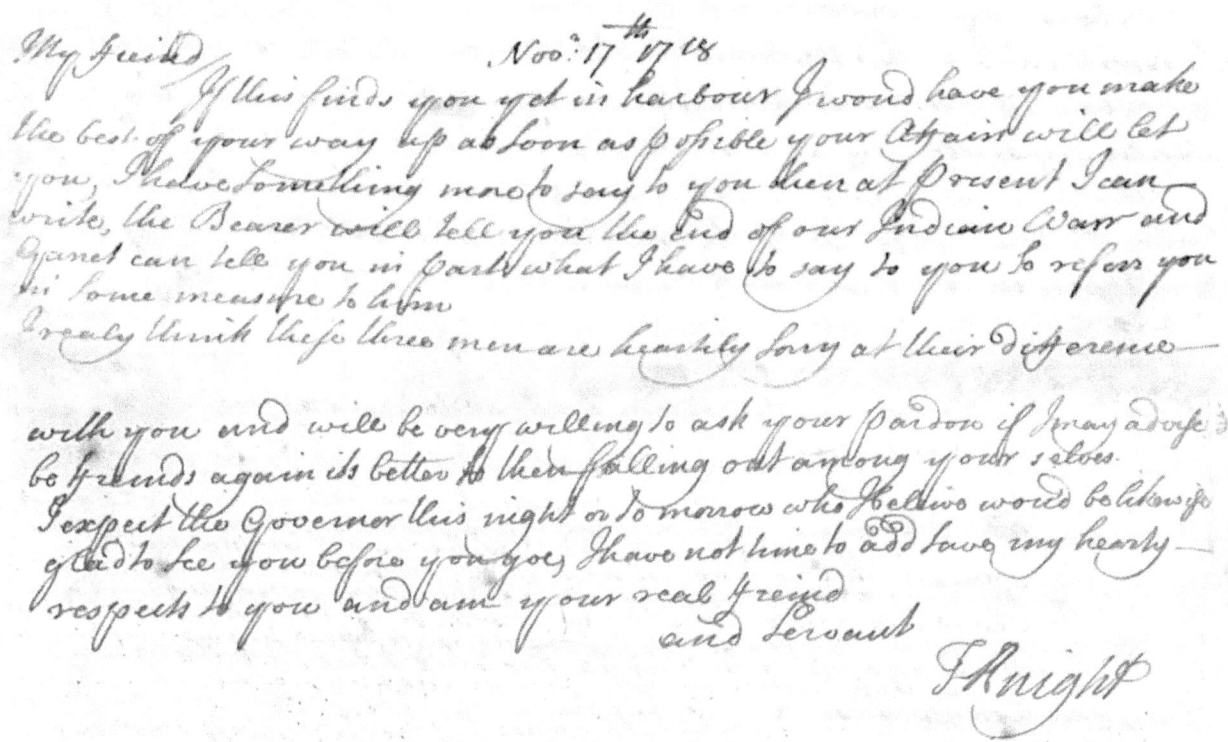

Figure 65: *Photo of Knight's Letter from the Minutes of NC Governor's Council, May 27, 1719.*

In reading the letter, there really isn't much there to incriminate anyone for anything. From the text in the letter, one can deduce that there was some sort of argument or falling out between Blackbeard and three of his crew. Other than that, the letter is purely social. But that didn't matter. It was enough to implicate Knight with Blackbeard, especially since it was signed with the closing comment, "your real friend."

Meanwhile, as Maynard's men dug graves and rummaged through Blackbeard's effects, Captain Brand made his way to Bath. He had reached the town called Queen Anne's Creek the same day that the Battle of Ocracoke had occurred. Today, that town is named Edenton. Upon arrival, he was greeted by Colonel Moseley, Colonel Moore, and Jeremiah Vail.[12] As you may recall, Moseley was the bitter political foe of Governor Eden and the key figure in prompting Spotswood to move against Blackbeard. He also met with a few other associates of Moseley, who shared his political goals. The next day, November 23, 1718, Brand and his entourage, including Moore and Vail, started out for Bath. Moseley remained behind.[13] After crossing Albemarle Sound, Brand and his entourage easily covered the fifty-mile distance to Bath and arrived there at about 10 o'clock that evening. Brand immediately went directly to Governor Eden's house, politely introduced himself, and explained the purpose of his visit. Brand wrote:

Chapter Twenty-Six

The "Sound" Brand mentioned was Albemarle Sound, which is between three and seven miles wide. Brand most likely used a ferry boat to cross the Chowan River at the north end of the sound, which is only one and a half miles wide.

I reached within 50 miles of Bath Town, on the 22nd. I got my self and horses over the sound with the assistance of Cols. Moseley and Cols. Moore two Gent. That have been much abused by Thach on the 23d. I sett forward again Cols. Moore and some of their friends went in with me as did severall of the Inhabitants when they heard I was Come in to take Thach, about ten a Clock att night I gott within three Mile of Town and desired Col. Moore to go in and Learn if Thach was there, he soon return'd to let me know he was not yet Come up but Expected Every minute. I parted from Col. Moors and went to the Governor and applied my self to him and let him know I was come in Quest of Thach.[14]

Governor Charles Eden was no country bumpkin from the backwoods. He was part of the English aristocracy, who was living in England at the time of his appointment. As mentioned in Chapter Eighteen, the decade before his appointment, North Carolina had experienced a great deal of political turmoil which came to a head in the Cary Revolution in 1711 and it was Moseley who led the opposition against the established government during that conflict. Among Eden's strongest political allies was Tobias Knight, his Chief Justice and Secretary of the Province.

Historian authors disagree as to Eden's first reaction when he opened the door and saw a naval captain standing on his porch in the darkness of the night. Some say he was shocked and perhaps frightened. Others say he was angry and resistant. **I believe that Eden was calm and controlled when he answered the door because he was well informed by loyal North Carolina residents who were keeping track of Brand's movements.** At any rate, Brand handed Governor Eden the letter that Lt. Governor Spotswood wrote earlier that month on November 7, 1718. It was a very polite letter that discusses Virginia's authority by saying, "I know not whether you and your people be well apprised of the extent of the Govt. of Virginia's power and Authority with Relation to Pyrats." Spotswood also informs Eden that Brand was there to apprehend the pirates and to seize their effects by adding, "I hope you will put a favourable Construction upon the friendly advise he brings you," referring to Brand. Then Spotswood closes the letter by saying. "It is very possible that I may have been misinform'd as to severall extraordinary passages that I hear with Respect to the Pyrats in your province."[15]

Reading between the lines, the gist of Spotswood's politely written letter was that Virginia has more power than you think. We plan to capture the pirates and take all their possessions and there's nothing you can do about it; I hope you assist Brand but it really doesn't matter, and believe you have intentionally done nothing about the pirates in your province, but I hope I'm wrong.

At that point, neither Eden nor Brand was aware of the fact that Blackbeard had already been killed along with twelve members of his crew and that his sloop and possessions had already been seized. Wanting a status report, Brand sent two small boats down the Pamlico to look for Maynard.[16] Captain Brand couldn't possibly have imagined the whirlpool of legal entanglements that awaited him upon his return to Virginia.

Twenty-Seven
Politics and Legalities

The central figure in the political intrigue that erupted in Bath was Tobias Knight. As mentioned earlier, he was the Secretary of the Colony, Collector of Customs, Chief Justice, and serving as a member of the Council, Knight was the second most important politician in the colony.[1] However, Knight was virtually bedridden for well over a year due to a prolonged illness.[2] His large brick house on the west side of Bath Creek provided him with waterfront access. In addition to Knight's loving wife, Katherine,[3] a man named Mr. Edmund Chamberlayne had been living there since August 1718. Chamberlayne eventually played a vital role in clearing Knight of the charges that were levied against him.[4]

There are two versions of the interactions between Brand and Knight in Bath. Both versions come from the same source, the transcript of Knight's hearing in May the following year. The first version is the one told by Brand, who was testifying against Knight. This is the version that most historian authors believe and is represented as fact in their narratives. Some of the locals told Brand that Knight had received some goods from Blackbeard, perhaps as payment for services. The first time that Brand went to Knight's house to inquire about the report, Knight was totally uncooperative and denied that any such goods were on his property. However, Brand went back the next day with the letter Knight had written to Blackbeard and a few other of Blackbeard's documents that linked him to Knight, all of which were found by Maynard on the *Adventure*. Brand referred to those documents as a "pocket book" that proved that Knight indeed had those goods. After seeing the evidence against him, Knight admitted his guilt and allowed Brand to search his property. Brand found twenty barrels of sugar and two bags of cotton in Knight's barn hidden under a pile of fodder. The following is from Brand's testimony in the *Minutes of NC Governor's Council*, May 27, 1719.

> After which Captain Ellis Brand Comander of his Majestys Ship the Lime declared that having received Information of Twenty Barrels of Sugar and two bags of Cotton lodged by Edward Thache at the House of Tobias Knight he asked the said Knight for those goods they being part of the cargoe piratically taken in the French ship and that the said Knight with many asseverations positively decryed that any such goods were about his plantation but yet the next day when the said Captain Brand urged the matter home to him and told him of the proofs he could bring as well by the persons concerned in sending the said goods as by Memdm: in Thaches pocket book he the T Knight owned the whole matter and the piratical goods afors'd were found in his Barne covered over with fodder.[5]

Knight denied any wrongdoing! In his rebuttal, he said that the goods belonged to Thache, and he was simply storing them temporarily for Thache because there was no other place in town where they could be housed. He also denied that those goods weren't hidden under fodder. Additionally, Knight categorically denied ever speaking to Brand about any such goods, adding that if Brand had visited his property, he would have given him full access. The following is from Knight's testimony in the *Minutes of NC Governor's Council,* May 27, 1719.

Knight's rebuttal to the charges

> The said Tobias Knight doth in the most solemn manner declare that the said pretended Evidence is every word false and that the said Brand never did at any time speak one word or mention to the said Knight in any manner whatsoever touching or concerning the sugars . . . if he had any other information against him he would be so civel as either to come himself or send his Lieut. to his house and every lock in his said house should be opened to him.[6]

Mr. Edmund Chamberlayne, Knight's house guest, totally corroborated Knight's rebuttal. As I shall discuss later on in this chapter and the next, there is compelling evidence that the entire case against Knight and Blackbeard was contrived by Moseley in collaboration with Virginia Lt. Governor Spotswood. The reasons will become quite clear after all the events revolving around this case are revealed. For now, based upon Knight's strong rebuttal, the collaborative testimony by Chamberlayne, and the continued support of Governor Eden, it suffices to say that I am convinced that Knight was telling the truth and that Brand was intentionally lying.

Chapter Twenty-Seven

Charles Johnson falsely implicates Governor Eden

The real scandal came when sixty hogsheads of sugar taken from the French ship were discovered in Governor Eden's storehouse. Many historian authors have repeated this incident in their books and articles. The problem is that it never happened. The primary source for this disgraceful tale was, you guessed it, Charles Johnson. Page 76 of *A General History of the Pyrates* says:

> When the Lieutenant came to Bath-Town, he made bold to seize in the Governor's Store-House, the sixty Hogsheads of Sugar, and from honest Mr. Knight, twenty; which it seems was their Dividend of the Plunder taken in the French Ship.[7]

The origin of this ridiculous claim came from Brand's February 6 letter. He mentions that a parcel of sugar consisting of sixty hogsheads (abbreviated as hheds.) and twenty barrels were among Blackbeard's effects along with the six black pirates who were captured at Ocracoke. They were considered as slaves, not pirates. Brand stated that they were delivered to him. However, he doesn't say where they were found or when they were delivered. Brand wrote:

> I sent to Lieut. Maynard to Come up to me at Bath Town, with the Sloops I having information that a parcell of Sugars which I found and were delivered to me, Sixty hheds. Twenty Barrells Six negroes all wth. were the effects of Thach.[8]

There is no other contemporary reference that those hogsheads and barrels were directly tied to Governor Eden or that he ever had any contraband goods. Based upon the volatile political climate and the number of Eden's enemies who were straining at the bit to obtain anything that could incriminate Eden, the total absence of any such comment serves as substantial proof that this never occurred. Yet, most modern accounts continue to repeat this absurd story.

Meanwhile, Maynard sent the *Ranger* back to Kecoughtan with the wounded naval personnel. The *Ranger* arrived on Sunday, December 1, 1718, as is evident from the logbook entry of the *Pearl* still anchored at Kecoughtan.

> Moderate gales & fair weather; in the afternoon the Ranger sloop that was man'd Arm'd & c by Capt. Brand of His Majesty's Ship Lyme on the Expedition against Edward Thach commonly call'd blackbeard a pyrate, anchored here from No. Carolina who acquainted us of the

destruction of the afore said Thach & most of his men, and the seizure of their Effects, by our men.⁹

There isn't any mention of when the *Jane* returned to Virginia, but it must have sailed there around this time. Maynard kept Blackbeard's *Adventure* to use for transportation back to Bath and then to Virginia. By mid-December, Maynard's search of Blackbeard's vessels and camp was finally complete. He, too, was ready to leave. Maynard sailed to Bath aboard Blackbeard's sloop *Adventure*. A letter written by Captain Gordon, which chronicled the event, reads that Maynard:

> Stayed about a fortnight or thre weeks there, to get of the Island some of Thatchs goods he had there in a tent; having found but very little in the pirats sloop, and then went up the river some forty or fifty leagues to receive orders from Brand.¹⁰

Upon hearing the details of Maynard's actions at Ocracoke, Governor Eden reached the limits of his tolerance. He wrote a letter to Colonel Thomas Pollock, who was a highly respected attorney, close friend, and leading politician in the province. Prior to Eden's arrival, Pollock had served as the acting governor. Pollock lived very close to the Virginia border and had many friends in that colony. The exact contents of Eden's letter are unknown, as no copy still exists, but fortunately, Pollock's response has survived. From Pollock's reply, one can piece together the gist of Eden's original letter. ¹¹

Apparently, Eden questioned the legality of Virginia's invasion of his province and discussed the royal statutes that expressly forbade the recent actions of the Virginia governor. Pollock's response, dated December 8, 1718, addressed all of Eden's concerns and included the following:

> I declare that I never heard any thing of any applications to Virginia concerning Captain Thach, nor nothing of any intended expedition out of Virginia, until I heard that Captain Brand was come in, and that he and Colonel Moore and Captain Veal were gone to Pamplico."¹²

After reading Pollock's legal opinion, Eden instructed Pollock to begin legal action against Virginia. Pollock entered a plea, questioning Spotswood's authority, and requested the return of all goods confiscated and sent to Virginia. Additionally, Pollock initiated a lawsuit against Brand for trespassing on the lands of the Carolina Lords Proprietors and for taking the goods mentioned above.¹³

Thomas Pollock (1654–1722) was a wealthy plantation owner, lawyer, and a member of the Provincial Council. He served as acting governor twice, once from 1712 to 1714, after the death of Edward Hyde, and again in 1722, after the death of Charles Eden.

Chapter Twenty-Seven

This action set in motion many other legal actions between North Carolina and Virginia that continued throughout 1719. I could write an entire chapter on the intricacies and complexities of these actions and a lawyer could probably write two complete volumes. However, in the name of expediency, I shall sum them up.

Spotswood's problems began with his *Act Against Pirates* that he drafted on November 13, 1718. That document was similar to a modern-day arrest warrant, and it specifically named Thatch. Unfortunately, the House of Burgesses didn't sign it until November 24, 1718, two days after Blackbeard's death, making the engagement illegal regardless of all the other issues.[14]

Compelled to answer Eden's complaints expressed through Pollock, Spotswood wrote him a letter dated January 28, 1719, in which he stressed that he did have the authority to invade North Carolina and that he thought that no pirate case could be tried in North Carolina. Selected excerpts from that letter are:

> I am impowered to take Pyrates in or out this Dominion which I hope will satisfy you that I have not exceeded the power given me by sending a Force to suppress Thatch and his crew . . . I have Authority to issue Warrants for apprehending Pyrates in any of those Provinces. That power was so fully restricted in the Warrant I gave Capt. Brand at his going into your province, and whereof I presume he gave you a Copy . . . I perceive no pirate case can be tried in your Province, but must be sent hither for that purpose.[15]

To ward off problems from his boss in London, Spotswood also sent a letter to Secretary Craggs dated May 26, 1719:

> I hope the Lords proprietors themselves will give little Credit to such Clandestine Testimonials when they shall know how dark a part some of their Officers have acted, particularly one who enjoyed the post of Secretary, Chief Justice, one of the Lordship's Deputys and Collector of the Customs held a private Correspondence with Thach, concealed a Robbery he committed in that providence, and received and concealed a considerable part of the Cargo of this French Ship which he knew Thach had no right to give or he to receive."[16]

The "Chief Justice" and "Collector of Customs" refer to Knight and the "private Correspondence" is of course the letter from Knight to Thache that was found by Maynard.

Meanwhile, Spotswood had other legal troubles. Back on November 20, 1718, the House of Burgesses drafted a recital of grievances against Spotswood addressed to King George I and sent William Byrd to England to deliver it. They intended to have Spotswood removed from office.

Figure 66: *Portrait of Alexander Spotswood. The image is dedicated to the public domain under CC0.*

> Having Duely Considered several Attempts of your LIEUTENANT GOVERNOR towards the Subversion of the Constitution of OUR GOVERNMENT, the depriving us of our Ancient Rights and Privileges, and many Hardships which he dayly exercises upon your Majestys Good Subjects of this COLONY, think we should not Discharge the Duty we owe to our SOVEREIGN, or the Trust reposed in us by our CONSTITUENTS, if we any longer forbear to lay them before your SACRED PERSON.[17]

Among their primary concerns was Spotswood's violation of the King's instructions that required him to assist other governors if they had any problems, not to invade their province. The following are quoted from King Georg's written instructions.

William Byrd (1674–1744) was a Virginia landowner, lawyer, surveyor, and a member of the governor's council from 1709 to 1744, and is considered the founder of Richmond, Virginia.

> You are forthwith to Communicate unto our Said Council such and so many of these our Instructions wherein their advice and Consent are mentioned to be requisite, as likewise all such others from time to time as you shall find Convenient for your Service to be Imported to them . . . You are not to Grant Commissions of Mark or Reprisals against any Prince or State, of their Subjects in Amity with us, to any person whatsoever without our Especial Command . . . In case of Distress of any other of our Plantations, you shall upon the Application of the respective Governors thereof to you, Assist them with what Aid, the Condition of our Colony under your Government, can Spare.[18]

The House of Burgesses knew these instructions, and so did Eden. This became part of the House's charges against Spotswood. He replied in a letter to the Commissioners for Trade and Plantations on December 22, 1718, in which he wrote:

> I did not communicate to the Assembly nor Council the Project then forming against Tach's Crew for fear of his having Intelligence there being in this Country & more especially among the present Faction an unaccountable Inclination to favour Pyrates of which I begg leave to mention some Instances.[19]

In response, the House of Burgesses published a complaint that read, "As to the destroying of Thache and his Crew that Story had better be kept in silence than told for if all the Circumstances of it were known they would make little for his Reputation."[20]

Spotswood finally explained his ultimate legal justification for his invasion of North Carolina and the meaning of the statement he wrote to Eden in his January 28, 1719 letter: "I have Authority to issue Warrants for apprehending Pyrates in any of those Provinces." He was referring to an order that was issued to the governor of Virginia in 1697 by King William III.[21]

This order gave the governors of Virginia, Maryland, New York, and Massachusetts the authority to "Seize, or cause to be Seized, the Person Commanding such Ship or Vessel, and any one or more Persons belonging to such Ship or Vessel."[22] Spotswood even cited that order on William Howard's order to charge. It begins with, "Pursuance of an Act of Parliament made in the Eleventh and twelfth years of the Reign of King William the third Entitled an Act for the more Effectual Suppression of Piracy."[23]

At that time, there really wasn't any government in North Carolina, and South Carolina didn't have a substantial militia or navy. Virginia was the only colony in the south with the capability to pursue and apprehend pirates. That's why William III issued an order giving the Virginia governor military authority in the Carolinas. But William died in 1701, and the laws at that time dictated that royal decrees expired six months after the king's death. Updating the edicts from King William III had been discussed by the Council of Trade, but this was "held up, possibly to avoid confusion with the extended offer of pardon." At one of the meetings of the Council of Trade, they commented on Spotswood's actions saying, "Spotswood, too, in Virginia, in spite of protests from some of the Councillors, tried a pirate under the Commission of William III, and the extended pardon only arrived just in time to save his unworthy neck from the gallows."[24] Spotswood didn't have a leg to stand on.

Pollock's suit against Brand was also proceeding during this time. It prompted Brand to write a letter dated March 12, 1719, to his boss at the Admiralty. He wrote:

King William III (1650–1702) ruled England and Scotland from 1689 to 1702. He ruled jointly with his wife, Mary, from 1689 to her death in 1694, as William and Mary. His successor, Queen Anne, was his cousin and his wife's sister.

> I cannot Omitt given there Lordships an acc. of the very extrodinary proceedings of the Goverers of North Carolina, when the court of Admiralty in Virginia was sett to proceed Against the pirates goods, a Lawyer appear'd for them and put in a plea against the juridiction of this court . . . nor doe I know of any Violence or ill treatment either himself or any of the inhabitants did meet with during the time I was in there government from any person under my command; when I left Bath Town the goverer went into the next County with me, I thought aft our parting my conduct had been such the time I was in there government that it would not allow'd of such base and scandilous exports as they are now pleas'd to give out.[25]

Meanwhile, back in December 1718, there was more intrigue brewing in North Carolina. While Maynard was in Bath, another highly charged political event occurred at the hands of Moseley. All the records for the colony were housed in the home of Captain John Lovick, a former naval officer who was the deputy secretary to the province. His home was located just four miles from the home of Moseley on Sandy Point along the north shore of Albemarle Sound, about five miles southeast of modern Edenton. Lovick was a friend and supporter of Knight and Eden, and Moseley suspected that there might be some sort of documents stored in Lovick's home that could be used to incriminate one of them. So, on December 26, 1718, Moseley, Moore, and three other armed men invaded Lovick's home at gunpoint. He was forcibly evicted from his home and Moseley and his men spent the next twenty hours searching through the records, looking for evidence that Eden was in collusion with Knight and Beard. They found nothing.[26]

Moseley's arrest in North Carolina

Governor Eden quickly learned about this outrage and ordered the arrest of Moseley and Moore on the grounds of unlawful entry. Outside of Lovick's home, as Moseley was being led away, he made a speech saying, "They could easily produce armed men to come and disturb quiet and honest men but could not raise them to destroy Thack but instead of that he was Suffered to go on in his villainies."[27] That speech gave Eden grounds to add more charges, which included discord and sedition, and making slanderous statements against the government.

While in Bath, Brand had arrested six of the former members of Blackbeard's crew. As stated in his February 6 letter, "I likewise had information of several Pyrates lurking there, I took six of them from the shoar."[28] They were loaded aboard the *Adventure,* and both Brand and Maynard prepared to return to Virginia. Brand traveled back by horse and Governor

Eden accompanied him part of the way. In a letter he wrote on March 12, 1719, Brand wrote:

Figure 67: *Woodcut Illustration of Blackbeard's Head from the 1837 The Pirate's Own Book. The image is dedicated to the public domain under CC0.*

When I left Bath Town the governer went into the next County with me I thought aft our parting my conduct had been such the time I was in their government that it would not allow'd of such base and scandilous exports as they are now pleas'd to give out.[29]

Maynard sailed back to Kecoughtan aboard the *Adventure*, with Blackbeard's head hanging from the bowsprit. This dramatic spectacle was captured in a woodcut image in the 1837 publication of *The Pirates Own Book: Authentic Narratives of the Most Celebrated Sea Robbers*, by Charles Ellms.[30]

After sailing through Hampton Roads, the *Adventure* rounded the point at Kecoughtan, and Maynard saw his ship, the *Pearl*, for the first time in forty-eight days.

It was Saturday, January 3, 1719. His actions were much more than simply being successful in accomplishing his mission. He was returning as a victorious hero. Forevermore, he would be known as the man who killed

Figure 68: *Close-up of Kecoughton (Kecoughtan) from a 1720 Map, Courtesy of Barry Lawrence Rudman.*

Blackbeard. He must have felt an enormous sense of pride and accomplishment. His commanding officer, Captain Gordon, likely experienced similar emotions. After all, it was his officer who made this tremendous naval victory. In what would be a mixture of respect and admiration, Maynard fired a nine-gun salute to his commander. Gordon returned the salute. The *Pearl's* logbook entry succinctly summarizes the events of that historical day:

> This day the Sloop Adventure Edward Thach formerly Master (a Pyrate) anchor'd here from No. Carolina commanded by my first first Lieut. Mr. Rob't Maynard who had taken the aforesaid Sloop, & destroy'd the said Edward Thach & most of his men; he also brought Thach's head hanging under his bowsprit in order to present it to the Colony of Virginia; he saluted me with 9 guns, I rendered the like number.[31]

Twenty-Eight
The Answers Lie with Cracherode

The definitive list of all of Blackbeard's men who were killed or captured that has been referenced for 300 years comes from Charles Johnson's *A General History of the Pyrates*.[1] This is exactly how it appears in his book.

That list was, of course, for a commercial publication. Within the British government, an official list was filed with the treasury board in England shortly after August 1719. Comparing the two lists, one sees that there are only three discrepancies in Johnson's account. Based upon its accuracy, it is apparent that Johnson must have seen an original report that was sent to England shortly after the engagement but before the amended treasury list that was prepared during the second half of 1719.

Before we go further, I would like to remind the readers of Captain Brand's report in which he states that ten pirates were killed and nine captured at Ocracoke, then adds six more arrested at Bath. "Pyrate had nineteen men . . . ten white men kill'd and the reset of the Prisoners were all wounded . . . I took six of them from the shoar."[2] That's a total of ten killed and fifteen captured.

The first discrepancy in Johnson's list is simple to explain. Without looking at any other source outside of Johnson's, *A General History of the Pyrates*, one can detect a glaring error. Johnson's captured list, which he labels as "were wounded and afterward hanged in Virginia," shows 16 names. But just four pages earlier, Johnson states that there were fifteen captured, writing, "and five-teen Prisoners, thirteen of whom were hanged."[3] Obviously, Johnson accidentally placed one of the "killed" pirates on his captured list. I think that Johnson needs to find a new editor! But which name was incorrectly identified as killed?

90   Of BLACK-BEARD.

The Names of the Pyrates killed in the Engagement, are as follow.

Edward Teach, Commander.
Phillip Morton, Gunner.
Garrat Gibbens, Boatswain.
Owen Roberts, Carpenter.
Thomas Miller, Quarter-Master.
John Husk,
Joseph Curtice,
Joseph Brooks, (1)
Nath. Jackson.

All the rest, except the two last, were wounded and afterwards hanged in *Virginia*.

John Carnes,	*Joseph Philips*,
Joseph Brooks, (2)	*James Robbins*,
James Blake,	*John Martin*,
John Gills,	*Edward Salter*,
Thomas Gates,	*Stephen Daniel*,
James White,	*Richard Greensail*.
Richard Stiles,	*Israel Hands*, pardoned.
Cæsar,	*Samuel Odel*, acquited.

Figure 69: *List of Killed and Captured at Ocracoke from A General History of the Pyrates, 2nd Edition, Page 90.*

Always follow the money! Great Britain offered a financial reward for the capture of pirates with higher payments made for officers and captains. To document the payments to the naval officers who made those arrests, a treasure report was drafted, signed, and filed. The man who signed that report was Anthony Cracherode, the Treasury Solicitor for Great Britain. The Cracherode report is absolutely invaluable to Blackbeard researchers,

Anthony Cracherode (1674–1752) was a lawyer and politician who sat in the House of Commons and was the Treasury Solicitor for Great Britain in 1719.

as it provides an enormous amount of information. This report includes all the details of the arrests, imprisonments, and deaths of Blackbeard's crew as well as those of Stede Bonnet's crew and a few other pirates.[4] The Cracherode report also separates the names of those who were killed and those who were captured. By comparing the two lists, it is clear that the misidentified pirate was Joseph Phillips. All nine of the names on Johnson's "killed" list appear on Cracherode's list, but Cracherode adds a tenth name, "John Philips—Sailmaker."[5]

The below list of pirates killed is precisely the way it appears in the Cracherode report.

Edwd. Thach Capt.
Philip Morton Gunner
Garrot Gibbons Boatswain
Owen Roberts Carpenter
John Philips Sailmaker
John Husk
Joseph Curtis
Joseph Brookes
Tho: Miller
Nath: Jackson

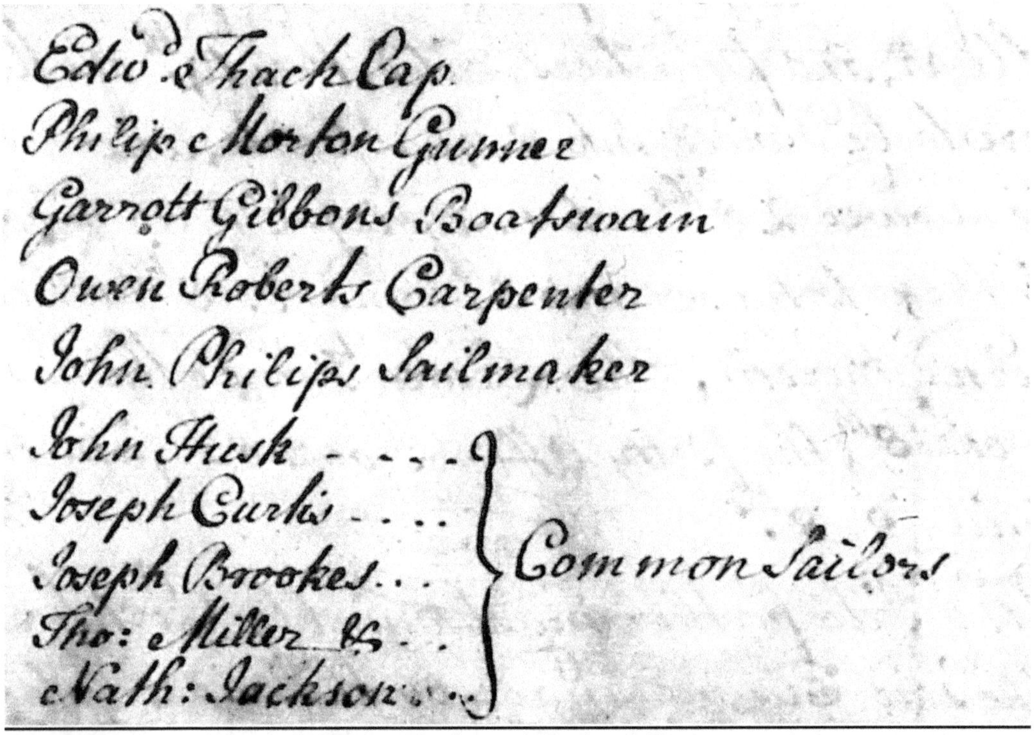
Edwd: Thach Cap.
Philip Morton Gunner
Garrott Gibbons Boatswain
Owen Roberts Carpenter
John Philips Sailmaker
John Husk
Joseph Curtis
Joseph Brookes
Tho: Miller
Nath: Jackson
Common Sailors

Figure 70: *List of Pirates Killed from the Original Cracherode Report.*

A comparison of the lists of those killed:

Johnson's List	Cracherode's List
Edward Teach (Commander)	Edward Thatch Capt.
Philip Morton (Gunner)	Philip Morton Gunner
Garret Gibbons (Boatswain)	Garret Gibbons Boatswain
Owen Roberts (Carpenter)	Owen Roberts Carpenter
Thomas Miller (Quarter-Master)	Tho. Miller
John Husk	John Husk
Joseph Curtice	Joseph Curtice
Joseph Brooks (1)	Joseph Brooks
Nathaniel Jackson	Nathaniel Jackson
	John Philips Sailmaker

Total: 9   Total: 10

By moving Joseph Phillips up to the "killed" list on Johnson's list, the total number of ten agrees with both Maynard's numbers and the Cracherode list. Notice that the first name was changed from Joseph to John. Apparently, the scribe who drafted the Cracherode report changed Joseph to John. I imagine this error was due to sloppy handwriting on one of the earlier reports.

The paragraph below this list is most interesting. It reads, "I am most humbly of Opinion, that in regard all the said last named persons appear by the said Certificate to have been killed in the said Engagement, and not to have been taken and convicted of Pyracy, no Reward is due in Respect of them, or any of them, by the words of his Majestys said proclamation."[6] Apparently, the British government didn't want to make any payments for any pirates who were killed. In order to receive any payment, the pirate had to be tried and convicted for piracy.

By reading Cracherode's list of those having been captured and convicted for piracy, a positive identification of all the rest of Blackbeard's pirates can be made. The list is preceded by a statement certifying payment to Gordon and Brand.

> And I do further most humbly Certifye to your Lordpps. That as to the said Claime made by the said Captains Gordon & Brand, by virtue of the said Certificate from the said Lieutent. Governour of Virginia dated the 28th of May 1719, there is only due to them & the officers and sailors concerned with them, in the said Captures, the sume of 280£. in respect of the 8 Pyrates taken in the said Engagement with

> Thach on the 22th. of Novm. 1718, and of the other 5 pyrates taken on Shoar.[7]

The below list of pirates captured is precisely the way it appears in the Cracherode report.

Hezekiah Hands Master (as appears by the Sd Lieutent Govern's Certificate)

John Carnes
Joseph Brookes Junr
James Blake
John Giles
Thomas Gates
James White
Richd. Stiles
John Martyn
Edwd. Salter
Steph: Daniel
Rich.d Greensail
Cesar

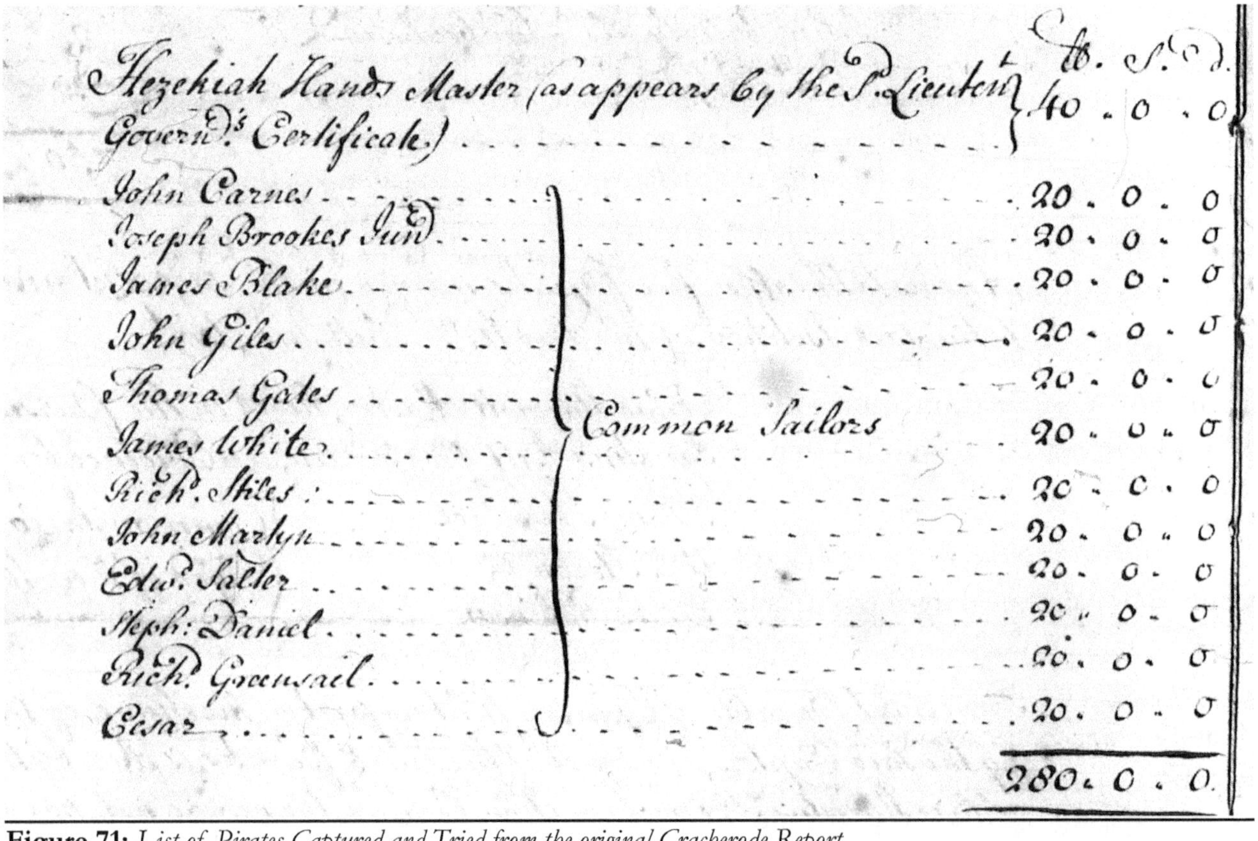

		£. s. d.
Hezekiah Hands Master (as appears by the Sd Lieutent Govern.d's Certificate)		40 . 0 . 0
John Carnes	Common Sailors	20 . 0 . 0
Joseph Brookes Jun.r		20 . 0 . 0
James Blake		20 . 0 . 0
John Giles		20 . 0 . 0
Thomas Gates		20 . 0 . 0
James White		20 . 0 . 0
Rich.d Stiles		20 . 0 . 0
John Martyn		20 . 0 . 0
Edw.d Salter		20 . 0 . 0
Steph: Daniel		20 . 0 . 0
Rich.d Greensail		20 . 0 . 0
Cesar		20 . 0 . 0
		280 . 0 . 0

Figure 71: *List of Pirates Captured and Tried from the original Cracherode Report.*

Cracherode added, "Which said sume of 280£ I am most humbly of opinion ought to be divided between the Sd. Captains Gordon, and Brand, and the Officers & other persons concerned with them in the said Engagement and Captures according to the proportions herein before Specifyed."[8]

The other two discrepancies on Johnson's list are the names of James Robbins and Samuel Odell. Neither of those names is on the Cracherode report. Here is a comparison list (note that I have removed the name of Joseph Phillips as he should have been on the dead list.)

Johnson's List	Cracherode's List
John Carnes	John Carnes
Joseph Brooks (2)	Joseph Brookes, Junr.
James Blake	James Blake
John Giles	John Giles
Thomas Gates	Thomas Gates
James White	James White
Richard Stiles	Richd Stiles
Caesar	Caesar
James Robbins	*Not listed*
John Martin	John Martyn
Edward Salter	Edwd Salter
Stephen Daniel	Steph: Daniel
Richard Greensail	Rich.d Greensail
Israel Hands (Pardoned)	Hezekiah Hands Master
Samuel Odell (acquitted)	*Not listed*
Total: 15	Total: 13

This discrepancy of 15 verses 13 is far more complex than just two additional names, or missing names on the Cracherode list. The answer can be found in Brand's report. As mentioned earlier in this chapter and the last, Brand states that 9 men were taken at Ocracoke and 6 in Bath for a total of 15. It appears that Johnson was correct when he listed James Robins and Samuel Odell on his list. So why aren't they on Cracherode's list? The only answer is that they weren't convicted, they were released.

Samuel Odell and James Robins can be explained after considering two sources. The first is Spotswood's letter to the Commissioners dated December 22, 1718, which states:

Samuel Odell and James Robins

> His [Blackbeard's] orders were to blow up his own vessel if he should happen to be overcome, and a Negro was ready to set fire to the

Powder had he not been luckily prevented by a Planter forced on board the night before & who lay in the Hold of the sloop during the actions of the Pyrats Tach.[9]

Charles Johnson recounts this same incident, saying:

> ... for before that, Teach had little or no Hopes of Escaping, and therefore had posted a resolute Fellow, a Negroe, whom he had bred up, with a lighted Match, in the Powder-Room, with Commands to blow up, when he should give him Orders, which was as soon as the Lieutenant and his Men could have entered, that so he might have destroy'd his Conquerors: and when the Negro found how it went with Black-beard, he could hardly be persuaded from the rash Action, by two Prisoners that were then in the Hold of the Sloop.[10]

Spotswood mentions one prisoner, a "Planter," and Johnson mentions two prisoners. These two could only have been Samuel Odell and James Robins. I can find no reference to Samuel Odell in any of the source documents or official reports, just Charles Johnson's *A General History of the Pirates* when he wrote, "Samuel Odell, was taken out of the trading sloop, but the night before the engagement. This poor Fellow was a little unlucky at his first entering upon his new Trade . . . "[11]

Johnson most likely saw the original report, which contained the names of Samuel Odell and James Robins. All the other names were correct. Therefore, we must conclude that Johnson was correct when he said Samuel Odell was "acquitted." This is one of the few instances that Johnson got right!

Many historian authors attribute Samuel Odell as the man who prevented the destruction of the *Adventure*, but Odell was a captain, and Spotswood clearly described this man as a "Planter." If it wasn't Odell, who was it? The only other person not on Cracherode's list was James Robbins. (Johnson spelled his name as Robbins, but other primary source documents spell his name as Robins with only one "b." I will use that spelling for the rest of this chapter.)

There is convincing evidence that James Robins was the "Planter" Spotswood referred to in the letter. Allen Norris published a book containing a transcript of all the deeds in Bath between 1696 and 1729. This book holds the answers to the identity of several of Blackbeard's crewmen. A deed transcribed in Norris' book names James Robins as the man who purchased lot #13 on September 9, 1718, from John Lillington.[12] That property had been the property of Governor Charles Eden just five months before. In late 1718, James Robins owned Eden's 400-acre property, including Eden's large house.[13] As mentioned in Chapter Eighteen, the property next to Robins' land belonged to Tobias Knight, who, of course, was the writer of the letter dated November 17, 1718, that was found by Maynard onboard the *Adventure* after the battle at Ocracoke. As was discussed in Chapter Twenty-Six and will again be discussed in Chapter Thirty, this letter played a vital role in the charges against Tobias Knight for his involvement with Blackbeard and was transcribed in the minutes of the council meeting of the North Carolina Governor's Council, May 27, 1719.

Blackbeard obviously had this letter in his possession before November 21, 1718, as he was killed on the morning of the 22nd and the letter was found on his sloop. How did he get this letter in just three days? The only answer is that James Robins, Tobias Knight's neighbor, brought it to him. Robins was most certainly the "Planter forced on board the night before," who Spotswood mentioned. As such, he would have been acquitted and would not be listed by Cracherode.

There is more conclusive evidence that James Robins must have been acquitted. There are several deeds recorded in Bath naming James Robins after 1718. Robins witnessed a deed in 1721,[14] and he sold all 400 acres of his property to Robert Campaine for £445 on July 9, 1724.[15]

With the mystery of the two acquitted visitors to Blackbeard's sloop solved, there aren't any other discrepancies, are there? *OR ARE THERE?* Well, unfortunately, the answer is yes.

Earlier in this chapter, I quoted Captain Ellis Brand's February 6, 1719 letter which reads, "Pyrate had nineteen men, Thirteen White and Six negroes, ten white men kill'd and the reset of the Prisoners were all wounded ... I likewise had information of several Pyrates lurking there [Bath], I took six of them from the shoar." I also quoted a section in Cracherode's report which reads, "in respect of the 8 Pyrates taken in the said Engagement with Thatch on the 22th. of Novm. 1718, and of the other 5 pyrates taken on Shoar."

Perhaps you noticed that Brand said he took six at Bath while Cracherode mentioned five. What is the reason for this discrepancy? Additionally, which ones of the captured pirates were the six "negroes" that Brand mentioned?

But most importantly, what happened to the remaining thirteen pirates? Johnson said that they were all hanged. But were they?

Twenty-Nine
Chained in Williamsburg

The thirteen pirates captured in both Bath and Ocracoke were eventually taken to Williamsburg and placed in the *goal* (the early eighteenth-century term for jail). Maynard returned to Kecoughtan on Saturday, January 3, 1719, aboard Blackbeard's captured sloop with all fifteen of his prisoners onboard.[1] Captain Gordon, Maynard's commanding officer and captain of the *Pearl*, would have ordered all the prisoners to be transferred to the ship and held for a short time prior to being sent to Williamsburg, just as he had done a few months earlier with William Howard, Henry Man, William Stoke, and Adult Van Pelt.[2]

The accommodations were abominable. There was no means of heating. Ventilation was almost nonexistent, and sunlight was extremely limited. The only toilet facility was a large wooden box with a hole in the top. One can only imagine the horrible smell that permeated through the stifling air. The pirates would have arrived sometime in January and were held there during the coldest months of the year. In addition to Blackbeard's pirates, other prisoners from the colony, including William Howard, and

Figure 72: *Exterior of the Actual Cells in Williamsburg where Blackbeard's Men were Held. Photo by Robert Jacob.*

Figure 73: *Interior of the Actual Cells in Williamsburg where Blackbeard's Men were Held. Photo by Robert Jacob.*

the three men arrested with him, would have been jammed into the small cells. Howard had been languishing there since October 30, 1718.[3]

With the addition of Howard, the complete account of all of Blackbeard's pirates held in chains in Williamsburg goes up to fourteen. I shall discuss him in greater detail a little later on, but first, let's discuss those captured in North Carolina. All contemporary reports about Blackbeard's pirates either mention the total numbers of those captured at Ocracoke and Bath or list the names of all those who were captured, but none of them put the two together. They do not indicate the names of the pirates taken at either of those specific locations. They also don't identify the race of the individuals captured. However, solving both of those questions is vitally important in solving the mystery of precisely what occurred in Williamsburg.

At the end of the previous chapter, I discussed the discrepancy in the number of those pirates taken at Bath. Brand lists six while Cracherode lists five. This can be easily explained by a simple error. Brand personally arrested those in Bath and should know the count. Additionally, there isn't anyone missing. The total numbers of all those arrested are the same in both reports. It seems likely that an error occurred, and one of those taken in Bath was accidentally added to the Ocracoke list by the time Cracherode received it.

There seems to be a discrepancy in the total number of Africans taken too. Brand mentions six Negroes, while Spotswood's address lists only five. Once again, I shall refer to Brand's February 6 letter, which reads, "Pyrate had nineteen men, Thirteen White and Six negroes ... "[4] Spotswood addressed the council in March of 1719 and requested their opinion as to how to proceed. The five negroes were problematic from a legal standpoint. Since slaves weren't legally responsible for their actions if they were following the orders of their owners, trying African pirates was always somewhat challenging. Spotswood's address can be found in the *Executive Journals of the Council*.[5] He said:

> The Governor acquainted the Council, that five Negroes of the crew of Edward Thack & taken on board his Sloop remaine in Prison for Pyracy ... desired the opinion of this Board, whither there be any thing in the circumstances of these Negroes being taken on board a Pirate Vessel, and by what yet appears, equally concerned with the rest of the Crew in the same diversity appears in their circumstances the same may be considered on their Tryal.

The difference between the two can also be easily explained if one accepts that Brand simply made a mistake in his identification of one of the captives.

In the eighteenth century, jails were called *Gaols*. The *Gaol* in Williamsburg was built as a two-cell debtor's prison in 1711. These two cells are among the original structures in Colonial Williamsburg and are the only documented surviving structures to have housed any of Blackbeard's crew.

Chapter Twenty-Nine

Brand wasn't at the battle of Ocracoke and may not have ever seen any of those who were captured there. Weeks after the battle, Maynard sailed the sloop containing the captured pirates to Bath, but they would have been imprisoned below deck. In any event, as the total numbers of all the pirates captured add up nicely, so there isn't anyone missing. The positive identity of four of those African pirates comes from the Minutes of the North Carolina Governor's Council, May 27, 1719, which says, "Evidences called by the Names of James Blake, Richard Stiles, James White, and Thomas Gates were actually no other than foure Negroe Slaves."[6] The reasons for their names being read in the minutes will be discussed shortly.

As for the fifth pirate of African descent, his identity can be ascertained through the process of elimination. He was the man named Caesar. The following chart compares Johnson's list with Cracherode's and specifies which ones were of African descent as well as giving the most likely location of their capture.

Johnson's List	Cracherode's List	African	Location of Capture	
John Carnes	John Carnes		Ocracoke	
Joseph Brooks (2)	Joseph Brookes, Jun.r			Bath
James Blake	James Blake	X	Ocracoke	
John Giles	John Giles			Bath
Thomas Gates	Thomas Gates	X	Ocracoke	
James White	James White	X	Ocracoke	
Richard Stiles	Rich.d Stiles	X	Ocracoke	
Caesar	Caesar	X	Ocracoke	
John Martin	John Martyn			Bath
Edward Salter	Edw.d Salter			Bath
Stephen Daniel	Steph: Daniel			Bath
Richard Greensail	Rich.d Greensail		Ocracoke	
Israel Hands (Pardoned)	Hezekiah Hands Master			Bath

Total: 7   Total: 6

Richard Greensail and John Carnes were hanged in Hampton.

It is certain that Spotswood would have spoken to Captain Gordon of HMS *Pearl* about his concerns over the five African pirates. Gordon would have separated them from the others, at least in terms of their treatment. The ones who were peacefully taken in Bath may have been treated differently, too. They weren't directly responsible for the deaths of any of Gordon's men.

That leaves the two captured white men who were combatants and participated in the action against Gordon's men, Richard Greensail and John Carnes. A logbook entry from HMS *Pearl* dated January 28, 1719, may reveal their fate. Captain Gordon wrote, "Yesterday in the afternoon the longboat came ... This morning sent 2 condemned pyrates ashore to Hampton to be executed, which about ½ past 11 was done accordingly."[7] There is no direct evidence that the two pirates Gordon executed were indeed Richard Greensail and John Carnes, but the idea is most compelling. Captain Gordon had the authority to convene an Admiralty court on board his ship and to carry out the sentence. He would have wanted to proceed as quickly as possible and not wait for a civilian court in Williamsburg that might turn them free on a technicality. Additionally, Gordon would have wanted his crew to see the execution of those responsible for the killing of their mates.

Everything changed for all the pirates who were being held in the dismal cells in Williamsburg when news of the new extension of the king's pardon arrived. On December 21, 1718, King George extended his proclamation. The entire text is available in the *British Royal Proclamations Relating to America 1603–1783*. The following is an excerpt of the proclamation:[8]

> We have thought fit, by and with the Advice of Our Privy-Council, to Issue this Our Royal Proclamation; And We do hereby Promise and Declare, That in case any the said Pirates shall, on or before the First Day of July, in the Year of Our Lord One thousand seven hundred and nineteen, Surrender him or themselves to One of Our Principal Secretaries or State in Great Britain, or Ireland, or to any Governors or Deputy-Governors of any of Our Plantations or dominions beyond the Seas, every such Pirate and Pirates, so Surrendering him of themselves, as aforesaid, shall have Our Gracious Pardon of and for such his or their Piracy or Piracies.

Of course, the entire proclamation is much longer, but this paragraph is the important one. Unlike the original proclamation, which had a cut-off date of January 5, 1718, after which any acts of piracy would render the person ineligible, this new one didn't mention any such date. Everyone was eligible for the pardon as long as they asked for it before July 1, 1719. Now, all of Blackbeard's pirates held in Williamsburg were eligible for a pardon.

William Howard must be considered the luckiest of all of Blackbeard's crew. He was freed the night before his scheduled execution. In fact, he lived a long life afterward. He purchased Ocracoke Island in 1759, and died there in 1794, at the age of 108.[9] Proof of his release can be found in

William Howard released

two documents. In a letter dated July 14, 1719, Captain Brand wrote that Howard "was found guilty and received sentence of Death Accordingly and his life is only owing to the ships arrival that had his Majesties pardon on board, the Night before he was to have been executed."[10]

Lt. Governor Spotswood also mentioned this event. However, he wasn't sure of what to do with the money that the authorities took from Howard at the time of his arrest and what to do with his two slaves.[11] He wrote:

> I received some days ago the Honor of your Lordships of the __ of August with his Majestys Commission for pardoning Pyrates which came very seasonably to save Howard their Quartermaster then under sentence of Death, but by his Majestys extending his Mercy for all Piracys committed before the 18th of August is now set at liberty . . . what I am therefore in doubt of is, whether by the remitting all forfeitures His Majesty intends only to restore the Pyrates to the Estates they had before the committing their Pyracies or to grant them a Property also in the Effects which they have Piratically taken. There is besides the two Negro Boys about £50 in money and other things taken from the aforementioned Howard & now in the hands of the officer who seized it on His Majestys behalf of which an inventory is lodged in the Secretarys office here. I therefore pray your Lordships advice & commands how these Effects are to be dispersed, where the person for whose possession they were found is pardoned.

At least two of the pirates taken in Kecoughtan by Captain Gordon the previous November were also granted their freedom. They were William Stokes and Adult Van Pelt. The minutes of a council meeting that was held on March 11, 1719, include, "It is the unanimous opinion of this Board that the said Stokes and Van Pelt are fit objects of his majesties mercy."[12]

In Chapter Twenty, I discussed the attack on William Bell incident that occurred in Bath on September 14, 1718, near Tobias Knight's house. Late that night, William Bell was traveling by periauger on the Pamlico River when he was attacked by several men in another periauger. The next morning, Bell identified his attackers as "one Thomas Unday and one Richard Snelling commonly called Titery Dick so be two of them and the others to be Negroes or white men disguised like Negroes."[13]

In the North Carolina political arena, Edward Moseley and Tobias Knight were bitter enemies. It went back to the Cary Revolution when they were on opposite sides. After Moseley heard about the Bell incident, he devised a plan to link Blackbeard to Knight and thus discredit Knight and possibly even Eden.[14] It has already been stated in an earlier chapter that Moseley, or at least some of his agents, were in close communication with Spotswood, and Moseley himself met Captain Brand when he traveled to Bath. At some point, Moseley heard about the Bell incident. It may have meant nothing to him at the time, but when news of Blackbeard's death reached Moseley, he realized that the situation could be turned to his advantage. He devised a complicated scheme aimed at proving Knight's involvement with Blackbeard. If Blackbeard was the attacker, it would place Blackbeard at the Knight house on the evening of September 14. Moseley could then proclaim that Blackbeard was there to make secret arrangements with Knight for the disposal of the goods he had recently taken from a French sugar merchant ship.

The most critical part of this scheme was to get Bell to change his story and to name Blackbeard and four of his crew members as his attackers. Without that, he wouldn't have any case. Apparently, Moseley was successful because Bell eventually testified against Blackbeard at the trial of the four crewmembers in Williamsburg.[15] The entire scheme's success depended upon getting collaborative testimony from some of the members of Blackbeard's crew still being held in Williamsburg. The only way for Moseley to obtain that testimony was to get help from Spotswood. Once again, Moseley's agents worked closely with Spotswood, providing the Virginia prosecutors with the details of Bell's assault and even some physical evidence. Financial arrangements were made for Bell to travel to Williamsburg and personally give testimony. Bell lost two horses along the way, which were eventually paid for by the Virginia Council as shown in their records.[16]

> That your Petitioner was at the charge of supplying the Government with Horses particularly for one Bell and his Son Evidences against Blackbeards Crew of Pirates taken in North Carolina who were tried here by a Court of Admiralty, in which service your petitioner lost two horses which cost him twenty pounds Currant money and hath received no satisfaction for the same.

The four pirates identified by the Virginia prosecutor as the ones who assisted Blackbeard during Bell's attack were James Blake, Thomas Gates, James White, and Richard Stiles. They were the ones who agreed to plead guilty, perhaps in return for their release. They stood trial on March 12, 1719. The four accused pirates all pled guilty and provided details about the attack that they most likely were given to them by the prosecutor. William Bell also added his testimony. In addition, Israel "Hezekiah" Hands testified as a witness for the prosecution. Hands was at Ocracoke

James Blake, Thomas Gates, James White, and Richard Stiles tried in Williamsburg

when the alleged assault occurred; however, he testified that Blackbeard and the other four all went to Bath the night of the 14th and returned with stolen goods taken from Bell.

The records of that trial, which were stored in Richmond, were destroyed in 1865 when the city was burned. However, a copy of the transcript was sent to North Carolina and became part of the minutes of the May 27, 1719, North Carolina council meeting. Fortunately, that document still exists. All of the information concerning this trial comes from that source.[17]

The trial wasn't only about Blackbeard, it involved Knight to the greatest extent possible. The *icing on the cake* was a personal letter from Knight to Blackbeard that Maynard had found on Blackbeard's sloop *Adventure* shortly after the battle at Ocracoke. The letter was dated November 17, which was just five days before Blackbeard's death. This letter was introduced during the trial as evidence of Knight's involvement with Blackbeard. Included in the transcript of the trial was a recommendation for Knight to be tried in North Carolina. Spotswood sent a transcript to Governor Eden, which launched a Council hearing accusing Tobias Knight of conspiracy.

> Whereas it has appeared to this Court Mr. Tobias Knight Secty of North Carolina hath given Just cause to suspect his being privy to the piracys committed by Edward Thache and his crew and hath received and concealed the effects by them piraticaly taken whereby he is become an accessary ... Its therefore the opinion of this court that a Copy of the Evidences given to this Court so farr as they relate to the said Tobias Knights Behaviour be transmitted to the Governor of North Carolina to the end he may cause the said Knight to be apprehended and proceeded against pursuant to the directions of the Act of Parliament for the more Effectual Suppression of Piracy.

It is obvious that the five pirates hoped to be released if they agreed to testify against their dead captain and pled guilty to the charges. It didn't work for the four Africans. They were hanged in Williamsburg. The document that Spotswood sent to North Carolina states that the four pirates were "Condemned and since Executed."

If the extension of the King's Pardon was available, why didn't these four pirates qualify? I believe it was because they were tried for assault and theft at their trial, not piracy. It's subtle, but within the legal parameters. It is ironic that if the four pirates had refused to confess to the assault, they probably would have qualified for the pardon and would have been released.

This leaves seven pirates still held in Williamsburg. Caesar and the six pirates taken in Bath, Hezekiah (Israel) Hands, John Giles, Joseph Brooks Jr., John Martin, Stephen Daniel, and Edward Salter. Each of them would have qualified for the extension of the pardon. Apparently, upon accepting the pardon, the pirate had to pay five shillings. Spotswood wrote a letter to Secretary Craggs dated May 26, 1719, that complained about some of the pirates not paying their fee. The most intriguing aspect of this letter is that it mentions seven pirates who have received their pardons, but only one, a "Condemned Negro," has paid the fee.[18]

The seven remaining pirates

> . . . having never received the value of one penny from any of the Pyrates that have either Surrendered, and had Certificates under the Seal of the Colony, for w'ch the Clerk was allowed to demand five Shillings a piece, yet I am well assured that no more than five paid any thing at all; And of seven that have rec'd their pardons, only one has paid the Attorney-Gen'l the common fee he receives for making out the like pardons even for a Condemned Negro, and he, too, was a person of a very notorious Character for his Piracys, and had his Money restored to him after he had been condemned, because there was no proof of its being piratically taken

The "Condemned Negro" was most certainly Caesar. It appears that he paid his fee and had his property returned to him. The other six must have eventually paid since there is evidence of some of them living long after their imprisonment. Hands, of course, would have been released for testifying for the prosecution and qualifying for the extension of the pardon. This is another instance when Charles Johnson got something right. He wrote, "The other person that escaped the Gallows, was one Israel Hands, Master of Black-beard's Sloop, and formerly Captain of the same."[19]

Unfortunately, Caesar, John Giles, Joseph Brooks, and Stephen seem to have disappeared from the historical record. But John Martin and Edward Salter were quite active in Bath after 1719. John Martin was the son of Joel Martin, who was among the first residents of Bath arriving in 1706.[20] Joel died on October 24, 1715. In his will, his son, John, inherited 220 acres from his father, the plantation north of Glebe Creek. John Martin sold the property on July 11, 1720, and James Robins witnessed the deed.[21] As you may recall, James Robins was the planter who was arrested at Ocracoke and later released.

Edward Salter is the most prominent of the pirate survivors. His name first appears in Bath's historical records in 1721, when he purchased two town lots from Henry Rowell. The deed was witnessed once again by fellow

Edward Salter in Bath

pirate James Robins.[22] In 1723, Salter bought 640 acres, and on November 12, 1726, Salter bought Governor Eden's 400 acres of property and his mansion from Robert Campaine for £600.[23] By 1727, Edward Salter was referred to as a "Merchant and gentleman" and owned the largest periauger with sails in the county along with a brigantine named the Happy Luke. In 1728, Salter bought 6 deeds for 3371 acres on the south side of the Pamlico River. He died in January 1734, and his periauger and brigantine are both mentioned in his will.[24]

Summary of the fate of all of Blackbeard's crew who were captured, based upon Charles Johnson's list	
William Howard	Accepted extension of pardon
John Carnes	Probably hanged in Hampton
Joseph Brooks (2)	Accepted extension of pardon
James Blake	Hanged in Williamsburg
John Giles	Accepted extension of pardon
Thomas Gates	Hanged in Williamsburg
James White	Hanged in Williamsburg
Richard Stiles	Hanged in Williamsburg
Caesar	Accepted extension of pardon
James Robbins	Acquitted
John Martin	Accepted extension of pardon
Edward Salter	Accepted extension of pardon
Stephen Daniel	Accepted extension of pardon
Richard Greensail	Probably hanged in Hampton
Israel Hands	Accepted extension of pardon
Samuel Odell	Acquitted

With dozens of twentieth- and twenty-first-century historian authors blindly following Johnson and writing that 13 pirates were hanged in Williamsburg, one nineteenth-century historian author got it right. The historian author was Shirley Hughson, who wrote Blackbeard & the Carolina Pirates, published in Hampton, Virginia in 1894.[25] Hughson wrote: "The trials were ordered to proceed immediately. They were held at Williamsburg, and four of the accused were condemned and afterward hanged."

The source Hughson cites is most interesting. This is exactly how it reads.

> An attempt to secure some details of these trials from the Virginia Admiralty Court Records proved fruitless. The clerk writes: "The

earlier records of this Court are in such a condition that I fear that I cannot give you the information asked for. I cannot even tell whether they go back as far as 1719; they are piled up in heaps in an upper room of the custom house building, and have been in that condition ever since the war. At the evacuation of Richmond, in the Great Fire, a large quantity of papers and records of the United States Courts, as well as of the General."

If all the records were truly lost, how did Hughson arrive at the number of four pirates hanged? Perhaps there were a few scholars still alive who had seen the documents before their destruction.

As for Blackbeard's goods that were confiscated, they were all sold at auction in Virginia along with the *Adventure*. Spotswood wrote a letter to James Craggs, Secretary of State for Britain, mentioning the sale of those goods.[26] Spotswood wrote:

> Having in my last taken Notice of some Goods in the possession of Thach and his Crew of Pyrates in North Carolina, w'ch were brought hither by Capt. Brand, of his Maj'ty's Ship, the Lyme, and the other Officers of the Sloops sent for Suppressing that Gang of Villains, I think it necessary now to informe you that these Goods, being proved to be piratically taken by this Thach in a french Ship bound home from Martinico, have, by a Decree of the Court of Vice Admiralty of this Colony, been condemned as such, and being perishable, have been sold at publick Auction the produce whereof in ye same Species for w'ch it was sold, amounts to 447 ounces ... penny weight of Spanish Gold, w'ch is of Virg'a Currency, £2,238; and is ready to be paid, after the necessary Changes of Transportation from Carolina, and of the Storage and Sale, and deducted, to the Owner, if they claim the same, or to whomsoever his Maj'ty shall appoint.

The *Adventure* and other of Blackbeard's goods were auctioned off in Virginia for £2,238.

Captain Gordon also mentioned the sale of Blackbeard's goods in a letter to the Admiralty dated October 2, 1719, in which he said, "I acquaint you that those goods, Viz: Cocco & Suggar haveing received damage, by salt water, which rendered them perishable, they were adjudged, upon survey most proper to be sold, to the highest bidder, by way of auction, which accordingly was done."[27] In another letter dated September 14, 1721, Gordon once again mentioned the sale of the goods as well as the *Adventure*. "Maynard got upon the Island or in the pirate sloop, all which put together; he tells his Majesties was sold for 2500 in Virginia: which if true he had a much better account of the sale than Capt. Brand or I had, because Mr. Chiswell the Agent Vectualer there for his Majesty was the person appointed to dispose of the vessel and cargoe."[28]

All the above testimonies, letters, council minutes, and documents completely close the book on the fate of all of Blackbeard's crew as well as his possessions and his sloop *Adventure*. There is only one question remaining. After the Virginia trial records were sent to North Carolina, what ever became of Tobias Knight?

Thirty
Tobias Knight's Ordeal

Governor Eden must have been stunned when he received the transcript of the trial of four of Blackbeard's crew. That trial had been held in Williamsburg on March 12, 1719. Within the testimony, there was substantial evidence given that Tobias Knight, Eden's friend and Chief Justice, was heavily involved with the illegal actions of the notorious pirate Blackbeard. Even more shocking was the strongly worded recommendation from the Virginia court that Knight should be arrested and tried.

> Whereas it has appeared to this Court Mr. Tobias Knight Secty of North Carolina hath given Just cause to inspect his being privy to the piracys committed by Edward Thache and his crew and hath received and concealed the effects by them piraticaly taken whereby he is become an accessary . . . Its therefore the opinion of this court that a Copy of the Evidences given to this Court so far as they relate to the said Tobias Knights Behaviour be transmitted to the Governor of North Carolina to the end he may cause the said Knight to be apprehended and proceeded against pursuant to the directions of the Act of Parliament for the more Effectual Suppression of Piracy.[1]

As mentioned in the previous chapter, the original Virginia records were destroyed by fire. Fortunately, the entire transcript was read into the minutes of the North Carolina Governor's Council meeting that was actually a hearing for Tobias Knight. Those minutes still exist and are housed in the North Carolina State Archives.[2] Unless otherwise noted, all the quotes in the rest of this chapter come from those minutes.

After receiving this transcript, Eden's council drafted an order dated April 4, 1719, which gave notice to Knight of the upcoming proceedings and

Chapter Thirty

ordered him to attend the meeting of the Board "to make answer to the several Depositions and other Evidences mentioned in the aforesaid ord." The council meeting was held on May 27, 1719. As no courthouse had yet been built in the province, the meeting was held at the house of Fred Jones Esq. Presiding over the council was Governor Charles Eden himself. Thomas Pollock, the province's most skilled lawyer and close friend of Eden's, also attended. Other members were "William Reed, Fra. Hosler, Fred. Jones, and Ruthd Landermon."

The members of Blackbeard's crew on trial were James Blake, Alias Jimmy, Richard Stiles, James White, and Thomas Gates. "Hesikia Hands late Master of the Sloop *Adventure* Commanded by Edward Thache," also provided testimony against his former captain. As discussed in the previous chapter, the entire trial revolved around the alleged attack on William Bell, who also provided testimony and was brought to Virginia at Lt. Governor Spotswood's expense.

Hands confessed to being part of Blackbeard's crew when he took "two French ships in the Month of August last past . . . that Thache plundered one of the ships of some Cocoa and the other brought in with him to North Carolina." After returning to Ocracoke, Hands remained there with the *Adventure* and the captured French ship while Blackbeard and the four crewmen on trial took sailed to Bath on a periauger. The purpose of that trip was to secretly meet with Tobias Knight to discuss how to legally dispose of the cargo that was obviously taken through piracy. Hands also testified that Blackbeard took some "chocolate, Loaf Sugar and sweet meats being part of what was taken on Board the French Ships" to give to Knight as a present. When Blackbeard returned to Ocracoke, Hands saw that he had "divers goods" that "Thache said he bought in the Country." However, after "hearing that one William Bell had been robbed" Hands came to the conclusion that the goods he saw matched the description of the goods that were stolen from Bell.

The four prisoners on trial corroborated Hand's testimony, saying that they were with Blackbeard when he when he visited Knight in September, arriving at his house sometime between midnight and one o'clock in the morning. Blackbeard delivered the gifts of three or four caggs of sweet meats, some sugar, and a bag of chocolate, and stayed there in conference with Knight until an hour before daybreak. The following is from the transcript.

> In September they went from Ocacock in a periaugoe with Edward Thache to the house of Tobias Knight lately of North Carolina and carried in the said periaugoe "three or foure Caggs of Sweet meats,

some loaf sugar, bagg of Chocolate" ... that they got to the said Knights house about twelve or one aclock in the Night ... that the said Knight was then at home and the said Thache staid with him til about an hour before the break of day and then departed.

After leaving Knight's house, they sailed about three miles to Chester's landing, where they encountered another periauger lying close to shore. On board were William Bell and an Indian boy. They came alongside and "Thache asked them for a dram and immediately jumped into the periaugoe and after some dispute plundered her carrying away with him, some money one cask pipes a cask of rum or Brandy some linen and other things."

William Bell testified that while at the "Landing of John Chester in Pamlicough river in North Carolina," a little before daybreak on the night of September 14, 1718, "that a white man that he since understands was Edward Thache," rowed his periauger alongside and "asked him if he had anything to drink." Bell replied that it was too dark for him to find anything, "whereupon the said Thache called for his sword which was handed him from his owne periaugoe and commanded the deponent to put his hands behind him in order to be tyed swearing damnation." Bell added that Blackbeard said that "he wou'd kill the deponent if he did not tell him truly were the money was." When Bell asked him who he was, his mysterious attacker answered that "he came from hell were he woud carry him presently."

The two men began to struggle and Blackbeard "called to his men to come on board to his assistance" and that "Thache demanded his pistols." In the course of the fight, "Thache in Beating the deponent broke his sword about a quarter of a yard from the point." Soon after, Bell ceased his resistance and told his attacker that all his valuables were locked in a chest and consented to open it up. He "took away £ 66–10 in Cash one piece of Crape Containing 58 yards a bin of pipes half a barrel of Brandy and several other goods" which included "a silver cup of a remarkable fashion being made to screw in the middle be the upper part resembling a chalice." Additional testimony was introduced into the record that such a cup was found by Lt. Maynard about the *Adventure* after the battle of Ocracoke. After taking what they wanted, the attackers "tossed his sails and oars overboard, and so rowed down the river." At the trial, Bell produced some physical evidence, his attacker's broken sword point Bell found in his periauger after the incident.

Bell testified that within five hours after the attack, he went to Governor Eden's house to report the incident. Eden sent him to the house of Tobias

Chapter Thirty

Knight, who was the Chief Justice and senior legal officer in the province. Knight immediately gave him a "warrant or hue & cry" for the arrest of his attackers, "which he now produces in court." Bell added, "yet the said Tobias Knight did not discover to the deponent that any such periaugoe had been at his house or that he knew of Thaches being in the Country."

Fortunately, after Bell's testimony, the contents of the infamous letter were read into the record. This was the letter that "the aforenamed Tobias Knight—directed to Capt. Edward Thache on board his sloop *Adventure* which Letter was proved to have been found among Thaches papers after his death." As you may recall, that letter was found by Lt. Maynard after the battle. For a complete transcript of the letter, see Chapter Twenty-Five.

Brand's testimony

Next, it was Captain Ellis Brand's turn to testify against the pirates. He "declared that having received Information of Twenty Barrels of Sugar and two baggs of Cotton lodged by Edward Thache at the House of Tobias Knight he asked the said Knight for those goods." Knowing that these goods were part of the cargo taken from the French ship, finding them would be proof positive of Knight's collusion with Blackbeard. Brand added that upon his first visit to Knight's house, Knight was very evasive and denied that "any such goods were about his plantation." But when Brand returned the next day, he pressed Knight on the issue and then produced "Thaches pocket book he the said Knight owned the whole matter and the piratical goods afors'd were found in his Barne covered over with fodder."

Brand restated his testimony in a letter he wrote to the Admiralty on July 14, 1719. It must be noted that Brand wrote this letter to his boss because he was in the middle of a lawsuit brought against him by the government of North Carolina. Brand wrote:

> [I] lay before there Lordships some part of the conduct of Tobias Knight Collecter of No: Carolina of his making abundance of difficulty, and adviseing the Governor not to assist me, and he constantly justifying the pyrates, I had information of a parcell of sugars and cotton hid in a barn of his and put on shoar by Thach the pyrate; which I found and were deliver'd up to me, when ever there shal be directions given to inquire into his behavior there will appear some very extrodinary proceedings, I having by me a letter of his to the Pirats.[3]

The next day, the Court of Admiralty gave the opinion that Knight knew of Blackbeard's piracies and was the receiver of stolen goods and as seen above in their recommendation to North Carolina, recommended that he be arrested and tried under the Act of Parliament for the more Effectual Suppression of Piracy.

Based upon all the evidence from the trial in Virginia, it appears that Blackbeard did visit Knight for a clandestine meeting on September 14, 1718, and then proceeded to attack Bell. It also appears that Knight was totally guilty, especially after the evidence of him hiding the stolen goods in his barn under a pile of fodder. Charles Johnson believed it, writing, "one Mr. Knight, who was his Secretary, and Collector for the Providence, twenty, and the rest was shared among the Pyrates."[4] As far as modern historian authors are concerned, some don't mention this incident at all, while others, like Kevin Duffus, author of *The Last Days of Black Beard the Pirate*, totally buy into the Virginia account as completely true and accurate.[5]

A few modern historian authors, including Kevin Duffus, go even further, offering a reason for Blackbeard's attack on Bell. Since it was vital for his meeting with Knight to remain a secret one, it was totally irrational for him to risk the entire scheme by foolishly attacking a stranger in a passing vessel. The only logical explanation was that Blackbeard was suffering from the advanced stages of syphilis and was no longer capable of rational thought. It would also explain why the fierce pirate didn't have the strength to confront Bell without the assistance of his crew.[6]

This is total speculation on the part of those modern biographers. There is absolutely no contemporary evidence that Blackbeard had contracted syphilis. It seems that this idea was introduced into the narrative sometime in the early twenty-first century.

After reading Knight's rebuttal and the further testimony given by Edmund Chamberlayne as well as Bell himself, I personally don't believe a word of the testimony given in Virginia. Since Knight was given advanced notice on April 4, 1719, he had almost two months to prepare a response to all the charges. That response was read into the record at the council meeting under the heading, "The humble remonstrance of Tobias Knight, Esq Secretary of this province and a Member of this Board in answer to the several Depossitions and other pretended Evidences taken against."[7]

Knight's rebuttal

There is one fairly modern biographer who seems to have been able to put all the facts of this case into proper perspective. This individual is historian author Robert E. Lee, who wrote *Blackbeard the Pirate* in 1974. As a general historian, Lee's skills are severely lacking. Most of his book is highly inaccurate, as it is a simple restatement of Charles Johnson. But when the trial records enter the story, suddenly Mr. Lee bursts forth with extremely detailed information and insightful legal opinions. You see, Mr. Lee was a lawyer.

Chapter Thirty

From his biography on the back of his book, we learn that Lee was a lawyer with degrees from Wake Forest, Columbia, and Duke Universities and was the author of sixteen law books and many articles written for the newspaper, *This is the Law*. Lee effectively dissected the evidence given against Knight as well as Knight's reply, writing, "Knight's broadside was fully supported by the evidence."[8] Lee added that "He successfully stated the principles of law involved and sought to contradict, point by point, each of the terms of evidence contained in the charges."[9] Apparently, Knight's debilitating illness had no negative effect on his cognitive reasoning.

For the documentation of Knight's response and supportive testimony of Edmund Chamberlayne, we go back to the minutes of the council.[10] He opens by saying "that he is not in any wise howsoever guilty of the least of those Crimes which are so slyly Maliciously, and falsely suggested and insinuated against him." One by one, he addresses the charges with sound legal reasoning and gives good reasons as to why they should be thrown out.

Beginning with Hands, Knight points out that he was at Ocracoke and wasn't present at Bath, so his testimony about what transpired between Blackbeard and Knight would be inadmissible as hearsay, or as Knight said, it was "out of the reach of his knowledge he being all the time at the said Inlet." Knight also pointed out that Hands had been held in prison for months before the trial "under the Terrors of Death" and that his evidence is "more of act, malice and designe against the said Tobias Knight than truth."

From there, Knight attacked the evidence given by the four pirates who were actually on trial. Knight began by saying that their testimony was "utterly false," then drew attention to a legal point that, according to the "Laws and customs of all America," individuals of African descent cannot give "Evidence of any validity against any white person whatsoever." Some historian authors have focused on this one aspect of Knight's defense as his only reason as to why their sworn statements should be excluded. This is incorrect. One must consider that a good lawyer always throws in every argument possible, not knowing which ones will stick. Knight added that their testimony was most likely "in hopes of obtaining Mercy" and therefore given under duress. Most importantly, Knight argued that he had been denied the opportunity to cross-examine them, because "they were then Condemned and since executed." The right to cross-examine one's accuser existed in early eighteenth-century America, just as it does today.

Dismantling Bell's testimony came next. Knight was a first-hand witness to the fact that Bell identified his attackers as "one Smith, Unday, Titery Dick, and others," when he reported the incident to Knight just a few hours after the assault. After Bell's return from Williamsburg, he recanted his earlier testimony in a statement made on April 25, 1719, in which he said, "that he doth verily believe that the said Thache never was at that time at the said Tobias Knights house." Knight then went on the offensive stating that his political enemies "cunningly suggested that said Edward Thache was at the said Tobias Knights house that night in which he was robbed which the said Tobias Knight has good reason to believe was rather an Artfull and Malitious design of those that drew the said Deposition."

As for Ellis Brand, Knight continued his legal attack on his accusers, calling Brand a liar. Knight said, "that the said pretended Evidence is every word false, and that the said Brand never did at any time speak one word or mention to the said Knight in any manner whatsoever touching or concerning the sugars mentioned in the said Evidence." Knight added that when he was told that "Brand had been informed that the said sugars had been caniveingly put on shore for the said Knights use," and that Brand was planning on searching his home, he spoke with Brand and explained that the goods were being temporarily stored for "the said Thache only til a more Convenient store could be procured by the Governor." Knight continued by saying that he "never did pretend any claim or right to any part thereof," of those goods and that if Brand or any of his men had requested to search the property, "every lock in his said house should be opened to him." Getting directly to the heart of the matter, Knight accused Mr. Maurice Moore, Jeremiah Veal, and others of that family as the ones who went to Virginia to give "the false and Malitious storys there suggested against him."

Regarding the infamous letter written back in November to Thache, Knight confirmed that the letter was indeed his and that it was accurately read, saying that he "doth believe to be true for that he did write such a letter by the Governors orders he having advised him by Letters that he had some earnest business with the said Thache." Then added that there was no evil intent to the letter. At the time, he believed that "Thache was as free a subject of our Lord the King as any person in the government." Reminding the council that he "had been confined to his bed by sicknes," Knight commented that he "never was able to goe out from his owne plantation," and that the only dealing he had with Thache or any of his crew was when "they had business at his office as secretary or Collector of the Kings Customs."

Knight concluded by stating that he did not see Blackbeard from the time he left in August until he returned "on or about the 24th of September last," and that he had no knowledge of Blackbeard or any of his crew being anywhere near his "house on or about the 14th day of September last past as is most falsly suggested in the aforesaid evidence given against him in Virginia." Then he turned it over to his house guest, Edmund Chamberlayne.

Chamberlayne had been living in Knight's house since August 1718. On the date in question, September 14, 1718, "for several days before and since he never was absent from the Tobias Knights house either by night or by day." He testified that "Capt. Edward Thache nor any of his Crew neither was any of them to his knowledge at the said Tobias Knights house either by night or by day until on or about the 24th day of the said last September." He added that it would have been impossible for "Tobias Knight to have had such communication with any person either within or without his said house without his knowledge."

When William Bell came to Knight's house on the morning of September 14, Chamberlayne was there and heard and saw everything. He testified that Bell initially identified his assailants as "one Thomas Unday and one Richard Snelling commonly called Titery Dick to be two of them and the others to be Negroes or white men disguised like Negroes." He added that Knight had a suspicion that "Wm. Smith and others" were also responsible.

His testimony finished by adding that William Bell personally told him about the 25th of April last, that he thought that "Tobias Knight was a very Civel Gent and his wife a very civel Gentle woman and he did not think or believe that the said Thache was there or that he knew any thing of the matter or words to that Effect."

After a short conference, the council concluded that the evidence from the Virginia trial against Knight was false and intentionally malicious. Knight was found not guilty and was acquitted.

> Evidences so far as the relate to the said Tobias Knight are false and Malitious and that he hath behaved himself in that and all other affairs wherein he hath been intrusted as becomes a good and faithful officer and thereupon it is the opinion of this board that he is not guilty and ought to be accquited of the said crimes and every of them laid to his charge as aforesaid.

Unfortunately, Knight's legal victory was of little use to him. His long illness finally got the better of him and Tobias Knight died just a few weeks after he was acquitted.[11]

Thus far, there is one piece of physical evidence that remains unexplained. What about Bell's "silver cup of a remarkable fashion . . . resembling a chalice" that was found by Maynard onboard the *Adventure* after the battle of Ocracoke? There is a logical explanation that makes sense. As you may recall, back in November 1717, Blackbeard took the *Great Allen*, just north of St. Vincent.[12] Among the goods taken by Blackbeard was a silver cup of such an impressive and exquisite design that another of Blackbeard's victims, Henry Bostock, felt compelled to mention it in his deposition.[13] Perhaps this was the cup that Maynard found and was later attributed to Bell's cup to help support the false case against Blackbeard.

As has been expressed in small bits and pieces throughout the past several chapters, the compelling evidence has caused me to conclude that Blackbeard did not secretly visit Knight around September 14, 1718, nor did he attack William Bell. Both events were concocted by Moseley and his allies in order to drive a political wedge into the Eden administration. Bell was somehow convinced to change his testimony and identify Blackbeard as his assailant. The cup found by Maynard was falsely identified as Bell's. The Knight letter, which was totally innocent by nature, was blown out of proportion and turned into evidence of conclusion. And finally, Brand intentionally lied to discredit Knight. As you may remember, at the time, a lawsuit against Brand had been filed by the government of North Carolina.

Lt. Governor Spotswood was happy to go along with the plot. When all this began, Spotswood was in significant trouble with his own council on a variety of charges, especially his illegal invasion of North Carolina, and William Byrd was in London speaking to the king in an attempt to have Spotswood removed. Spotswood must have thought that if he could prove the government in North Carolina was weak or even corrupt, it would help justify his decision to go into another sovereign province to attack Blackbeard. The Moseley plan was perfect. It alleged that a pirate had attacked a citizen within the capital city itself with no response on the part of the government and that the Chief Justice was financially involved with the very same pirate.

Their scheme didn't work in 1719. Knight's brilliant reply dismantled every legal point of the argument. However, it became successful in 1724, and for the next 300 years. Charles Johnson repeated the false charges as if they were facts and added a few new allegations against Eden. Other historian authors over the years followed suit, not even considering Knight's response.

Governor Eden sold his Bath property in March 1720 and moved to a large house on the west side of the Chowan River. The nearby town of Queen Anne's Creek was eventually renamed Edenton in his honor. He died of yellow fever on March 26, 1722. Part of the inscription on his tombstone reads, "he brought ye Country into a flourishing condition, and died much lamented."[14]

Governor Charles Eden was never under any suspicion of wrongdoing during his lifetime; however, that changed shortly after his death. Only eight days later, on April 3, 1722, Spotswood was finally removed as Lt. Governor.[15] Spotswood never mentioned Eden in any of his letters. Instead, Blackbeard became linked directly to Eden in the writings of Charles Johnson.[16] Subsequently, many historian authors have used Charles Johnson's books as primary sources to further discredit this highly respected statesman.

Thirty-One
Will the Real Israel Hands Please Stand Up?

One of the best-known of all of Blackbeard's crewmen is Israel Hands. He was immortalized in literature as one of the characters in Robert Louis Stevenson's classic 1883 novel, *Treasure Island*. Israel Hands was Long John Silver's master gunner. Toward the end of the novel, Hands is dramatically killed by the story's main hero, young Jim Hawkins. Later editions of Charles Johnson's book, *A General History of the Pyrates*, were still being released during Stevenson's lifetime. It is reasonably certain that Stevenson read a copy of Johnson's book and thought that Israel Hands, a real pirate, would be an excellent name for one of his pirate characters. After all, according to Johnson, Israel Hands was the only one of Blackbeard's crew to survive.

Most of what was written about Hands was in Johnson's book. He identifies Hands as Blackbeard's sailing master on the *Queen Anne's Revenge* writing, "*Israel Hands*, Master of *Teach*'s Ship."[1] According to Johnson, when the pirate ship was run aground, it was Israel Hands that came to Blackbeard's assistance as the commander of the sloop *Adventure*. Johnson wrote:

> From the Bar of *Charles-Town*, they sailed to *North-Carolina*; Captain *Teach* in the Ship, which they called the Man of War, Captain *Richards* and Captain *Hands* in the Sloops ... Accordingly, on Pretence of running into *Topsail* Inlet to clean, he grounded his Ship, and then, as if it had been done undesignedly, and by Accident; he orders *Hands*'s Sloop to come to his Assistance, and get him off again, which he endeavouring to do, ran the Sloop on Shore near the other, and so were both lost.[2]

Chapter Thirty-One

The historical record contradicts Johnson's account. According to David Herriot, Hands had been in command of the *Adventure* before the *Queen Anne's Revenge* ran aground, but at the last moment, Blackbeard sent his quartermaster, William Howard, to take command and come to his assistance. Herriot said:

> They arrived at Topsail-Inlet in North Carolina, having then under their Command the said Ship Queen Anne's Revenge, the Sloop commanded by Richards, this Deponent's Sloop, commanded by one Capt. Hands, one of the said Pirate Crew and a small empty Sloop which they found near Havana ... the said Thatch's Ship Queen Anne's Revenge ran aground off of the Bar of Topsail-Inlet, and the said Thatch sent his Quartermaster to command this Deponent's Sloop to come to his Assistance;[3]

Johnson chose Israel Hands as the main character in one of the more colorful events in his chapter on Blackbeard. He wrote:

> The aforesaid *Hands* happened not to be in the Fight, but was taken afterwards ashore at *Bath-Town*, having been sometime before disabled by *Black-beard*, in one of his savage Humours, after the following Manner.—One Night drinking in his Cabin with *Hands*, the Pilot, and another Man; *Black-beard* without any Provocation privately draws out a small Pair of Pistols, and cocks them under the Table, which being perceived by the Man, he withdrew and went upon Deck, leaving *Hands*, the Pilot, and the Captain together. When the Pistols were ready, he blew out the Candle, and crossing his Hands, discharged them at his Company; *Hands*, the Master, was shot thro' the Knee, and lam'd for Life; the other Pistol did no Execution.—Being asked the meaning of this, he only answered, by damning them, that *if he did not now and then kill one of them, they would forget who he was.*[4]

This story is so popular that modern pirate reenactors portraying Hands often wear a bloody bandage about their knee and walk with a crutch. The reliable historical record remains silent. There are no contemporary accounts that mention this or any other similar event. After Blackbeard's death, Hands was arrested in Bath and taken to Williamsburg to stand trial. As discussed in Chapters Nineteen and Twenty, he testified against Blackbeard and four of his other former shipmates, who were on trial for piracy. Throughout the trial, Hands made no mention of ever being shot by Blackbeard. He also made no mention that he ever witnessed Blackbeard shooting anyone. It seems highly probable that if this incident had actually occurred, Hands would have enthusiastically retold it to further discredit his former captain.[5] In my opinion, this anecdote never occurred at all. I believe it was completely spawned from the colorful imagination of Johnson.

Johnson continues Hand's story with his dramatic last-minute pardon. Johnson wrote:

> *Hands* being taken, was try'd and condemned, but just as he was about to be executed, a Ship arrives at *Virginia* with a Proclamation for prolonging the Time of his Majesty's Pardon, to such of the Pyrates as should surrender by a limited Time therein expressed: Notwithstanding the Sentence, *Hands* pleaded the Pardon, and was allowed the Benefit of it.[6]

Once again, Johnson seems to have confused Hands with Howard. As seen in Chapter Twenty-Nine, it was Howard who received this miraculous pardon the night before his death. Captain Brand wrote that Howard "was found guilty and received sentence of Death Accordingly and his life is only owing to the ships arrival that had his Majesties pardon on board, the Night before he was to have been executed."[7]

The final sentence written by Johnson regarding the life of Israel Hands was that he "is alive at this Time in London, begging his Bread."[8] As expected, there is nothing in the historical record to support this. What is even more interesting is that Israel Hands wasn't even his real name.

Hezekiah Hands was his real name. As mentioned in Chapter Twenty-Eight, Great Britain offered a financial reward for the capture of pirates with higher payments made for officers and captains. In order to document the payments made to the naval officers who made those arrests, a treasure report containing the names of all the pirates was signed by Anthony Cracherode, the Treasury Solicitor for Great Britain. One of the names on this report was "Hezekiah Hands Master (as appears by the Sd Lieutent Govern.s Certificate.)"[9] Lieutent is an abbreviation for lieutenant, in reference to Lt. Maynard who is mentioned on the governor's certificate as the arresting officer. Payment for Hands is listed as £40, the rate "For every Lieutenant, Master, Boatswain, Carpenter & Gunner."[10]

Anthony Cracherode's report on the arrest of pirates in North Carolina and South Carolina was based on financial accounting and showed the reward money paid for their arrest and trial.

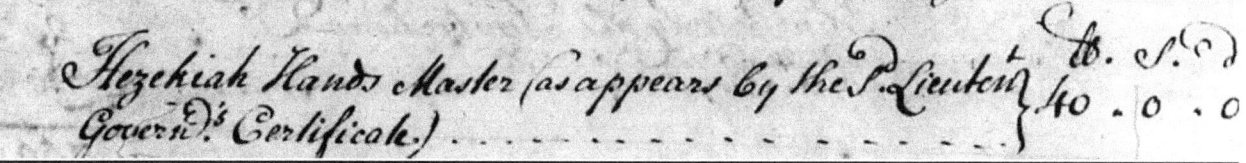

Figure 74: *Hezekiah Hands Entry on Cracherode's Report dated June 20, 1719.*

Chapter Thirty-One

The trial in Williamsburg of four of Blackbeard's crew took place on March 12, 1719. A man listed as Hesikia Hands was the star witness for the prosecutor. A short excerpt of Hand's testimony reads:

> Hesikia Hands late Master of the Sloop Adventure Commanded by Edward Thache being sworne and examined deposed that he was on Board the said Sloop Adventure at the takeing of two French ships in the Month of August last past and that all the prisoners at the Barr were on Board the said Sloop and bore Armes under Thache at the time of the said piracy.[11]

Although spelled differently, it is clear that this was the same Hezekiah Hands listed in Cracherode's report. It is doubtful that Charles Johnson ever saw either the Cracherode report or the transcript from the trial in Williamsburg. If he had, he may have had more accurate information in his book. There is no doubt, however, that Johnson saw *The Tryals of Major Stede Bonnet and Other Pirates*, published in 1719. The second half of Johnson's chapter on Stede Bonnet is taken almost word-for-word from that book. Hands is only mentioned once in the text of *The Tryals of Major Stede Bonnet and Other Pirates*. In David Herriot's deposition, he is only referred to as "Capt. Hands."[12] Herriot doesn't mention any first name.

Charles Johnson had chosen Hands as the central figure of his imaginative episode where Blackbeard shoots him under the table. However, such an important character needed a first name. Johnson decided it should be Israel.

Thirty-Two
Meet Edward Thache–A Bristol Man Born

Edward Teach was a Bristol Man born, but had sailed some Time out of Jamaica in Privateers, in the late French War; yet tho' he had often distinguished himself for his uncommon Boldness and personal Courage, he was never raised to any Command, till he went a-pyrating.[1]

This is how Charles Johnson began the chapter on Blackbeard in his second edition of *A General History of the Pyrates* and all subsequent editions. For almost three hundred years, this statement encompasses everything most people knew about Blackbeard's early life. Johnson's first edition, released only a few months earlier, places Blackbeard's birth in Jamaica. The original chapter on Blackbeard begins with this statement:

> Edward Thatch, (commonly called Black-beard) was born in Jamaica, and was from a Boy bred up to the Sea; in the late War he sail'd for the most Part in Privateers, yet, tho' he had often distinguished himself for his uncommon Boldness, and personal Courage, he was never raised to any Command.[2]

The first edition of *A General History of the Pyrates* was released on May 14, 1724. The second edition was released three months later.

It appears that after the release of the first edition, someone who claimed to have first-hand knowledge of Blackbeard's birthplace told Johnson that he was born in Bristol, not Jamaica. The copy was edited and "A Bristol Man Born" became the literary gospel of Blackbeard's origins.

The first historian author to question Johnson's version was John F. Watson in his book, *Annals of Philadelphia and Pennsylvania, in the Olden Time*. Watson introduced the name of Edward Drummond as Blackbeard's real name. His source seems to have been a person living in Hampton, Virginia, who told the author that he was a relative of Blackbeard and that his real name was Drummond. Watson also mentions a doctor named Cabot, who allegedly was Blackbeard's surgeon and was the direct ancestor of another person living in Virginia who verified the Drummond account. In Watson's 1844 edition, he wrote:

> I happen to know the fact that Blackbeard, whose family name was given as Teach, was in reality named Drummond, a native of Bristol. I have learned this fact from one of his family and name, of respectable standing, in Virginia, near Hampton ... His surgeon, for a part of his time, was a Doctor Cabot, who became the ancestor of a family of respectability settled in Virginia. The name of Teach, it may be observed, seems to be a feigned name, because no such name can be found in the Philadelphia or New York Directories.[3]

Keeping with Blackbeard's Jacobite tradition, Drummond is an old Scottish Clan name that dates back to the thirteenth century.

Although Blackbeard is discussed at length in Watson's first edition of *Annals of Philadelphia and Pennsylvania*, released in 1830, there is no mention of Drummond in that edition. It appears that the Drummond legend was told to Watson sometime later than 1830. Perhaps after reading his book, this unnamed Virginia resident from Hampton wrote Watson and relayed the Drummond account. Without verifying any of the alleged facts, Watson added a new Blackbeard legend to the historical record.

As this new idea gained traction during the twentieth century, several historian authors wrote in their publications that Edward Drummond was indeed Blackbeard's real name. One of those historian authors was Hugh Rankin. In his 1960 book, *The Pirates of North Carolina*, he wrote, "His name originally, it seems, was Edward Drummond, and he began his career as an honest seaman, sailing out of his home port of Bristol."[4]

The first notable historian author to discredit the Drummond name was Robert Lee in his 1974 book, *Blackbeard the Pirate*. He quotes the Watson paragraph and then writes, "The above, in the opinion of the present author, is merely a rash statement picked up by Watson a hundred and twenty-odd years after the death of Blackbeard."[5] Lee even mentions two reputable historians who refuted Watson's claim just a few years after it was written. They were Robert R. Howison, *History of Virginia*, 1846, and Charles Campbell, *Introduction to the History of the Colony and Ancient Dominion of Virginia*, 1847. Both authors mentioned that Blackbeard's name was Teach and that they could find no reference to him having any relatives in Virginia.[6]

As late as 2012, while researching my first book, I noticed that the official Colonial Williamsburg website identified Blackbeard's real name as

Edward Drummond. This has now been removed, as the Drummond legend gradually fell out of favor. Modern historian authors generally agree that there is no truth to this story. With the Drummond distraction finally resolved, twenty-first-century historian authors refocused their efforts on the only other clue to Blackbeard's true identity, Edward Teach, a Bristol man born.

The logical starting place in the search for Blackbeard's origins would seem to be Bristol. In the mid-twentieth century, historian Edmund Harding was unable to find any record of any person with the surname of Teach, or any variation, living in Bristol during the late seventeenth century.[7] In researching his 2006 book, *Black Beard: America's Most Notorious Pirate*, Angus Konstam searched the 1698 Bristol census and found no reference to anyone with the surname of Teach, including all the many other spellings.[8] It appeared that the Bristol connection was a dead end. That all changed in 2016 when Baylus Brooks published *Quest for Blackbeard: The True Story of Edward Thache and His World*. Thanks to *Ancestory.com*, a wealth of church and county records became searchable through the internet. Records that were overlooked or difficult to get in years past were suddenly available.

Everyone was looking in Bristol for someone named Teach. They didn't think about the surrounding communities. Records show that a Reverend Thomas Thache lived with his wife, Rachel, in Stonehouse, about thirty miles to the northeast of Bristol.[9] Church records show that Thomas and Rachel gave birth to a boy named Edward Thache on June 14, 1659. He was baptized two weeks later at St. Cyril's Church in Stonehouse, where Thomas Thache served as rector.[10] Young Edward grew up and married a woman named Elizabeth. They had two children, a boy named Edward, and a daughter named Elizabeth. Naming both son and daughter after the parents was common at that time. It is this young Edward Thache Jr. who Brooks believes to be the notorious pirate Blackbeard.[11]

The son of a respected country reverend, Edward Sr. was a "member of a gentleman's family." The four of them, Edward Sr, wife Elizabeth, and their two children, the other Edward and Elizabeth, left for Jamaica sometime in the 1680s.[12] There are no christening records of their children in Jamaica, indicating that they were both born in England. Additional proof that the Thache family left England around that time comes from the 1696 census. If they were still living in England after May 1, 1695, their names would have been recorded.[13] The port they would have used was the one closest to their home. It was also the largest seaport in England. This was the port of Bristol. It appears that Charles Johnson was correct. Edward Thache did leave Bristol and settle in Jamaica.

Chapter Thirty-Two

Jamaica was the most profitable English colony in the Americas

Precisely when the Thaches arrived in Jamaica still remains a mystery. Their place of residence was Spanish Town, also known as St. Jago de la Vega. It was the capital of Jamaica during the Spanish rule and the English unofficially named it Spanish town. Only about ten miles west of the main harbor, it was the administrative center of the government and the location of St. Catherine's Church, the largest church on Jamaica in 1699. It is still an active church today, where the bodies of Anne Bonny and Mary Read are supposedly buried. Since the destruction of Port Royal in the 1692 earthquake, Spanish Town was the most important city on the island. By now, Edward Sr. was a ship's captain. He was listed as such in several documents of his time. Spanish Town would have been an ideal place for a person of his status to settle.

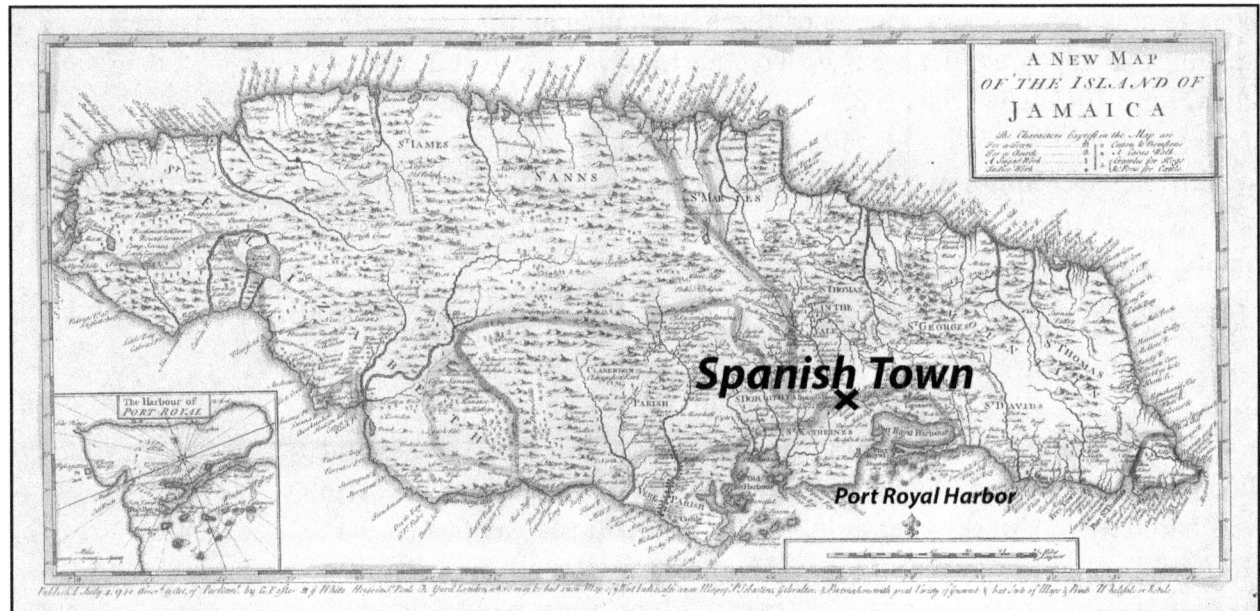

Figure 75: *1740 Map of Jamaica. The image is dedicated to the public domain under CC0.*

The Theach family first enters the historical records of Jamaica with the death of Edward's wife, Elizabeth. She died in January 1699. Burial records of St. Catherine's church show that in 1699, Elizabeth Theach, wife of Captain Edward Theach, was buried there. At that time, it was fairly common for people to remarry rather quickly after the death of a spouse. St. Catherine's church records also show that six months after Elizabeth's death, Edward remarried a local named Lucretia Poquet Maverly Ethell in July 1699. A further look at the church records shows that Edward Sr. and Lucretia had three children: Cox, Rachel, and Thomas. These records also mention his first two children, Elizabeth and Edward Jr.[14]

Note the spelling of their names above. Most of the records from Spanish Town spell their name Theach, whereas the records from England spell their name Thache. Early eighteenth-century records also indicate that the Theach family lived on the south side of the city. Examination of all the other Jamaican records reveals that no other family with the name Theach or any other spelling exists anywhere else on the island.[15] Edward and Lucretia had three children—Cox, born on July 8, 1700, Rachel, born February 6, 1704, and Thomas, Born November 17, 1705.[16]

Meanwhile, young Edward Jr. was following in his father's footsteps. He joined the English Navy and was stationed aboard HMS *Windsor*. Earlier in the War for the Spanish Succession, HMS *Windsor* harassed French and Spanish shipping near Petit-Goave and San Domingue. In 1704, it sailed to Newfoundland and then to Ireland. Captain Tudor Trevor took command and sailed for the Caribbean on May 18, 1706.[17] The *Windsor* arrived at Jamaica on July 28, 1706, according to an article in *The Boston News-Letter Boston* edition of September 30 to October 7, 1706. It reads:

> Jamaica, August 9th. One the 28th July Arrived here Commadore Kerr from England with 10 Sail of Men of War, and a fire Ship, with the Ships under his Convoy. Admiral Whetstone, on the 8th Currant hoysed his Flag on board Her Majesties Ship Windser and sailed for Cathageen Portobel on the Spanish Coast.[18]

Precisely when Edward Thache Jr. joined the navy is unknown. Documents showing his naval service have yet to be found. He may have been serving onboard the *Windsor* when it was raiding ships at Hispaniola in 1703 or he may have signed on while it was at anchor in Kingston Harbor in 1706. What is certain is that he left Jamaica as one of the crew onboard the *Windsor* in 1706. Proof of this comes from the will of his father, Edward Sr.

Edward Thache, Sr. died on November 16, 1706, at age forty-seven. He was buried beside his first wife, Elizabeth. According to the laws of the time, the oldest son inherited everything. That would leave Lucretia and her three small children with absolutely nothing. Cox was six, Rachel was two and a half, and Thomas was only one. There would be no pleasant way for the Thache family to survive. Edward Jr. was a man of honor. He would not let that happen and signed a document turning his inheritance over to his Stepmother and his three half-siblings.[19]

This single document is the most important item in establishing Edward Thache Junior's life. This all-encompassing deed places him on the *Windsor* as a member of the navy, establishes him as the only son of Captain Edward Thache, Sr., and links him to all of his half-siblings. It was signed and dated December 10, 1706, and was filed with the Registrar General's Department at Spanish Town. Miraculously, it still exists. The following is an excerpt of the highlights of the letter:

In October 1720, Anne Bonny and Mary Read were tried for piracy in Spanish Town, along with Jack Rackham and his crew.

> To all to whom these presents shall Come or may Concern Edwd. Thache on Board her Majesties Shipp Windsor Sends Greetings . . . that I the said Edwd. Thache Son and heire of my said late Father Edward Thache soe deced as aforesaid for and in Consideration of the love and ffection I have for and bear towards my Brother and Sister Thomas Theache and Rachell Theache Infants and for divers other good Causes and Considerations . . . doe Grant remise release and for ever quitt claime unto my Mother in law Lucretia Thache Widdow all my right title interest estate claime.[20]

We now have a clear and direct line that begins with an English family living near Bristol. They migrated to Jamaica in the late seventeenth century with their son Edward. In 1706, as the war with the French was on, Edward sailed as a member of the Royal Navy. Perhaps Charles Johnson was correct when he wrote, "Edward Teach was a Bristol Man born, but had sailed some Time out of Jamaica in Privateers, in the late French War."

The final confirmation comes from Edward's half-brother, Cox, who became a captain in the local artillery and claimed to be the brother of Blackbeard.[21] This was recorded in a collection of letters published by Charles Leslie in 1739 under the title, *A New and Exact Account of Jamaica*.[22] One of the passages in this collection says, "At this time, the famous Edward Teach, commonly known by the Name of Black-beard, infested the American Seas . . . his Mother is alive in Spanish-Town to this Day, and his Brother is at present Captain of the Train of Artillery."[23]

If it seems like it just couldn't be that easy, you are correct. It isn't that easy. This account fits the traditional version perfectly, but it doesn't answer all the questions that arise from the other known facts throughout Blackbeard's mysterious life.

Three aspects of the traditional version of Blackbeard's origin have always bothered me. The first is the ease with which Governor Eden granted Blackbeard a pardon so soon after he took vessels off Charles Town Bar. The second was the quick way in which Blackbeard formed a partnership with Stede Bonnet. It isn't logical for Bonnet to hand over control of his sloop to a perfect stranger. It seems that there had to be some sort of prior relationship. Third is the Jacobite aspect of Blackbeard's life. Thache isn't Scottish, and their hometown of Stonehouse, England, was not known to have any Jacobite sympathies.

Upon researching Blackbeard's origins, a fourth element arose, which is the connection of some of his crew to the town of Bath. Both John Martin and William Howard had wealthy families who owned property there. They weren't typical of pirates, yet they both held the position of quartermaster under Ben Hornigold, and Howard was also Blackbeard's quartermaster. This Bath connection ties into the easy way in which Blackbeard received his pardon. Perhaps there is an alternative explanation of Blackbeard's origins that addresses these issues.

Thirty-Three
Meet Black Beard

There have always been two aspects of the traditional version of Edward Thache's early life that bothered me. The first is the rapid and unchallenged way in which he established a friendly relationship with the government of North Carolina in Bath. He and his pirates had just left Charles Town, where they took seven vessels and terrorized the local citizens ashore. Those actions clearly made all of them ineligible to apply for the king's pardon. It seems that Thache was fairly certain that the North Carolina governor would approve his request. This implies that Thache already had some sort of prior relationship with someone who held a fairly high office within the province. The second aspect of the traditional version of Edward Thache's early life is the fast and unquestioned way in which Stede Bonnet handed over the total command of his sloop to him at Nassau in 1717. Until relatively recently, no mainstream publication on Blackbeard offered any sort of explanation for these two concerns. This changed in 2008, when a book was released that offered a rational and very believable alternative for Edward Thache's origins, which effectively explained these two problematic events. This book was *The Last Days of Black Beard the Pirate* by Kevin Duffus. The answer was fairly simple. Blackbeard wasn't Edward Thache from Jamaica, he was Edward Beard from Bath.

This astounding theory began with genealogists John Oden, Jane Bailey, and Allen Norris.[1] It started with a research project to study the deed records in early Bath. As they pored over the records and found connections between Bath residents and Black Beard's pirates, one name emerged as a person who could actually be this famous pirate's father, James Beard. Allen Norris put her thoughts down in 2003 in an obscure publication titled *Beaufort County, North Carolina Deed Book I, 1696–1729: Records of Bath County, North Carolina*.[2] Eventually, Kevin Duffus became aware of their work and introduced their theory to the world. The story begins with Captain James Beard of Barbados and South Carolina.

The Beard family originally settled in Barbados. There is a marriage record in St. John Parish, Barbados, dated April 13, 1664, between William Beard and Susannah Murrey. Although unproven in the historical record, it is believed that William Beard was James's father. There is also a marriage record in Barbados dated June 1, 1693, between Jane Beard and George Bell. Jane was likely a close member of James' family. Beginning in 1688, there was a mass migration of settlers from Barbados to Goose Creek, South Carolina, which is about fifteen miles north of Charles Town. James and his three brothers, Matthew, George, and William came to the Goose Creek area sometime in the late seventeenth century. James and his brothers are named in local records throughout the early eighteenth century. Captain James Beard was mentioned in a warrant for land in South Carolina dated August 29, 1699. He is also identified as a witness in a legal document concerning the estate of Philip Dewitt of Nevis Island

Figure 76: *1690 Map of Charles Town. The image is dedicated to the public domain under CC0.*

Chapter Thirty-Three

and Henry Symonds of South Carolina in November 1700. This implies that Beard sailed from Charles Town to Nevis Island regularly. There is also a Charles Town document dated February 20, 1702, which names James Beard as appraiser of the estate of Colonel John Berringer, formerly of Barbados. Berringer's sister was married to Governor James Moore of South Carolina.[3]

James Beard was a sea captain and master of the sloop *James*. Virginia export records list a "Captain James Beard, master of the James of New York, a 10-ton vessel that cleared Virginia on September 18, 1704, bound for New York." He arrived safely, as his name and sloop are listed twice in the 1704 customs records of New York. The first entry was made on May 29, 1704, and the second was made on October 14, 1704. These read, "New York merchants Jno. Scott & Caleb York who imported goods from Virginia on the sloop *James*, whose master was James Beard."[4]

The Beard family had strong Scottish ties. Their name was originally spelled Baird and, according to the historians and genealogists who authored the official Baird Clan website, it was among the oldest clans in Scotland. This clan seemed to always be in trouble with the government. The Baird Clan history website states:

> During the 17th century, the clan system slowly broke down, and the Bairds were no different. By the Late 17th century, we read how the Sir James Baird of Saughtonhall oppressed the Highlanders in their region.
>
> The Bairds of Auchmedden shared a strong Jacobite sentiment that led them into the uprising of 1715, and 1745. William Baird of Auchmedden, 7th of Auchmedden carried the responsibility to raise funding or men in Aberdeenshire for Lord Lewis Gordon. His efforts resulted in raising two regiments for the Jacobite cause, far exceeding efforts in other parts of Scotland.[5]

The first paragraph above offers two interesting facts. The first is "Sir James Beard." This wasn't our Captain Beard, of course, but he might have been his namesake. Second, it describes the clan system breaking up in the late seventeenth century, precisely when the Beards seem to have migrated to Barbados. Considering their strong Jacobite ties, it is likely that Captain Beard's sloop *James* was named for James II, the British king living in exile in France.

Charles Town was heavily involved militarily during the War for the Spanish Succession. The English port city was unsuccessfully attacked by

War of the Spanish Succession (1701–1714) was essentially between England, Scotland, the Dutch Republic, and the Holy Roman Empire against France and Spain.

the Spanish in 1702, which prompted Governor James Moore to launch a combined land and naval attack on St. Augustine. Charles Town was attacked again by a naval force of Spanish and French ships in 1706. In the years between those two attacks, privateer letters of marque were issued by English governors to assist in suppressing enemy attacks.[6] No documents have been found that name Captain James Beard as one of the captains who participated in the 1702 attack on St. Augustine, or as a privateer before 1706, but it is very likely that he was. As mentioned above, in 1702, James Beard was linked to Colonel John Berringer, whose sister was married to Governor James Moore. When Governor Moore gathered his attack fleet that same year, he would have used all available sea captains, and James Beard was in the right place at the right time. Evidence that James Beard was a privateer is circumstantial. In 1706, James Beard suddenly had a great deal of money and began buying large tracts of land in North Carolina. What other explanation accounts for a sea captain suddenly finding riches during times of war?

Figure 77: *Close-up Showing Beard Creek and Bath from a 1709 Map of North Carolina, Courtesy of the University of North Carolina at Chapel Hill.*

James Beard's first North Carolina property was on the north side of the Neuse River, just across from where the Marine Corps base of Cherry Point is today. He was issued a patent for the 640 acres of property by Governor Nathaniel Johnson on June 1, 1706.[7] Perhaps this was for service as a privateer in the war. A creek ran through his property, which eventually became known as "Beard Creek," a name that remains today. But he was active in Bath, too. In May of 1706, James Beard was called to jury duty at a General Court proceeding.[8] Shortly afterward, he bought a lot in Bath, the capital of the province, about 30 miles north of his Beard Creek property. A deed dated September 23, 1706, shows Capt. James Beard purchased the half-acre lot number 13 on Front Street in the town of Bath.[9] A few weeks later, he purchased 375 acres west of town, just south of Glebe Creek. This property became his permanent home, and was known as "Beard Land."[10]

James Beard had a wife named Elizabeth and two children, a daughter and a son. The daughter's name and age were determined through the record of her marriage. Court records state that Susannah Beard was sixteen years old when she was married to John Martin Franck in 1711. This places

James Beard's family

Susannah's birth date sometime in 1695.[11] The name and birth date of his son is more problematic. Captain James Beard died in South Carolina on October 21, 1711. He hadn't had time to make out a formal will, so he left a nuncupative will which read:

> We the under written on Sunday Night last ye 21st Instant about seven of ye Clock, In the Evening, Capt. James Beard then very weak butt then very sensible did hear him declare that if it pleased God to call him out of this world before next day when he designed to make his will, in such a case then he desired Mr. Jno. Morgan to take care of all his effects such of them were left to remit to his wife & son.[12]

Upon his death, his son automatically inherited the property with no need for a new deed. Unfortunately, that meant that no records were filed which would provide us with his name. He certainly assumed ownership, as there are no records that anyone else owned the property. Since a person had to be at least twenty-one years old to own property, young Mr. Beard must have been born in 1690 or before. Shortly after James' death, his wife, Elizabeth, remarried Captain William Masten. They remained on the Beard Land property until 1718. This makes logical sense. Young Mr. Beard had gone to sea. He naturally would have allowed his mother to remain in their house. James' daughter, Susannah, inherited his property at Beard Creek on the Neuse River.[13] It is interesting to note that North Carolina records show that the owner paid his taxes on the property until the end of 1718, precisely when Black Beard was killed.[14]

Over the centuries, there have been three common spellings of our favorite pirate's nickname—*Blackbeard*, *Black-beard*, and *Black Beard*. Before proceeding, I must take a moment to discuss these spellings. In all the previous chapters, I used the most common spelling: Blackbeard. However, in this chapter, we are assuming that his last name was actually Beard, the son of James Beard. The first name of Edward is a guess and is solely based on the fact that he became known as Edward Thache. During that time, the word "Black" was often added to pirate's names. Examples of this are Black Sam Bellamy and Black Bart Roberts. If that held true for Edward Beard, his name would be Black Beard. This makes for a very amusing play on words, Black Beard referring to both his name and the long black beard he wore on his face.

Making a direct connection between the son of James Beard and the pirate is the next step. Even though all the evidence is circumstantial, it is strong and most compelling. It begins with one of the community's earliest settlers, Joel Martin.[15] He owned a 220-acre plantation on the west side of Bath Creek.[16] When James Beard first arrived in Bath, he bought a town lot from Joel Martin,[17] then he purchased his Beard Land property on the west side of Bath Creek, which was very close to Joel Martin's.[18] One might say that they were neighbors. Martin was also one of the attorneys who processed James Beard's will.[19] Joel Martin died in 1715. In his will, he left all 220 acres of his land to his son, John Martin, who was present to take possession of the property.[20] This is the same John Martin who served as Hornigold's quartermaster and accompanied Black Beard back to Bath in 1718. He was arrested by Captain Brand and taken to Williamsburg for trial, but was released according to the extension of the pardon. Please refer to Chapters Fifteen and Twenty-Nine for more details.

Another one of Black Beard's pirates came from the Bath area. This was his quartermaster, William Howard. Records show that William's father, Philip Howard, arrived in North Carolina as an indentured servant in 1663. Beaufort County deed records show that Phillip Howard owned land on the north shore of the Pamlico River near North Divining Creek in 1702. They also indicate that Philip had a son named William, who was born in 1686. Philip owned land in Perquimans Precinct between 1702 and 1703, then he moved to Hyde Precinct where he purchased 320 acres on the Pamlico River east of Bath.[21]

A likely possible scenario is that at the end of 1715, while Blackbeard was in Bath visiting his mother, John Martin and William Howard decided to go with him to fish the Spanish wrecks off Florida. While in Florida, they met Ben Hornigold and joined his crew.

In addition to Martin and Howard, genealogists have found that Joseph Curtice, Joseph Brooks, Joseph Brooks Jr., Thomas Miller, Nathaniel Jackson, Stephen Daniel, and John Phillips all have surnames that appear in Bath County deeds, and Garret Gibbons and John Gills appear to share surnames with businessmen who traded with Bath residents.[22] These names should sound familiar. They were the names of the pirates killed or captured at Ocracoke with Black Beard. Please see Chapter Twenty-Eight.

So far, I haven't touched on the original two issues that troubled me in the first place: how Black Beard was able to receive the king's pardon so easily and why Stede Bonnet was so easily convinced to give Black Beard total command of his sloop. The answer to the first question is Tobias Knight. As Chief Justice and Secretary of the Province, Knight was the most trusted and respected legal authority in North Carolina. All requests for the king's pardon would have gone through him. If Knight approved the request, the governor would naturally issue the pardon without question. James Beard's "Beard Land" property was just one plantation to the north of Tobias Knight's property.[23] Black Beard and Knight were neighbors.

Chapter Thirty-Three

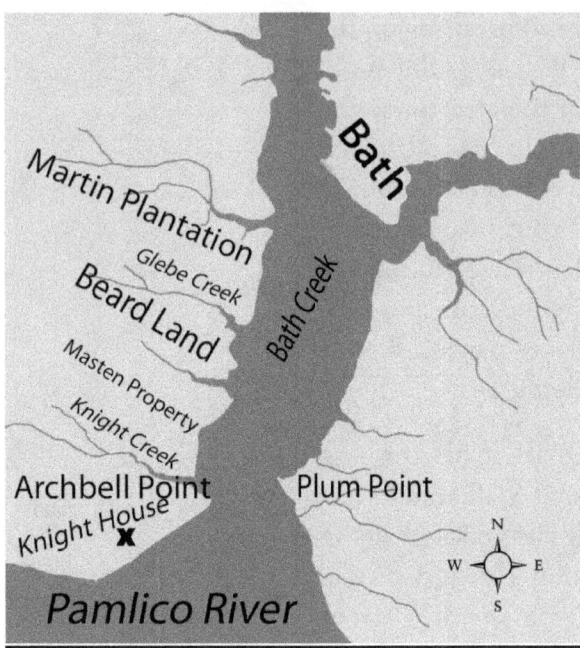

Figure 78: *West Bath Creek Plantation, Tobias Knight's property, Beard Land, and the Martin Plantation.*

Tobias Knight held his position since his appointment on January 24, 1712,[24] although he probably lived in the area for many years before that. As Chief Justice, Knight was the man who signed the final disposition on James Beard's will.[25] Black Beard must have dealt with him on many legal issues, especially since his house was so close by. After Black Beard left for a life as a pirate, his mother, Elizabeth, stayed on his property, next to Tobias Knight. It is logical to believe that when Black Beard first landed in Bath during the summer of 1718, his first stop wasn't at the nearest tavern. It was at his old house to see his mother.

Then there is Tobias Knight's letter to Black Beard, found by Maynard after the battle of Ocracoke. The letter that begins with the greeting, "My Friend." There is nothing in the letter to suggest any impropriety of collusion, but the nature of the words implies a friendly relationship. It doesn't seem like an important legal bureaucrat would write such a letter to a man he barely knew. The letter can be found in Chapter Twenty-Six, but I will provide it here again for convenience:

Nov. 17th 1718

My Friend

If this finds you yet in harbour I would have you make the best of your way up as soon as possible your affair will let you. I have something more to say to you than at present I can write, the Bearer will tell you the end of our Indian Warr and Garret can tell you in part what I have to say to you so refer you in some measure to him. I realy think these three men are heartily sorry at their difference With you and will be very willing to ask your pardon if I may advise be friends again its better so then falling out among your selves.

I expect the governor this night or tomorrow who I believe would be likewise glad to see you before you goe. I have not time to add save my hearty respects to you and am your real friend and servant[26]

T Knight

Knight was known as a man of the utmost integrity. It would be unlike him to approve Black Beard's request, knowing that he had just committed so many acts of piracy. As mentioned earlier, Knight was suffering from a prolonged illness. Knight wrote:

> That he did verily beleive at the same time that the said Thache was as free a subject of our Lord the King as any person in the government and the said Tobias Knight doth further say in his owne Justification that when the said Thache and his Crew first came into this government and surrendered themselves pursuant to his Majestys Proclamation of indemnity the said Tobias Knight then was and for along time had been confined to his bed by sickness[27]

I believe that when Black Beard arrived in Bath, he visited Knight at his home. After exchanging pleasantries, Black Beard told Knight that he had been a pirate earlier but now wished to be granted the king's pardon. Black Beard carefully omitted his actions of the past month. Knight was eager to help his old friend, as he had done for Stede Bonnet a few weeks earlier. I'm certain that Bonnet wasn't forthcoming with his piracies either.

If Tobias Knight knew Black Beard before, when his name was Edward Beard, why would he call him Thache? I believe that this was at Black Beard's request, as he had a mother and a sister still living in the area. He asked Tobias to continue using his assumed name for their sake. Knight may not have understood the necessity of the request when Black Beard first made it, but after his death, and once all of his crimes were exposed, Knight realized the need to continue with the deception.

How about Stede Bonnet? What was his connection to the Beard Family? Stede Bonnet's father was Edward Bonnet and his uncle was Thomas Bonnet, Jr. Stede's uncle Thomas died and left a will dated October 20, 1678. The will mentions several family members as well as their servants. One of their servants was named William Beard, the same William Beard who was James Beard's brother and who was mentioned in Charles Town in the early 1700s.[28] That's the connection! The Beard family was associated with the Bonnet family in Barbados.

Beard's link to the Bonnet family

This revelation is stunning. We knew that the Bonnets and the Beards lived on Barbados in the 1680s and 1690s, but now, we know that William worked for the Bonnets in some

Chapter Thirty-Three

capacity while his brother, James, was learning the trade as a sea captain. With William Beard working for the Bonnet family, it isn't difficult to imagine the Beards living in close proximity to the Bonnets. Taking this scenario to the extreme, it certainly is amusing to entertain the possibility that Stede Bonnet and Black Beard may have even played together as children. As the century came to a close, James and his three brothers left Barbados to seek their fortunes in South Carolina. James took his wife, Elizabeth, and his two children with him. Stede Bonnet remained on Barbados to run his plantation.

Both families were strong Jacobites. In 1714, the shocking news reached Barbados that Queen Anne had died, and the new king was George I. All loyal Jacobites were distraught at the idea of their monarch, James, living in exile while a German sat on the throne. The Bonnets hoped for a successful revolt in Scotland, but by 1717, they realized that such a revolt wasn't going to materialize. News had reached Barbados about a pirate base developing on Nassau with a great number of Jacobites. There was even talk of these pirates organizing themselves into a Jacobite navy. Stede Bonnet wanted to be part of that navy, so he purchased the sloop *Revenge*, hired a crew, and set sail.

Bonnet's intentions may have been to seek out his old friend, Black Beard, right from the start. When Bonnet arrived in Nassau, he learned that Black Beard was out at sea with Hornigold, so he sailed off on his own to give piracy a try. He had some success at Charles Town, where he took two vessels, but his sloop was met with disaster at the hands of a well-armed Spanish ship when sailing past the wrecks on the Florida coast. Meanwhile, Black Beard returned to Nassau and was told that Stede Bonnet had recently been in port and was asking where he might find the bearded captain. Delighted at the prospect of forming a partnership with his old friend, Black Beard waited patiently for Bonnet to return to Nassau. The two old friends were reunited when Bonnet's damaged sloop limped into port and Black Beard happily agreed to take command while his friend recovered from his wounds.

Of course, that scenario is a complete assumption based upon two facts: William Beard worked for Stede Bonnet's uncle and Black Beard took command of Bonnet's sloop *Revenge*, in Nassau in September 1717. But that likely scenario explains Black Beard and Bonnet's rapid partnership in terms that are more believable than any other accounts.

Assuming that Black Beard was the son of James Beard, when and where did he adopt the name of Edward Thache? Obviously, this was an alias that he took when he first turned to piracy to protect his sister and mother still living in North Carolina. They might have been safe if he was just becoming a pirate, but he planned on supporting the Jacobite movement. In England, family members of known Jacobites were being arrested, too. Concealing his true identity would have been imperative.

As with just about everything in Black Beard's life, the answer to one question seems to create several more questions. In this case, two unresolved concerns remain after we assume that Black Beard was James Beard's son. The first concern is his alias. Why choose Edward Thache? That's an odd name to pick out of the air as an assumed identity. The second issue is the fact that there was a man named Edward Thache from Jamaica who fit the profile of Blackbeard perfectly and who was identified by his half-brother, Cox Thache, as Blackbeard, in the 1720s. Only one scenario completely resolves both of these concerns at the same time.

This explanation is highly speculative and isn't supported by any documentation. It isn't even mentioned by any other historian authors to my knowledge. It is a theory that I personally developed based on the need to resolve these two concerns in a logical manner.

Edward Thache was serving as a privateer on board a ship operating sometime near the end of the war with Spain and France in 1714. Edward Beard happened to be on that same ship. The two men knew each other well enough to have shared their backgrounds. Edward Thache died while at sea. Shortly afterward, the war was over, and many privateers were thinking about turning to piracy. At that same time, the Jacobite revolt was beginning in Scotland. Edward Beard decided to become a pirate and to change his name for the protection of his family back home. In doing so, he chose to assume the identity of Edward Thache. This way, anyone looking to go after his relatives would be drawn to Jamaica rather than North Carolina. This explanation would also resolve the question about Black Beard being identified as Edward Teach in Philadelphia in 1715. He had already assumed his alias by the time he began sailing from Jamaica to Pennsylvania.

In summary, the Black Beard theory where he was the son of Bath resident James Beard makes a lot of sense. There are too many connections between Bath and Black Beard to be simply coincidental. This theory, which is grounded in fact, answers the questions about Black Beard's relationships with Tobias Knight and Stede Bonnet. On the other hand, this theory flies in the face of three hundred years of tradition, which is also grounded in historical fact. Until further information is revealed, the choice of which version of Black Beard's origins to believe is totally up to the individual.

Thirty-Four
Meet Mrs. Blackbeard

Every pirate needs a good woman by his side, and according to Charles Johnson, Blackbeard needed fourteen. This last wife was a sixteen-year-old whom Blackbeard married while in Bath in 1718. She was the daughter of a wealthy plantation owner, and Governor Eden even presided over the ceremony. Many of Blackbeard's other wives were still living somewhere. Johnson concludes his short narrative on Blackbeard's wives by adding that he shared his wife sexually with five or six of his crew. Johnson's sensationalized and highly provocative account was mentioned previously in Chapter Four, but it needs to be restated once again:

> Before he sailed upon his Adventures, he marry'd a young Creature of about sixteen Years of Age, the Governor performing the Ceremony. As it is a Custom to marry here by a Priest, so it is there by a Magistrate; and this, I have been informed, made *Teach*'s fourteenth Wife, whereof, about a dozen might be still living. His Behaviour in this State, was something extraordinary; for, while his Sloop lay in *Okerecock* Inlet, and he ashore at a Plantation, where his Wife lived, with whom after he had lain all Night, it was his Custom to invite five or six of his brutal Companions to come ashore, and he would force her to prostitute her self to them all, one after another, before his Face.[1]

Of all the falsehoods created by Johnson about Blackbeard, this one seems to have gained the most traction. Most of the general public, who cannot recall any other details of Blackbeard's life, can recount the story that he had fourteen wives. It is the ultimate version of a sailor having a girl in every port. The truth is that nothing exists in the historical record to suggest that Governor Eden performed such a wedding ceremony or that Blackbeard was ever married, other than an offhand comment by Captain

Chapter Thirty-Four

Captain Ellis Brand's letter mentioning Blackbeard's marriage.

Walter Hamilton was British Governor of the Leeward Islands from 1715 to 1721, which included Antigua, Montserrat, Nevis, St Christopher, and the Virgin Islands.

Brand. In his February 6, 1719 letter describing the events surrounding Blackbeard's time in Bath, Captain Brand wrote:

> . . . and gave out he design'd to be an inhabitant & leave of his Piraticall Life and the sword to put a life so to his designs he marryed there.[2]

Brand himself was in Virginia when this alleged wedding occurred. In the late summer of 1718, he hired a spy to gather intelligence on Blackbeard in Bath. Captain Brand didn't arrive in Bath until late November 1718. There is no way to determine if Brand was told about a Blackbeard's marriage by his spy or if one of the residents in Bath told him the story in person. The casual way in which Brand mentions it and the fact that no additional details are included causes this marriage to come under scrutiny. One point to consider is Brand's meaning of his statement, "he marryed there." In the eighteenth century, words had different connotations than they do today. That phrase may have simply meant that he became involved with the community. Today, a similar statement is used when people are said to be "married to their work."

No church or court records of such a marriage exist, but that fact alone is inconclusive. Many records have been lost over the centuries, especially considering the turmoil that North Carolina experienced during the Revolutionary War and the Civil War. Another point to consider is that no woman came forward after Blackbeard's death to claim any of his property or possessions. This can also be explained by the fact that he was a pirate and such a wife would naturally want to distance herself from such controversy.

There is one other comment in the historical record that I am compelled to mention in the name of thoroughness. One year before Blackbeard arrived at Bath, as his career as a pirate captain was just beginning, Governor Walter Hamilton of the Leeward Islands wrote a letter dated October 7, 1717, to the Council of Trade and Plantations, describing this new pirate captain. In that letter, he wrote:

> The ship is commanded by one Captain Teatch, the sloop by one Major Bonnett an inhabitant of Barbadoes, some say Bonnett commands both ship and sloop. This Teatch it's said has a wife and children in London, they have comitted a great many barbarities;[3]

With no specifics, this hearsay comment, "it's said has a wife and children in London" has no credence. Discarding Governor Hamilton's and Captain Brand's letters, the only information the world had on Blackbeard's wife or wives came from Johnson's *A General History of the Pyrates*. That changed in 1974 with the release of Robert E. Lee's book, *Blackbeard the Pirate*. Lee was the first historian author to suggest a name for this fourteenth wife that Blackbeard married in Bath. Her name was Mary Ormond. Lee wrote:

> There has not been found any authenticated record or any statement of the identity of the girl whom Blackbeard married. Tradition has it that the girl's name was Mary Ormond, and there exists a letter written by a relative of a Mary Ormond so stating.[4]

Lee adds that the Ormonds were prominent in Bath and lists the dates of their accomplishments, but all those dates he gave are after 1738.[5] Details about this letter do not appear within the text of that chapter, but Lee does provide additional information in the notes at the end of the book. Lee wrote:

> This letter, written from Route 1, Pinetown, N.C. (about five miles from Bath), on June 16, 1947, signed by Mrs. Ada S. Bragg, and addressed to a relative, Mrs. E. P. White of Buxton, N.C., stated that Mrs. White's great-great-aunt, Mary Ormond, became Teach's last wife. Mrs. Bragg is now deceased. During August, 1966, the author read this letter while interviewing Mrs. White at her home in Buxton, where, up until December, 1965, she had been postmaster for thirty-five years. Mrs. White's mother before her marriage to Lindley Tyer, Mrs. White's father, was Anne Elizabeth Ormond. Mrs. White stated that she was unable personally to trace her ancestors beyond great-grandparents.[6]

Simply stated, Ada Bragg, a resident of Bath, apparently had more knowledge of Mrs. White's ancestors than did she. In 1947, Ada felt compelled to write Mrs. White and inform her that she was a direct descendant of the brother of Mary Ormond, the woman who married Blackbeard. It is logical to assume that Mary Ormond's sibling was her brother, as the Ormond name would have only been passed down to Mrs. White's mother, Anne Elizabeth Ormond, through the male line.

The generations in this letter just don't add up. Mrs. White was born around 1900. Her great-great-aunt would have been born around 1785 at the earliest. I arrive at this date based on my own ancestry. I was born in 1955 and my great-great-grandfather was born in 1840, one hundred fifteen years before my birth. Blackbeard's wife was supposed to be sixteen years old in 1718, placing her birth in 1702. That would leave Ada's account at least two "greats" short. Perhaps Ada was generalizing when she wrote, "great-great-aunt." Beyond that discrepancy, is there any credibility to this claim?

The Ormonds were a prominent family in Bath, but not until the 1730s, about twenty years after Blackbeard's death. There were three Ormonds mentioned in the records: Wyriott, William, and Roger. It is believed that Wyriott and William were brothers. Researcher and historian author Baylus Brooks believes that Roger was Wyriott's son, but this is unlikely. As I shall explain a little later on, I believe that Roger was another brother. The earliest record of Wyriott Ormond in Bath was when he appeared as an attorney on October 2, 1739. William first appeared when he witnessed a land exchange in Beaufort County on August 7, 1737. Deed records also show that William bought a tract of land at Mallard Creek on August 29, 1737. The following year, he became a member of the Beaufort County Land Commission. Roger Ormond was first mentioned in connection to Bath when he was appointed as tax collector at Bath Town on July 23, 1734.[7]

Prior to those dates, Wyriott Ormond was mentioned as attending a meeting of Masons at the Devil Tavern near Temple Bar in London on December 15, 1733. William Ormond was listed as living in London's St. Anne Soho district from 1733 to 1734. The matriarch of the family was Wyriott's and William's mother, Mary Ormond, who died in England and was buried at Richmond Parish of Surrey on June 25, 1737.[8] It is logical to believe that after their mother's death, Wyriott and William followed Roger to Bath, who had been there as the tax collector since 1734.

Wyriott Ormond owned estates in London and his first wife, Elizabeth, was the daughter of Arthur Moore, a member of the Board of Trade and Plantations.

It would appear that Mary Ormond couldn't have been the wife of Blackbeard. There weren't any Ormonds living in North Carolina when Blackbeard was there. This doesn't mean that Ada Bragg was completely wrong when she suggested that Mrs. White was the ancestor of a sibling of Blackbeard's wife. Ada just got a few of the facts mixed up. A fascinating explanation of this relationship to Blackbeard's wife was put forth by historian author and researcher, Allen Hart Norris in her book, *Beaufort County, North Carolina Deed Book I, 1696–1729: Records of Bath County, North Carolina*.[9] All the information supporting that explanation comes from her book. By the way, this is the same author and book that first introduced the theory that Blackbeard was the son of Bath resident, James Beard.

Norris worked backward from the way ancestry is traditionally established. Instead of starting with Blackbeard, she started with Mrs. White. With no documentation concerning any wife of Blackbeard's, Norris sought to confirm Ada Bragg's letter by linking Mrs. White to the sibling of someone Blackbeard may have had the opportunity to marry. This was done primarily through wills, deeds, and some local books on ancestry. Norris' final conclusion may be incorrect, but there is no doubt that it is intriguing. I'll go through this as clearly as I can. If you have trouble following the details, there is a chart that follows to help put the pieces in their proper place.

Mrs. White's parents were Mamie Ormond Tyler and Christopher Miller. Mr. Miller was from Hatteras, and upon their marriage in 1897, his wife left Bath to be with her husband. Mrs. White's grandmother was Ann Elizabeth Ormond, not her mother, as Ada Bragg stated. It was the grandmother, Ann Elizabeth Ormond, who married Lindley Tyler in 1867, and was a direct descendant of Wyriott Ormond, the man who came to Bath in the mid-1730s. Ann's father was Henry Ormond, her grandfather was Thomas Ormond, her great-grandfather was Roger Ormond, and her great-great-grandfather was Wyriott Ormond. So far, so good. Mrs. White was the great-great-great-great granddaughter of Wyriott Ormond.

Jumping back to Wyriott Ormond, I mentioned earlier that Baylus Brooks identified Roger Ormond, the one mentioned in 1734, as Wyriott's son. According to Norris' research, Roger Ormond was Wyriott's son, but his son Roger was born in 1740. It is unlikely that Wyriott would have had an earlier son named Roger, too. Either way, Roger Ormond was the son of Wyriott Ormond and Ann Darden. This was a very interesting woman. In 1725, Ann Darden acquired ownership of James Beard's former plantation. Wyriott was Ann's third husband. Her first husband was named Abraham Adams and her second husband was William Ormond, Wyriott's brother.

Ok, so Mrs. White is a direct descendent of Wyriott Ormond. Where does Blackbeard enter the picture? It comes from Wyriott's son, Roger Ormond, born in 1740. Roger married a woman named Mary Barrow in 1765. At the conclusion of the ceremony, her name legally became Mary Ormond, just as today when the wife usually takes the last name of her husband. This is the Mary Ormond who has ties to Blackbeard. Not as his wife, but as his niece.

Mary Barrow was the daughter of John Barrow, who was born around 1700. The name of Mary's mother, John's wife, remains unknown, but documents from the time led Norris to conclude that Mary's mother was the daughter of Thomas Worsley. This is important because Thomas Worsley had the plantation right next to the Beards. He naturally would have been good friends with James Beard and his family. This is proven by the existing records that show that Thomas Worsley's son, John, married James Beard's daughter, Susannah Beard Frank. In the theory that Blackbeard was the son of James Beard, Susannah was his sister.

Chapter Thirty-Four

According to Norris, it is likely that Thomas Worsley's other daughter married Blackbeard, an old family friend. If this were the case, Mary Barrow Ormond's mother would have been the sister of the woman who married Blackbeard. The problem with this intriguing theory is that there is no proof that Thomas Worsley had any daughters at all. The records only show two sons. This isn't unusual. Most records come from deeds and inheritance. Women are rarely mentioned in those types of records.

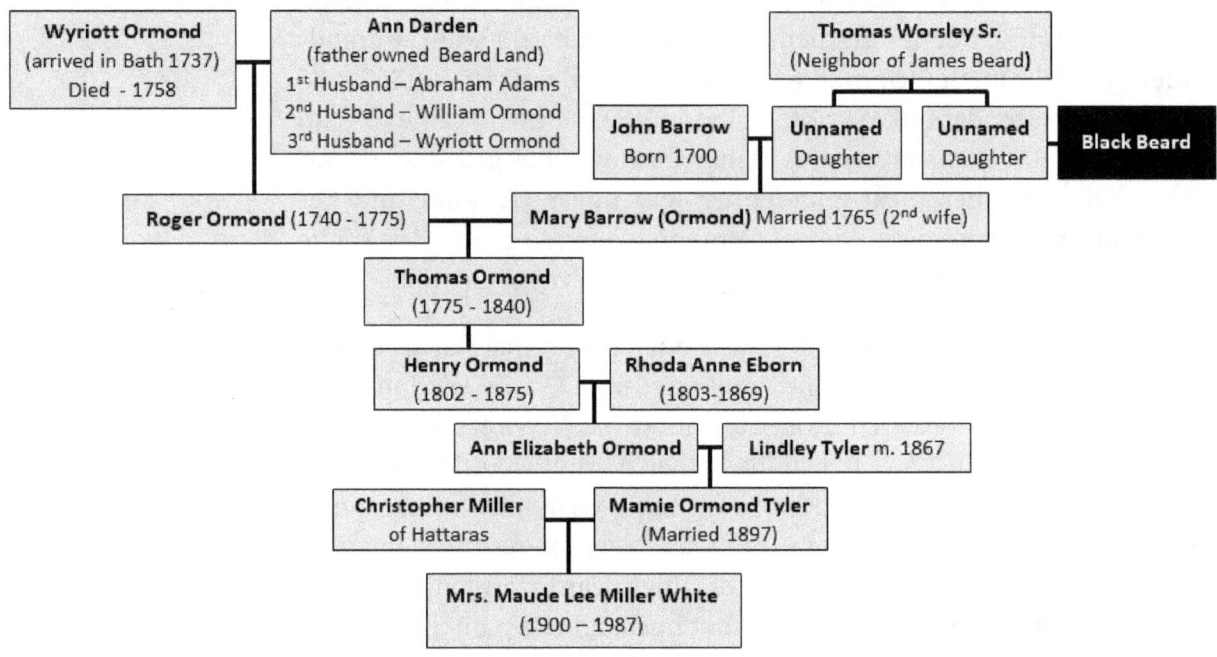

Figure 79: *Family Tree of the Ancestry of Mrs. White*

In my opinion, Allen Hart Norris worked extremely hard to find a link between someone named Mary Ormond and Blackbeard. After several questionable assumptions, she finally arrived at this conclusion. It may be true, but without any substantial evidence, or more believable circumstantial evidence, I hesitate to accept this as accurate. Furthermore, I don't think that Blackbeard took a wife at Bath, or anywhere else, for that matter. Of course, I'm always prepared to change my mind in light of some new piece of information previously unrevealed.

Thirty-Five
Twisted Tales

Sunrise was still an hour away, and early morning fog lay thick along a shore of reeds and tall grasses. The air was motionless and heavy. The only sounds came from awakening shore birds and the gentle lap of the tidal water. Through the fog, two young men in a small rowboat emerged from between the reeds. The combination of the fog and the dim light made visibility exceptionally poor, which was ideal for the purposes of these two men. They were planning a robbery. The boat edged out along the tidal floodplain and into the Hampton River. The men were careful not to make any noise as they glided through the water. Eventually, the bow of their tiny boat struck a tall wooden post jutting up out of the water. It had been put there intentionally two years earlier. With one of the men holding onto the post to steady their boat, the other carefully climbed up to the top, where a gruesome object awaited him. It was Blackbeard's head.

After Blackbeard's death at the battle of Ocracoke, Lt. Maynard cut Blackbeard's head off and brought it back to his home port of Kecoughtan, modern-day Hampton. As a warning to other would-be pirates, Blackbeard's head was placed atop a post near the entrance to the Hampton River, where it had been slowly rotting for two years. The young man climbing the pole came face to face with Blackbeard. His skin was dark and leathery, and stringy remnants of his tangled black hair and beard still clung to his decaying skull. Taking a deep breath, the young man hurriedly pried the grisly head off the post and tossed it to his friend in the boat. With their prize safely aboard, they quickly rowed away. This theft was more than a simple college prank or dare. These students wanted it for a very special purpose. Once back in Williamsburg, the two students secretly silver-plated it and took it to the Raleigh Tavern, where it was converted into a punch bowl.

Chapter Thirty-Five

This dramatic tale of how Blackbeard's head was stolen was totally the product of the author, Robert Jacob. Whereas the historical record does show that Blackbeard's head was placed on a post at the entrance of the Hampton River, there is no credible documentation that describes what became of it. Blackbeard's head may have been stolen, or it may have simply dropped into the water. However, the legend that Blackbeard's head was stolen and used as a punch bowl at the Raleigh Tavern is well documented in John Watson in his book, *Annals of Philadelphia and Pennsylvania*, in 1844. As for the theft being perpetrated by college students, this part of the legend was passed down among college students at William & Mary in the mid-twentieth century and relayed to me personally. Watson wrote:

> When the vessel which captured Blackbeard returned to Virginia, they set up his head on a pike planted at "Blackbeard point," then an island. Afterwards, when his head was taken down, his skull was made into the bottom part of a very large punch bowl, called the infant which was long used as a drinking vessel at the Raleigh tavern at Williamsburg. It was enlarged with silver, or silver plated; and I have seen those whose forefathers have spoken of their drinking punch from it, with a silver ladle appurtenant to that bowl.[1]

The story quoted above is perhaps the most interesting and even the most believable of all the twisted tales told about the world's most famous pirate. There are hundreds of fanciful legends about Blackbeard, which usually involve the location of his treasure. Some of them seem credible until the facts are scrutinized. Throughout this chapter, I shall examine some of the more intriguing stories passed down over the years. My purpose for retelling these accounts is not to inform, but rather to amuse.

Mulberry Island treasure legend

The earliest printed story to give a precise location for Blackbeard's buried treasure appeared in Clement Downing's 1737 book, *A Compendious History of the Indian Wars; with An Account of the Rise, Progress, Strength, and Forces of Angria the Pyrate*. Clement was a British naval officer who spent much of his career hunting pirates in the Indian Ocean. Upon his return to England, he wrote a book on his experiences. During his service, he met a Portuguese man named Anthony de Silvestro, who was a crewman aboard a vessel in the service of the Moors. Silvestro told Clement that before he came to the Indian Ocean, he was in the British Royal Navy and was aboard one of the two sloops that attacked Blackbeard at Ocracoke. The following is what Downing wrote in his book:

> This Anthony told me, he had been amongst the Pyrates, and that he belong'd to one of the Sloops in Virginia, when Blackbeard was

taken. He informed me, that if it should be my lot ever to go to York River or Maryland, near an Island called Mulberry Island, provided we went on shore at the Watering Place, where the Shipping used most commonly to ride, that there they Pyrates had buried considerable Sums of Money in great Chests, well clamp'd with Iron Plates. If any Person, who uses those Parts, should think it worth while to dig a little way at the upper End of a small sandy Cove, where it is convenient to land, he would soon find whether the Information I had was well grounded. Fronting the Landing-place are five Trees, amongst which, he said, the Money was hid.[2]

This fascinating story seems like it was written for Hollywood. It is the perfect tale of pirate treasure. The loot was buried in several great chests at the upper end of a small sandy cove near the place where ships land to take on water. The exact spot is marked by five trees. Mysterious and intriguing clues don't get any better than this.

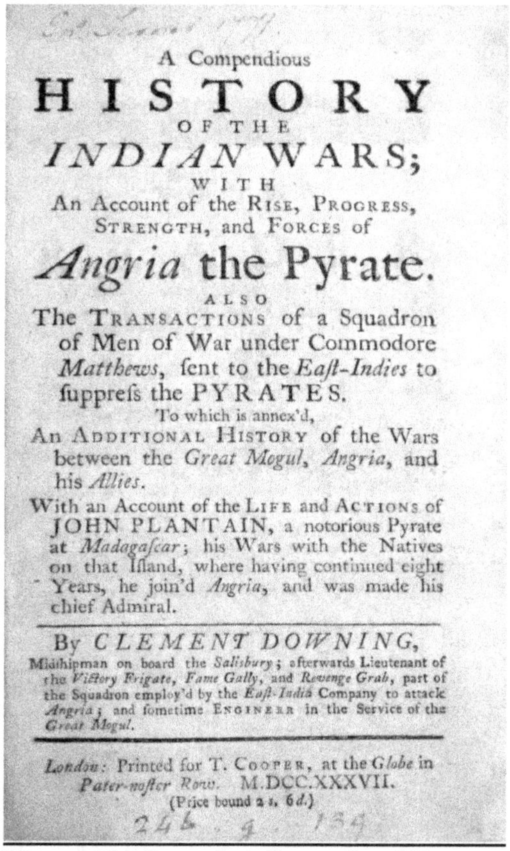

Figure 80: *Cover of* A Compendious History of the Indian Wars. *The image is dedicated to the public domain under CC0.*

The problems come from the location of Mulberry Island. First of all, the York River is in Virginia, not Maryland. Looking at Virginia maps, Mulberry Island can be located, but it is in the James River, not the York River. Ok, so Anthony de Silvestro got a few details wrong. After all, he wasn't from Virginia; he was Portuguese. Mulberry Island is only about ten miles up the James River from Hampton Roads, where HMS *Lyme* and HMS *Pearl* were stationed. For Blackbeard to bury treasure on that island, he would have had to sail right past the headquarters of the Royal Navy in the southern colonies. On top of this, the James River was the busiest river in Virginia. Its banks were lined with plantations. Not to mention that the James River is the water route to the capital of the colony, Williamsburg. This renders Mulberry Island among the most unlikely hiding spots for Blackbeard's treasure. In modern times, Mulberry Island is part of the U.S. Army's Fort Eustis.

Lunging Island, New Hampshire, has a Blackbeard legend too. This small, rocky, and remote island lies about six miles from the mainland of New Hampshire. With only a few summer homes on the island, it is uninhabited for much of the year. According to the legend, Blackbeard stopped there in 1715 and left part of his treasure in a cave near the shore. Before setting sail, he sent his fourteenth wife ashore and then left her there. Today,

Lunging Island, New Hampshire treasure legend

Chapter Thirty-Five

her ghost walks the shoreline, hoping for Blackbeard's return.[3] There are numerous internet articles about Blackbeard on Lunging Island and The History Channel even did a story on this legend in 2001. In this episode, several historian authors were brought to the location and commented on the story's feasibility. Needless to say, I was shocked to hear them giving this ridiculous story any credence.

According to the New Hampshire historian author, Dennis Robinson, the source of this legend is Prudy Crandall Randall, who owns the only summer home on the island that she calls "Honeymoon Cottage." She contends that her father told her the story in the late 1920s.[4] The date of 1715 for his visit was chosen because Blackbeard's activities are too well documented from the summer of 1717 to his death in November 1718. However, the 1715 date is inconsistent with Charles Johnson's story of Blackbeard's fourteen wives. In *A General History of the Pyrates*, Blackbeard's fourteenth wife was the girl he married at Bath in 1718. Another problem with this legend is that in 1715, Blackbeard wasn't yet a successful pirate captain. He wouldn't have accumulated much of a treasure to hide. Finally, the treasure was supposedly hidden in a rocky cave along the shore. This cave is completely underwater at high tide. At low tide, when the cave is visible, rough surf makes any attempt to enter this cave exceptionally dangerous. A small wooden rowboat would have been crushed against the sharp rocks and any treasure lost. **The Lunging Island legend must be regarded as total fiction.**

Blackbeard Island, Georgia treasure legend

Blackbeard Island, Georgia, is an obvious place to have a legend of buried treasure. A study of Georgia maps reveals that this island acquired its dubious name in the mid-nineteenth century. This 1886 map clearly shows Black Beard Island, but earlier maps identify this island as either Black Bend Island or Black Bear Island. Apparently, the name was intentionally altered to support a tale of buried treasure. The legend itself is also very vague, just that he used the island as a depository for some of his treasure.

A 1999 tourist publication titled *Longstreet Highroad Guide to the Georgia Coast* mentions that Blackbeard marked his buried treasure with a "spike in an old tree, and it is said practiced other

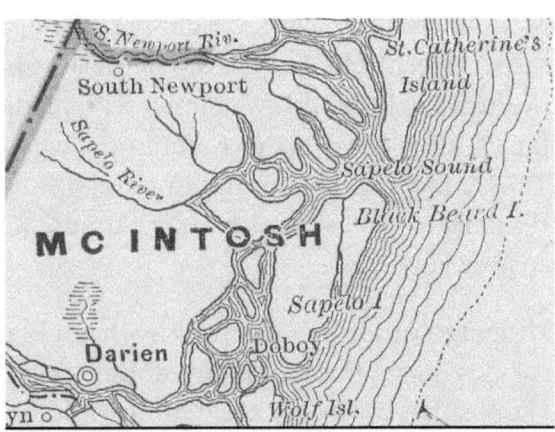

Figure 81: *Close-up of an 1886 map Showing Black Beard Island, Georgia. The image is dedicated to the public domain under CC0.*

342

nasty pirate-like activities such as beheading his 16th wife and six crew members and burying them in a mass grave on the island."[5] The legend of buried treasure goes back to the 1880s when fake treasure maps were being sold, indicating that the treasure was buried on the north end of the island.[6]

Some abandoned structures on the island add to the legend. These structures are easily explained when looking at the island's actual history. In 1800, the United States Navy bought the island from its original French owners. The island had lots of live oak trees that the navy wanted for ship construction. In the 1880s, about the same time the treasure maps appeared, the navy built a quarantine station to help combat an outbreak of Yellow Fever. A crematory was built in 1904, which is still visible today. This station was closed in 1909, and the island eventually became a national wildlife refuge.[7]

The Creeks sold Blackbeard Island to a group of French plantation developers in the mid-eighteenth century.

There is no doubt that Blackbeard sailed past this island every time he sailed along the coast of North America. The last time he passed that way was in May 1718, just before he laid siege to Charles Town. Whereas Blackbeard had the opportunity to stop and bury some treasure on the northern tip of the island, it is very unlikely that he did. In 1718, Georgia was technically part of Spanish Florida, although the English were constantly challenging Spanish authority by probing into the territory. In 1718, a rather large population of Creeks inhabited three of Georgia's barrier islands, St. Catherines, Ossabaw, and Sapelo. Blackbeard Island is actually part of Sapelo Island.[8] It would be very unwise to bury any treasure on an Island that was occupied by the Creeks. I believe that Blackbeard Island's treasure legend has no basis in fact.

Two ominous-looking towers stand on islands in the waters frequented by Blackbeard. The first is Blackbeard's Tower on Nassau and the second is Blackbeard's Castle on St. Thomas. There is no doubt that Blackbeard was on Nassau several times, from 1716 to 1718. The legend contends that he occupied this tower while ashore to watch for approaching vessels. There are also tales that the tower is haunted by Blackbeard's ghost and is the source of unexplained lights locally known as that "Teach's Light."[9] On St. Thomas, a similar story is told concerning a structure called Blackbeard's Castle. Blackbeard's Tower near Nassau was originally a water tower and Blackbeard's castle in the Virgin Islands was built in 1679 as a military watchtower by Danish colonists. The building was originally known as Skytsborg Tower, meaning "sky tower."[10] It appears that both of these legends originated in the twentieth century with the rise of tourism. On Nassau, there would be no logical reason for anyone to watch for approaching vessels from the east side of the island when the only approach to the

Treasure legends on Blackbeard's Tower, Nassau and Blackbeard's Castle, St. Thomas

harbor was from the west. Whereas Blackbeard sailed past St. Thomas, there was a very strong English presence on the island in 1717, and he would have been arrested the second he stepped ashore. I believe that both of these legends have no basis in fact and were created solely for tourists.

Blackbeard's base and treasure on Amelia Island, Florida

Tranquil, charming, and quaint, Amelia Island has amazing beaches, a wide assortment of hotels and Bed & Breakfasts, wonderful restaurants, and a long and rich history. One of the barrier islands, it is located on the east coast of Florida just below the Georgia border. It is the ideal spot for tourists who appreciate the finer things and are looking for an out-of-the-way spot. There are several museums and the remnants of an early nineteenth-century Spanish fort. There is also a large Civil War period fort, Fort Clinch, which was constructed in 1847 and offers an exciting look back into American history.

Locals throughout the island and even several visitor websites claim that the island has long been associated with pirates, specifically Blackbeard. The island has a pirate-themed museum that further touts this legend. It seems logical at first look. Amelia Island is in a great location, close to the main shipping lanes that follow the Gulf Stream along the Florida–Georgia coast. The legend contends that not only did Blackbeard use the island to ambush passing vessels, but he buried some of his treasure there, too. The most fascinating aspect of this legend is that the burial spot of his treasure is marked by a chain hanging from a big oak tree, rumored to be near downtown Fernandina Beach. Treasure hunters have claimed to have seen this mysterious chain hanging down from a branch on a tree, but when they go back to the same spot with a shovel, the tree, and chain are nowhere to be found.[11]

Blackbeard sailed past Amelia Island on a number of occasions, but there was no gap in his timeline that would have given him the opportunity to establish a temporary base there. Even though the old Franciscan Mission that had been established on the island had closed by 1718, small groups of farmers still inhabited the island. While researching my second book, *Pirates of the Florida Coast: Truths, Legends, and Myths*, I discovered that many local pirate legends originated with fishing guides, who told pirate stories to their clients in the early twentieth century to boost business. I believe that the Amelia Island Blackbeard legend originated as one of those fishing stories and has no basis in fact.

The Brick House legend of Blackbeard's Home on the Albemarle

Closer to Blackbeard's home in North Carolina stands a house rumored to have once been occupied by Blackbeard himself. It's a structure called "The Old Brick House" and is located along the Pasquotank River, close to Elizabeth City. Unlike some of the other legends, this one was described in detail in Catherine Albertson's 1914 book, *In Ancient Albemarle*. The story was not created by the author, merely recorded by her from local tales she may have heard or read.

The house still stands today and is on the National Register of Historic Places. It is a well-built two-story home with a chimney at each end and five gable windows across the front on the second floor. According to Albertson, "Teach and his drunken crew would come, seeking refuge after some bold marauding expedition, in the hidden arms of that lovely stream." From the river, one can occasionally see ghostly light emanating from the house and shimmering on the water. Locals call it "Teach's Light."[12] Inside the home, there were secret passages that led to an underground tunnel that came out near the river's edge. Albertson wrote:

> And should a strange sail heave in sight, or one which he might have cause to fear was bringing an enemy to his door, quickly to the secret closet near the great mantel in the banquet hall would Blackbeard slip, drop quietly down to the basement room beneath bending low, rush swiftly through the underground tunnel, slip into the waiting sloop and be off and away up the river or down.[13]

In addition to the basement escape tunnel, the legend contends that he kept some of his prisoners there. Albertson wrote, "It is said that the basement room of the Brick House served as a dungeon for prisoners taken in Teach's private raids and held for ransom."[14] The reason locals are so certain that this was indeed the home of Blackbeard was that a circular granite grinding wheel was found on the property, "sunken in the ground at the foot of the steps and bears the date of 1709, and the initials E.T."[15]

With evidence like that, how could anyone doubt the authenticity of this claim? A grinding wheel with the initials E.T. carved in it must have belonged to Edward Teach. That last comment was just a bit sarcastic. E.T. could have been the initials of millions of people. Second, why would a pirate mark his grinding wheel with his initials? I can't imagine Blackbeard going out to the mill to grind some corn for his breakfast. Homeowners wouldn't put their personal initials on tools and implements, the manufacturers would. Last, archaeologists date the house's construction to 1735.[16]

Authors who create stories like this imagine Blackbeard being in North Carolina for years. In reality, he only had about one month to live such a life in the large brick house. My final comment on this amusing story concerns the house's location. In her book, Albertson wrote that the house was "A few miles down the Pasquotank from Elizabeth City."[17] In reality, it is less than a mile up the Pasquotank from Elizabeth City.

Chapter Thirty-Five

Governor Eden's secret tunnel

Many people believe that the Governor Eden Tunnel legend is the most probable location for Blackbeard's hidden treasure. According to local tradition, a tunnel was constructed that led directly from Bath Town Creek to the cellar of Governor Eden's house, which was used by Blackbeard to smuggle some of his stolen goods directly into the Governor's cellar.[18] At first, this story gives the impression of being quite reasonable. Governor Eden's 300-acre plantation was located on the west side of Bath Creek next to Tobias Knight's property at Archbell Point. The record of Eden's purchase has been lost, but it is certain that he took possession sometime in early 1716.[19] There is no question that both Blackbeard and Eden were in Bath for about three of the last six months of 1718. Collusion between the two men was firmly established in the minds of all who read Charles Johnson's *A General History of the Pyrates*:

> When the Lieutenant came to *Bath-Town*, he made bold to seize in the Governor's Store-House, the sixty Hogsheads of Sugar, and from honest Mr. *Knight*, twenty; which it seems was their Dividend of the Plunder taken in the *French* Ship;[20]

This story appears in just about every book and website that discusses Blackbeard's lost treasure. An example of such a website is titled *7 Places Blackbeard's Gold Could've Been Stashed,* where the author wrote, "Legend has it that Blackbeard was able to slip easily in and out of Eden's estate (presumably to deliver the governor's cut of the loot) by using a special rock path or, in some versions, an underground passage between."[21] The tunnel legend goes back much further than the twentieth century.

The earliest known description of this tunnel was in a letter dated March 31, 1857, written by Joseph Bonner. He was a resident of Bath for many years and remembered playing on the property once owned by Eden when he was a young boy in the 1820s. The old, abandoned mansion was still standing and the ruins of the tunnel were visible. Bonner wrote:

> I was accustomed to visit this building in my early childhood. Its massive walls, capacious halls, rich workmanship of the interior, and palace like appearance, indicated that it had been a abode of wealth. A subterranean passage some 60 or more yards in length, communicated from a brick wall near the margin of the creek, with the cellar of the dwelling.[22]

The tunnel's next appearance in print was in 1894. A newspaper article recounted the statements of two men, Thomas Latham and Jno Burgess, who visited the property and saw the remains of the ruined mansion with an old cellar. Part of the subterranean passage was still visible.[23] Catherine Albertson included this tale in her 1914 book, *In Ancient Albemarle:*

> Nearly on a line with this, at the water's edge, is shown the opening of a brick tunnel, through which the Pirate Teach is said to have conveyed his stolen goods into the governor's wine cellar for safe keeping.[24]

All of this evidence seems fairly compelling except for one thing. Eden didn't own the property when Blackbeard was there. Deed records from 1718 show that Eden sold the tunnel land property to John Lillington on April 10, 1718, about two and a half months before Blackbeard arrived.[25] For the entire time Blackbeard was in North Carolina, Eden was living on a 320-acre plantation at Bridge Creek north of town on the east side of Bath Town Creek.[26]

We aren't quite finished with this tunnel land. There are a few more surprises to reveal. Eden sold his property to two men, Stephen Elsey and James Robbins.[27] As you may recall from Chapters Twenty-Eight and Twenty-Nine, James Robbins was listed as one of the pirates arrested on the *Adventure* after Blackbeard's death. He was probably the planter mentioned by Spotswood as the man who prevented the "Negro" from blowing up the sloop.

> His [Blackbeard's] orders were to blow up his own vessel if he should happen to be overcome, and a Negro was ready to set fire to the Powder had he not been luckily prevented by a Planter forced on board the night before & who lay in the Hold of the sloop during the actions of the Pyrats Tach.[28]

Was James Robbins more than a visitor on the *Adventure*? Did he purchase the plantation in order to secretly receive stolen goods from pirates? Imagine the real estate listing. Beautiful Two-story Mansion, three bedrooms, two dining rooms, no bath, has a long subterranean tunnel leading to the docs ideal for smuggling. No documentation exists to suggest that this house ever contained stolen contraband.

Finally, the property was sold to Edward Salter on November 12, 1726, for £600.[29] That name should sound familiar, too. He was the cooper who was forced into service aboard the *Queen Anne's Revenge* after Blackbeard took the *Margaret* near Puerto Rico in December 1717. Salter stayed with Blackbeard and was with him when he took the two French ships off Bermuda and was among those pirates arrested by Captain Brand in Bath in November 1718. He was taken to Williamsburg to be tried for piracy.

Chapter Thirty-Five

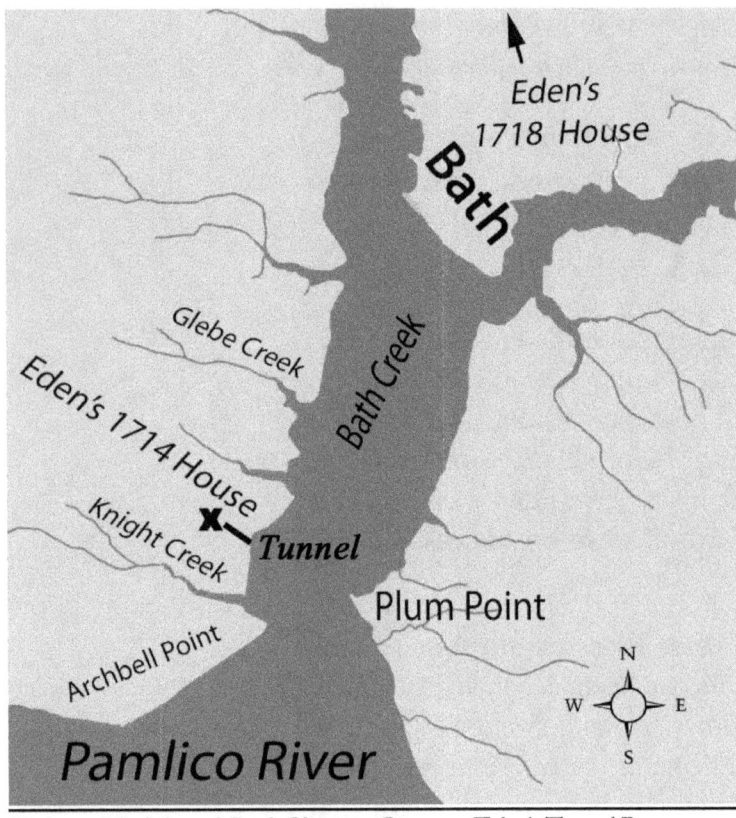

Figure 82: *Map of Bath Showing Governor Eden's Tunnel Property.*

Blackbeard's treasure legend at Cape Henry, Virginia

Apparently, he was also among the pirates who were released.

Visiting this property while researching this book, no signs of the old plantation remain and there is no trace of any tunnel. My thoughts on this subject are that it was never a tunnel at all, only a brick ramp leading from the dock to the mansion's cellar. One hundred years later, the walls of a collapsed and overgrown ramp would appear to have been a tunnel to the untrained eye. Brick ramps leading from the water's edge are common in eighteenth-and nineteenth-century homes. I have seen several over the years. It makes rolling barrels and carts with cargo much easier.

I have saved the best for last. The sole source for this story is a Virginia Beach tourism website titled *The Lost Treasure of Edward Teach*.[30] Tourism is big business, especially in vacation and beach towns like Virginia Beach. I lived there for many years and can testify to the fact that it is a terrific place to visit. It's also only about eighteen miles from Hampton, where Blackbeard's head was taken and placed on a post in the river. Hampton hosts a yearly pirate event called the *Blackbeard Pirate Festival*, which commemorates Maynard's victory over Blackbeard. It's also a huge block party where, unfortunately, most of the visitors don't even realize that Blackbeard was a real person. Further up the road from Hampton is Colonial Williamsburg, a large collection of restored and reconstructed eighteenth-century buildings that serve as a living history museum. It's also where Blackbeard's captured pirates were taken to stand trial. In light of this combination of history and tourism, it is surprising that someone would write a story about Blackbeard for their website that makes no logical sense and has no basis in fact.

In his final months, with his main base on Ocracoke, Blackbeard would sail to Virginia Beach regularly to hunt for ships. He would drop off a few men at Cape Henry to serve as lookouts, then continue west along the shore until he reached a mooring at the end of a dirt road named Pleasure House Road. There is no need to explain why Blackbeard would want to stop there or how that road got its name. While Blackbeard was enjoying the pleasures found there, his men at Cape Henry would keep a close watch for vessels. They were careful to note the direction the vessels were sailing in. If a ship was sailing from the east, it was a merchant vessel, but if it was sailing from the west, it probably belonged to the Royal Navy. The lookouts would signal Blackbeard at Pleasure House Road. If he received the signal for a merchant vessel, he would quickly set sail to intercept it. On the other hand, if the ship was identified as a naval vessel, Blackbeard would be able to escape through the inland waterways back to the ocean and then to his base at Ocracoke.

On one particular day, as Blackbeard and most of his crew were drinking and partaking in other pleasurable activities at the Pleasure House, they received the signal that a merchant ship was approaching from the east. The pirates quickly boarded and sailed out to meet the unsuspecting ship. The pirates were easily able to catch the ship because it was weighed down by an excessive amount of heavy treasure. Just as his men had finished transferring the treasure aboard his sloop, he heard the sound of naval guns being fired. To his amazement, two royal naval ships were rapidly closing in. He didn't want to be caught with the treasure onboard, so he took all of it to the nearby beach and buried it there among the sand dunes. Afterward, he headed back to Ocracoke. Unfortunately, this was only a few days before he was killed by Lt. Robert Maynard. To this day, no one has ever found the treasure, and it remains hidden beneath the sands of what is now First Landing State Park.

I shouldn't have to point out all the ridiculous facets of this incredible story, but I will. The ludicrous claim that merchant ships are always sailing from the east and naval vessels are always sailing from the west is the most obvious. Each naval vessel leaving port by sailing from the west would eventually have to return by sailing from the east. The same holds true for the merchant ships.

Pleasure House Road is a real place. It runs approximately north-south through a modern suburb and ends where the Chesapeake Bridge Tunnel begins. The origin of its name is unknown, but it probably acquired the name fairly early in the town's history. The problem comes from the signal between the lookout at Cape Henry and someone standing watch on the sloop where it lay at anchor at the end of the road, almost six miles through dense shoreline pine trees. There is no way a signal could be seen from that distance. To add to the problem, there is no inland waterway that Blackbeard could have used as an escape. The nearest waterway is Lynnhaven Bay, which dead-ends in a few miles.

Chapter Thirty-Five

The end of the story is the most fantastic. Blackbeard was in a rush to escape the two approaching English warships that were just about on top of him. Instead of making a fast escape, he spends time going ashore and burying his treasure. Since there is no way for a sloop to come near the shore, Blackbeard would have had to take the treasure ashore by rowboat. All this would have been done in plain sight of the two warships, which would have easily reached his position by then.

This last tale sums up the public's mindset when it comes to Blackbeard. Enthusiasm is strong among the masses, and many people revel and delight in hearing and repeating fascinating local tales of pirate treasure. Unfortunately, they seldom stop to consider the facts or even the logical circumstances which surround those legends. That being said, in the final chapter of this book, I will explore a few other stories of Blackbeard's buried treasure that just might be true.

Thirty-Six
Blackbeard's Treasure

Let it be understood that the scenario about to be set forth is based upon facts and opportunity. Nonetheless, it is all speculation that was completely imagined by the author, Robert Jacob. Let's turn back the clock to 1718, in the port of Nassau.

It's a dark night in early May 1718, and Blackbeard sits in his cabin pondering his future. A heavy matter weighs on his troubled mind. By now, Blackbeard has realized that his partnership with Stede Bonnet didn't grow into the collaborative and effective team that he had hoped for. Bonnet is not Hornigold. Instead, Blackbeard sees Bonnet as an insufferable and incompetent liability. Their friendship has soured to the point where Blackbeard cannot even stand the sight of him, but breaking up such a partnership isn't so easy. The men who remain loyal to Bonnet will have to be dealt with. And then, there's all their treasure. The enormous task at hand is now for Blackbeard to devise some sort of scheme that will force Bonnet and his men to leave peacefully without expecting their share of the loot.

The next day, Blackbeard's fleet of four vessels anchored at Nassau, and he and his men went ashore. Those who had been in Nassau before were delighted to visit with their old friends from other ships. The majority of Blackbeard's crew were relatively new and experiencing the pleasures that the pirate port had to offer for the first time. Blackbeard had a warm and comforting reunion with his old partner, Ben Hornigold. He also talked a bit of Jacobite treason with Charles Vane. More importantly, the details of the king's proclamation now became clear to Blackbeard. Contemplating his future, he decided that he would accept the pardon, at least as a pretense, and start a new life in North Carolina. However, this complicated

Chapter Thirty-Six

matters and accelerated Blackbeard's timeline. If he was going to deal with Bonnet, it would have to be done quickly.

Ultimately, Blackbeard knew that the *Queen Anne's Revenge* would have to be sunk. With the fleet's prime ship lost, the uneasy partnership would quickly dissolve. Bonnet and his men would go their separate ways, but only if they believed that the ship had sunk by accident. Otherwise, they might seek retribution. A plan began to formulate in Blackbeard's mind. Running the *Queen Anne's Revenge* aground as if by accident would be easy enough. The issue that had been troubling Blackbeard was how to keep the treasure. It would also be lost when the *Queen Anne's Revenge* went down. Somehow, all the treasure would have to be secretly offloaded before the ship was scuttled. Afterward, Bonnet and his men would believe that the treasure was lost along with the ship, and they wouldn't press Blackbeard for their share or come after him seeking vengeance later. The question that was burdening Blackbeard's mind was how to offload this treasure from the *Queen Anne's Revenge* right under the watchful eyes of his crew without raising suspicion.

Blackbeard sat all alone in a tavern, as his friends had retired for the evening. His head hung low and his shoulders slumped with the troubles that racked his brain. The proprietor wouldn't dare question his presence so late in the night. After all, this was Blackbeard. Glancing up, he stared intently through the window at his ship anchored in the harbor. With a solemn look on his face, Blackbeard sighed. It was a good ship that had brought him so much success. He detested the thought of sinking it, but he knew that it would have to be done. Another noggin of rum was placed on his table. Turning away from the window, Blackbeard smiled and his eyes twinkled. He kindly thanked the bartender for his service, then returned to his previous posture, his mind deep in thought. Suddenly, his head jerked up with anticipation. His eyes lit with excitement and passion. Blackbeard knew what to do.

The plan was brilliant! It called for a diversion of immense proportions. That diversion was Charles Town. His plan was both simple and effective. While most of his pirates were occupied, taking prizes scattered around the entrance of the busiest harbor in the south, Blackbeard could clandestinely have the treasure offloaded to the small Spanish sloop in his possession and sail it away unnoticed. Blackbeard could depend on about twenty-five loyal crewmen. Some of them would transfer the treasure to the sloop, while the others would make sure that the rest of the pirates were busy somewhere else and not watching the *Queen Anne's Revenge* too closely. To pull off this deception, Blackbeard would need to involve some of the leaders in his crew. Fortunately, two of Blackbeard's most trusted allies were William Howard and Hezekiah Hands.

This was the first phase of Blackbeard's plan. The second phase would require a little more imagination. For the plan to work, the small sloop would have to sail away loaded with the treasure and return empty without anyone noticing that it was gone. If he was able to pull that off, everyone would believe that treasure was still aboard the ship. Blackbeard knew that it would take several days for the sloop to make the round trip. He also knew that if the supply of prize vessels at the harbor's entrance dwindled, his men might want to leave the Charles Town area and sail on. This would be disastrous for his plan. If the pirates decided to leave before the sloop returned, they would notice that it was missing and become suspicious. Blackbeard needed something else that would compel his men to stay at the harbor's entrance willingly for about five days. The solution would be the most challenging part of his plan. He would have to convince his crew that they had contracted some sort of disease, perhaps yellow fever, and that they needed to remain near Charles Town for several days until they received the drugs required for treatment. Blackbeard knew that it would take the authorities in the city several days to gather the drugs he requested and put them all into a chest of medicine.

The small Spanish sloop was the key to Blackbeard's plan. Approaching Charles Town, he made certain that the sloop remained in tow with no one onboard. The first ship sighted was the *Crowley,* which they seized just as it cleared the bar. Blackbeard couldn't believe his luck. This was the perfect ship needed for his plan to succeed. It contained important dignitaries and passengers who could be used as hostages to bargain for this chest of medicine. This included Mr. Samuel Wragg, a member of the Council of South Carolina. Threatening Wragg would certainly force the governor to agree to his demand for such a chest. This also afforded Blackbeard the opportunity to take care of another concern he had: Captain Richards.

When Bonnet was removed from command, Richards was named captain of the *Revenge*. Of all of Blackbeard's men, Richards was the most likely man to realize Blackbeard's scheme. As captain, Richards would be sailing the *Revenge* around the harbor in search of ships. Observant, quick, and intelligent, Richards might notice that the sloop was gone. To make matters worse, Richards had been with Blackbeard long enough to know how his mind worked. Richards had to be distracted somehow, and in such a way that he wouldn't be watching the harbor. Blackbeard solved this problem as effectively as he did the others.

Chapter Thirty-Six

Blackbeard signaled the *Revenge* to send their captain over to the *Queen Anne's Revenge*. Once aboard, he explained to Captain Richards that one of the passengers, a Mr. Marks, was selected to go ashore and deliver a message to the governor with his demands for the chest of medicine. He also explained that someone would have to accompany him ashore. Gently grasping Richards by the shoulder and looking deep into his eyes, Blackbeard expressed his total confidence in Richards's abilities and told him that there wasn't anyone else he could trust with such a difficult and important task. Then, Blackbeard suggested that Hezekiah Hands go with them, just to make sure that nothing went wrong. A passing pilot boat was easily captured, and Richards, Hands, and Marks boarded. They pushed off and set sail for the docks.

Choosing Hezekiah Hands to accompany Richards ashore was perhaps the cleverest part of Blackbeard's plan. Hezekiah Hands was one of two of his officers who were in on the plan right from the start. Once ashore, Hands would ensure that Richards was preoccupied to the point where he wouldn't notice the missing sloop. Hands pulled off his mission with extraordinary discretion and cunning proficiency. Richards and Hands walked the streets of Charles Town, intimidating the local citizens to the greatest extent possible. They visited the taverns, drinking their fill without paying. There was no way for Richards to see Blackbeard's vessels from shore, especially considering that he was probably drunk for a substantial part of the time.

That night, while everyone slept, a few of Blackbeard's inner circle quietly lifted the treasure out of the hold of the *Queen Anne's Revenge* and lowered it into the sloop which was tied alongside. This wouldn't be as challenging as one might think. In today's market, one pound of gold dust is worth about $37,000. A single chest could easily hold fifty pounds, which would be worth $1,850,000 in today's money. Ten chests would be worth eighteen million dollars. It's easy to envision ten strong men secretly loading twelve or fifteen such chests into the sloop. John Martin went with them as the sloop's captain. He had been Hornigold's quartermaster and was highly experienced in such matters. Once the treasure was onboard, the sloop quietly pushed off and drifted with the outgoing tide into the current. When it was far enough away, the sails were raised, and it

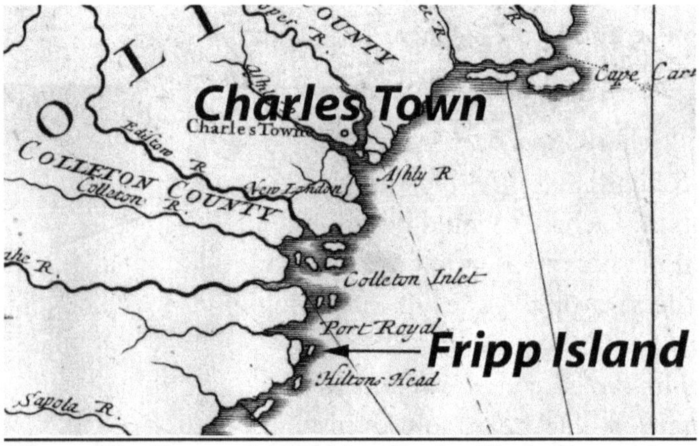

Figure 83: *Close-up of 1709 Map Showing Charles Town and Fripp Island, Courtesy of the University of North Carolina at Chapel Hill.*

sailed south to an island that Blackbeard had personally selected earlier, Fripp Island.

Fripp Island was a perfect choice. It was only about one day's sail from Charles Town. A sloop could easily get there and back in four days. It was also totally uninhabited. There would be no one to witness the pirates landing or burying their treasure. Most importantly, Blackbeard knew the island well. The island was named for John Fripp, who had been appointed sheriff of Colleton County in 1702. His father had been a privateer in the seventeenth century, and John followed in his father's footsteps. During the recent war with the Spanish and French, Fripp was heavily involved in the defense of Charles Town. He had a small fort constructed on his island that was used as a lookout station to warn Charles Town of any impending attack. Blackbeard

Figure 84: *Blackbeard Buries His Treasure by Howard Pyle. The image is dedicated to the public domain under CC0.*

had been a privateer in that war, operating in the Charles Town area, and part of his duties was to stop at Fripp Island a few times to deliver supplies to the men stationed there. After the war, the island was deserted. There is a long, deep channel running behind the island that is completely hidden by the tall sand dunes. Blackbeard could tell his men precisely where they could land safely without being detected by passing vessels sailing up the coast. Except for the fact that Blackbeard wouldn't have accompanied his men, this illustration drawn by Howard Pyle accurately captures Blackbeard's pirates burying their treasure on an island.

Meanwhile, back on the *Queen Anne's Revenge*, it was apparent to Blackbeard that none of the other pirates had seen his men spiriting the treasure away the night before. The empty boxes in the hold gave everyone the impression that the treasure was still onboard. If asked where the sloop was, Blackbeard was prepared to answer that he had sent it out to look for prizes sailing up from the south. For three days, Blackbeard maintained his composure as he anxiously awaited the return of the sloop. Finally, he saw the sail of the sloop approaching from the south. Casually glancing around, everything looked normal. There were no questioning looks from his men. As the sloop came closer, Blackbeard eagerly watched for a sign of success. John Martin walked to the rail and nodded. Blackbeard realized

Chapter Thirty-Six

that his deception had worked, but his plan wasn't over yet! He had to sink the *Queen Anne's Revenge*.

The next day, the chest of medicine was delivered, and the drugs were administered to his relieved men. Since they weren't really sick in the first place, the treatment appeared to work marvelously. The fleet set sail for Topsail Inlet, a location carefully chosen by Blackbeard. He knew the Carolina coast well and knew precisely where and how to sink the *Queen Anne's Revenge* that would make it look accidental. They arrived at dusk. Knowing that daylight would be needed to rescue the men from the sinking ship, Blackbeard waited outside of the inlet as his other sloops entered. Now for the final stage of his plan. As the sun came up, the sails were raised and the *Queen Anne's Revenge* headed straight for the channel. Blackbeard himself was at the helm. With a slight shift of the helm, the ship imperceptibly veered off to the port side. Only a few pirates onboard were prepared for what happened next.

The ship struck a sandbar with great force. The seams split open and water rushed into the hold. Panic ensued among most of his men, as they clamored to escape the sinking ship. Blackbeard and those who were part of the plan carefully watched the others to make sure that they didn't go into the hold containing the empty treasure boxes. The sloop that came to their assistance was the *Adventure*. Commanding that sloop were William Howard and Hezekiah Hands. This was also part of Blackbeard's well-conceived plan. It would appear to Bonnet, Richards, and all the others, that the *Adventure* was trying to help get the ship off the sandbar, but in reality, Howard and Hands were there to make sure that it sank. They also made sure that the *Adventure* sank too.

After taking control of the Spanish sloop that had served him so well, Blackbeard renamed it the *Adventure*, not to be confused with the other sloop *Adventure* that had just sunk. After marooning some of Bonnet's men on Harbor Island, Blackbeard sailed to Ocracoke, but he only remained there for a very short while. He sailed the *Adventure* back to Fripp Island and recovered the treasure. Martin knew exactly where he had buried the treasure a week or two earlier. The sloop maneuvered into the channel behind the dunes and the men came ashore. All the treasure was still there, precisely where they hid it. Once the treasure was onboard their sloop, the pirates sailed north to Bath to accept the king's pardon.

Now that this scenario of the events of May and June 1718 has reached its conclusion, I would be remiss if I didn't explain why I believe this narrative is feasible, and discuss the documented facts that support it. After my explanation, you may find that it isn't quite as fictional as it may have seemed. As I mentioned in Chapter Thirty-Two, when I first began researching Blackbeard around 2007, I was bothered by three things that just didn't make sense. In seeking answers to these nagging questions, I discovered that other historian authors were bothered by these things too, but none of them offered any suggestions or solutions.

From Blackbeard's actions in Bath, he obviously planned on accepting the king's pardon in early May 1718. Keeping that in mind, these three questions come into focus. Why would he risk getting that pardon by attacking Charles Town? What was his great need for a chest of medicine and why didn't he just purchase it and stay out of trouble? Why did he sink the *Queen Anne's Revenge*? The scenario put forth in this chapter neatly addresses all of those questions. Let's take a closer look at the facts that drove me to create this plausible story.

The small Spanish sloop was brought to Charles Town in tow, with no one on board. In David Herriot's deposition, he said that it was "a small empty Sloop which they found near Havana."[1] The names of Blackbeard's trusted pirates who could have been part of his plan were the men who remained with him after his split with Bonnet. These men included John Martin, Hezekiah Hands, and William Howard. John Martin and Hezekiah Hands are listed in Cracherode's report[2] and William Howard was later charged for piracy in Williamsburg.[3]

The *Crowley* was the first ship captured by Blackbeard, taken as it cleared the bar. Aboard was Samuel Wragg, who was held hostage until a chest of medicine could be delivered. One of the hostages, Mr. Marks, was sent ashore to deliver Blackbeard's message to Governor Johnson. Marks was accompanied by Richards and Hands, who "walked upon the Bay, and in our publick Streets, to and fro in the Face of all the People." All of these facts can be found in Stephen Godin's June 13, 1718 letter, in addition to many other source documents.[4]

Fripp Island was uninhabited at the time and is a one-day sail from Charles Town. I have visited Fripp Island and have seen the hidden channel that runs behind the sand dunes. Additionally, the visitor's center on Fripp Island used to tell a story about Blackbeard's use of that channel to ambush passing vessels and to bury his treasure. They even had a map of the island displayed on the wall that contained Blackbeard's image. John Fripp did own the island; however, the part of my story that tells the island was used as a lookout post during the war can't be verified. I remember reading that tidbit of history somewhere about ten years ago, but I have been unable to locate that article again. However, considering the fact that Charles Town

Fripp Island

was brutally attacked by the Spanish in 1706, it seems likely that a station would have been constructed there.

The prime reason I chose Fripp Island in my scenario as the location that Blackbeard used to bury his treasure is the long-existing local legend that this is precisely what he did. I lived in the area on and off for about seven years between 1984 and 1997. In addition to the visitor's center mentioned above, I recall talking to a man who was about ninety years old about Fripp Island. He told me that when he was a child, sometime around 1910, every time his parents took him to the beach on Fripp Island, he would take along a shovel to hunt for Blackbeard's treasure. This legend is strong among the locals and has endured for centuries. If this was indeed the place where Blackbeard buried his treasure, he must have recovered it. Just about every inch of the island has been excavated and developed in recent years. Today, the island is covered with a very exclusive gated community where a few Hollywood celebrities are rumored to have homes.

Most historian authors believe that Blackbeard was a privateer in the war against Spain and France. Most of the pirates from Nassau started out that way. There are no documents that Blackbeard was among the privateers who sailed near Charles Town, but it is reasonable to assume that he was. Charles Town was the center of most of the action during that war. Of course, since I can't verify that Fripp Island was used as a lookout base, Blackbeard's presence on Fripp Island delivering supplies can't be verified. That detail was added based on the local legend that he spent a lot of time there.

Identifying John Martin as the commander of the sloop that carried the treasure to Fripp Island was based on logic and facts. As mentioned in Chapter Fifteen, Martin joined Blackbeard in Nassau. Traveling from Nassau to Bath, Martin was more of a passenger rather than a member of the crew. As such, Martin wouldn't have had any assigned duties and the other crew members wouldn't expect to see him and his absence wouldn't be noticed. Although not a regular member of Blackbeard's crew, Martin was one of Blackbeard's trusted allies. He was among those who were arrested by Captain Brand in Bath, proving that he remained with Blackbeard after the split with Bonnet and his men. Obviously, someone had to command the sloop. It couldn't be Blackbeard, because his absence from the ship would arouse suspicion. Of all of Blackbeard's small group of loyal and trusted pirates, there were only three men who were qualified to handle such a task. One was Hezekiah Hands, but he was ashore with Richards. The second was William Howard, but as the quartermaster, his absence would have been missed as well. That left John Martin, Hornigold's former quartermaster.

The last and most controversial part of this narrative is the intentional sinking of the *Queen Anne's Revenge* and the sloop *Adventure*. At their trial, most of Bonnet's men testified that they believed that Blackbeard deliberately wrecked the ship in order to cheat them out of their share of the treasure.[5] In his deposition, David Herriot identified Hezekiah Hands as captain of the sloop *Adventure* but stated that Blackbeard sent William Howard over to command the *Adventure* at Topsail Inlet. It was Howard who steered the sloop to come to Blackbeard's assistance and wrecked that sloop in the process.

> That the next Morning after they had all got safe into Topsail-Inlet, except Thatch, the said Thatch's Ship Queen Anne's Revenge ran aground off of the Bar of Topsail-Inlet, and the said Thatch sent his Quartermaster to command this Deponent's Sloop to come to his Assistance; but she ran aground likewise about Gunshot from the said Thatch, before his said Sloop could come to their Assistance, and both the said Thatch's Ship and this Deponent's Sloop were wrecked.[6]

The strongest argument on the side of the sinking being intentional is the absence of the treasure itself. In 1996, the wreck of the *Queen Anne's Revenge* was discovered exactly where it was expected to have sunk, just outside of the entrance to Topsail Inlet, near Beaufort North Carolina.[7] Since then, underwater archaeologists have been scouring the wreck site in search of artifacts, especially any possible treasure items. So far, according to researchers Mark U. Wilde-Ramsing and Linda F. Carnes-McNaughton, only four coins and a tiny amount of gold dust have been found. Three of the coins were identified as a single *one reale* coin and two *half reale* coins. In today's money value, those coins would equal about $25. The gold dust comes to only 20.8 grams in weight. Archaeologists believe that this gold dust accidentally fell through the floorboards.[8]

If the *Queen Anne's Revenge* sank accidentally, what happened to the treasure? According to all of Bonnet's men, they didn't take it with them. None of Blackbeard's men could have taken any of the treasure, either. If they did, they would have been seen by Bonnet's men ashore. As you may recall from Chapter Seventeen, only twenty-five of his men were marooned on Harbor Island. The rest remained in town. The standard rebuttal most often used by those who are on the side of the accidental sinking theory is that there just wasn't any treasure to begin with. Their rebuttal generally includes comments that most of Blackbeard's prize vessels contained cargo that the pirates threw overboard, and there was very little money taken. Occasionally, someone will add that whatever treasure they did have would have been spent in ports on drink and women.

Chapter Thirty-Six

These arguments make no logical sense. Looking at the historical record, an estimate of the minimum amount of loot onboard the *Queen Anne's Revenge* can be made. Since Blackbeard didn't lose any vessels in battle or from bad weather, everything of value that was taken between September 1717 to June 1718 could have been stored on their ship. During that time, they captured a total of fifty-four vessels. Forty-seven of those were taken before they stopped at Nassau and seven were taken at Charles Town or on the way to Topsail Inlet. Disregarding the cargo those vessels carried, every one of them would have had a moderate supply of coins onboard that would be used to pay docking fees and taxes, and to buy supplies for the return trip, not to mention the money the passengers would have carried.

According to Captain Dubois' report, there was a significant amount of gold dust onboard *La Concorde* when it was taken. Wilde-Ramsing and Carnes-McNaughton estimate the total to weigh about twenty pounds.[9] This included fourteen ounces of powdered gold. But it doesn't include the five pounds of gold powder taken from M. Turgot.[10] Another French captain, whose vessel was taken near Saint Vincent, reported that Blackbeard's men took six thousand pounds sterling from an English ship.

> Half an hour before the privateer was taking an English ship anchored at Layou, another cove of Saint-Vincent and obliged the captain who had sent his crew ashore with all his money to go fourth on board. There, by dint of blows and threatening to have him hanged, he compelled him to send for his treasure which consisted of six thousand pounds sterling.[11]

Captain Christopher Taylor of the *Great Allen* reported that Blackbeard took his treasure which consisted of "silver and silverware, all for about eight thousand sterling livre."[12] That's worth about $152,000 in today's money.[13] In his deposition, Henry Bostock reported that "he believed they had much Gold Dust on board," and that he "heard about fifteen ounces of Gold Dust" that Blackbeard's men took from Robert McGill's vessel.[14] While cruising the Spanish waters off Mexico, the *Jamaican Dispatch* reported that the ship was "filled with much treasure."[15]

As for Blackbeard's pirates spending all their money ashore, that statement makes absolutely no sense. Blackbeard and Bonnet were continuously at sea from the time they left Nassau in September 1717 until they returned in May 1718. None of Blackbeard's vessels made a single stop at any port during that time. According to their theory, when they finally landed at Nassau, Blackbeard divided all the treasure into shares and issued everyone in the crew their percentage of the loot. The crew then spent everything in town. It's hard to conceptualize all of Blackbeard's crew spending all their money in just a few days, especially when the testimony of Bonnet's pirates states the opposite. They all said that they had been cheated out of their shares when the *Queen Anne's Revenge* sank. There was no mention of them receiving their shares in Nassau a few weeks earlier.

After leaving Charles Town, *The Boston News-Letter* reported that Blackbeard "plundred, and took all their Provisions, and some Rice, and about 4000 Pieces of Eight."[16] Richard Allein, the Attorney-General at Stede Bonnet's trial, estimated the amount of cash Blackbeard took at Charles Town at "about fifteen hundred Pounds Sterling, in Gold and Pieces of Eight."[17]

Now that the existence of a treasure on board Blackbeard's ship has been established, I offer one final comment concerning the believability of the story I have woven above in this chapter. If Blackbeard didn't smuggle his treasure off the *Queen Anne's Revenge* somewhere between leaving Nassau and the ship's sinking at Topsail Inlet, what happened to it?

I believe that after retrieving his treasure from Fripp Island, Blackbeard and his men returned to Ocracoke and then went on to Bath to accept the king's pardon. At some point, they divided the loot into shares. This event must have happened sometime before September 1718, because quartermaster William Howard was arrested in Kecoughtan (Hampton), Virginia, on September 16, 1718. His arrest was done on the orders of Lt. Governor Spotswood. At the time of his arrest, Howard had £50 in coins. That's about $5,000 in today's money.[18] If the value of the treasure was as high as I suspect, his share would be far greater than £50. Howard just took enough money to travel from Bath to Virginia. Apparently, Howard lived comfortably for the rest of his life. He purchased the island of Ocracoke in 1759. Many of the island's residents can trace their direct ancestry to him. Howard died in 1794, at 108 years old.[19]

By far, the most successful member of Blackbeard's crew was Edward Salter. He was arrested in Bath in November 1718 by Captain Brand, along with John Martin and several other members of Blackbeard's crew. He was tried for piracy in Williamsburg and then released in accordance with the extension of the proclamation. Salter returned to Bath and went into his trade as a cooper. In 1721, he first appeared in the Bath Deed records, when he bought two town lots from Henry Rowell. This purchase was witnessed by James Robins, the same man who owned the Beard property and who was on the *Adventure* when Blackbeard was killed and arrested by Maynard.[20] On November 12, 1726, Edward Salter purchased a 400-acre estate for £600. That property was the same estate that Governor Eden had owned in 1714, the same estate that was later known as the tunnel land

Chapter Thirty-Six

and rumored to be the place where Blackbeard may have buried his treasure. In addition to having a copy of the 1726 deed of purchase, Edward Moseley's 1733 Map shows Edward Salter as the owner of that property at Archbell Point.[21] When he died in 1734, he was the largest landowner in North Carolina. In his will, his occupation was listed as a cooper.[22] Where did a cooper get that kind of money?

Now for the part of this book that everyone has been waiting for since they read the first chapter. What became of Blackbeard's share? Where did he bury his treasure?

The first issue that must be tackled is the dispute about buried treasure itself. Many historians refute the idea that pirates buried treasure at all. These historians contend that pirates spent their money as fast as they stole it. They also contend that there are no source documents that support any pirates burying treasure except the well-known treasure that Captain Kidd had buried in New York. This, of course, is completely incorrect. There are source documents that describe pirates burying treasure, such as William Dampier's book, *A New Voyage Round the World*. Historians also overlook the fact that everyone buried treasure. There were no banks in colonial America and no safe place to store valuables. Farmers, plantation owners, and tradesmen in the cities concealed their valuables in secret compartments within their homes or buried somewhere on their property. Why should pirates be any different?

It was July 1718. Blackbeard and his crew were on Ocracoke. They had just returned from Fripp Island, where the treasure had been temporarily buried. Now it was time to divide the treasure into shares. As the captain, Blackbeard's share would have been far larger than the others. Unlike the usual tales of buried treasure portrayed in the media, Blackbeard wasn't planning on hiding everything in one place, only to be retrieved at some later date. Blackbeard was planning on diversifying his portfolio. Looking ahead, he knew that he would need available funds at each location where he planned on doing business. Without banks in which those funds could be deposited, Blackbeard would have to bury portions of his treasure in several locations. There are hundreds of places rumored to be the one and only spot where Blackbeard buried his treasure. This number can be significantly reduced by tracing Blackbeard's movements. Obviously, if Blackbeard hadn't sailed there after July 1718, he couldn't have buried any treasure there. This eliminates anywhere in the Caribbean, Nassau, Georgia, Florida, and New England. In fact, there are only three locations that Blackbeard was known to visit during that time: Bath, Ocracoke (including the surrounding rivers of Pamlico Sound), and Philadelphia.

Dampier's book, A New Voyage Round the World, was released in 1697 and was an autobiography recounting his journeys around the world as a member of a pirate crew.

Blackbeard's August 1718 visit to the Philadelphia area is documented in Chapter Nineteen. The more detailed descriptions come from historian author John R. Watson in his book *Annals of Philadelphia and Pennsylvania, Being a Collection of Memoirs, Anecdote, and Incidents of the City and Its Inhabitants from the Days of the Pilgrim Founders*. Watson mentioned Governor Keith's warrant for his arrest,[23] described Blackbeard's visit to High Street, and retold the "traditional story that Blackbeard and his crew used to visit and revel at Marcushook at the house of a Swedish woman,"[24] But that's not all Watson has to say about Blackbeard. He included several accounts of Blackbeard's treasure being buried in the Philadelphia area. The more interesting accounts include:

Philadelphia founded in 1682 by William Penn

> Schuylkill waters, that the pirates of Blackbeard's day had deposited treasure in the earth. The fancy was, that sometimes they killed a prisoner and interred him with it, to make his ghost keep his vigils there and guard it.[25]

> They believing that Blackbeard and his accomplices buried money and plate in numerous obscure places near the rivers; and sometimes, if the value was great, they killed a prisoner near it, so that his ghost might keep his vigils there and terrify those who might approach.[26]

> Robert Venables, the old black man who died in 1834, aged 98, told me that he knew personally an old black man, and Carr, a drayman, in Gray's alley, both of whom had been with Blackbeard. He had heard that Blackbeard had dealings with "Charles," the owner of a shallop packet to Burlington—who used, when about to start, to go around the little town, crying, "ho! Burlington, ho!" He supplied the pirate with flour, &c. Heard often of pirates' money. He knew that Murdock, Riley, Farrel and others, went to Point-no-point to dig—success not known—some said they were frightened off.[27]

These local tales of Blackbeard's treasure are very interesting, but they lack any real substance. Every inch of early Philadelphia has been excavated many times since 1718 for the construction of new buildings. If anyone found Blackbeard's buried treasure at any of those locations, they didn't report it. Watson's accounts do confirm that just two generations after his visit in 1718, the local population of Philadelphia held a strong belief that Blackbeard buried some of the treasure there. The last account from Watson's 1850 edition mentions the location of Burlington, New Jersey. That's where we are headed next.

Chapter Thirty-Six

Burlington was first settled by Quakers in 1677 and was the capital of the province until 1702. The oldest building still standing in town was built in 1685.

Burlington was an important city in 1718 when Blackbeard was in Philadelphia. It's just fifteen miles up the river and Blackbeard could have easily sailed there in a few hours. Two exceptionally intriguing stories appeared in the *New York Times* in October 1926. These articles focus on Miss Florence E. Steward and the legend that Blackbeard's treasure was buried on her 217 Wood Street property. The first article was in the Friday, October 8, 1926 issue and was a short item buried on page 25.[28] The next day, Saturday, October 9, 1926, a follow-on feature article appeared on page one.[29]

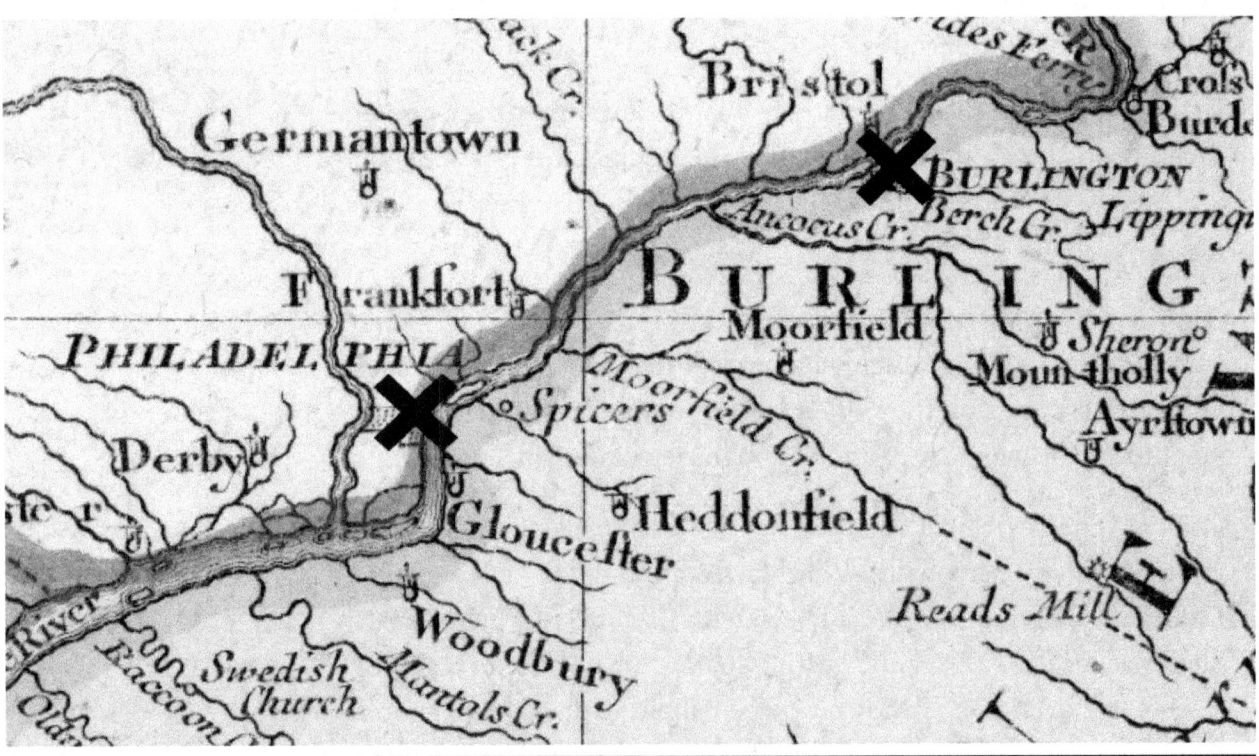

Figure 85: *Close-up of Philadelphia to Burlington from Eighteenth Century Map. The image is dedicated to the public domain under CC0.*

The property had been in Miss Florence E. Steward's family for several generations. In her backyard stood an old black walnut tree. There was a very old legend told by locals and Florence's family members that Blackbeard used the tree as a marker and had buried his treasure at its base. Known by then as the "Pirate Tree," the legend also contends that Blackbeard buried a Spaniard on top of the treasure whose spirit would serve as a guard. This tale seemed feasible, as the tree was only about two hundred yards from the river's edge. It was close enough for Blackbeard to carry his treasure, yet far enough away from those traveling on the river to see him burying it.

Florence had heard this story many times but never attempted to dig for it. That changed when she decided to sell her home. She began thinking about the legend more and more, and came to realize that if it were true, she may lose out on a fortune. On October 7, 1926, she hired some workmen to dig up her yard. Apparently, she didn't watch them too closely. In the afternoon, Florence noticed that the workmen had gone, leaving a large hole in her property. Neighbors told her that they had seen these workmen loading a heavy object into their truck before they left. Some inquisitive schoolchildren found a skull and some bone fragments in a corner of the excavation. The police were called in and a fence was placed around her property.

These two thought-provoking articles raise some interesting questions. Did those workmen find part of Blackbeard's treasure buried at the base of the old black walnut tree? If they did, they didn't tell anyone. It's amusing to imagine two impoverished New Jersey laborers suddenly discovering enormous wealth and living out the rest of their lives in luxury somewhere in Argentina. As for the skull found at the digging site, most buried treasure pirate legends include a dead body buried along with the treasure. The police didn't do any forensic examination of the skull. According to the article, it was believed that the skull was part of an Indian burial site from a time before European occupation. I believe that there is a strong possibility that Blackbeard did bury part of his treasure under the black walnut tree that later was owned by Miss Florence E. Steward and that this treasure has already been found.

At first, Ocracoke seems like the best place to look for Blackbeard's treasure. He had established a base there and spent much of his time either on the beach or sailing nearby. In reality, it is a terrible place to look for Blackbeard's treasure. Blackbeard was burying his funds at locations where he intended to use them for business. There was nothing to buy on Ocracoke, no merchandise, not even land. Deeds for land purchases were recorded at Bath. Consequently, the financial transactions also took place in Bath, the business center of the province. But just to be thorough, let's take a look at Ocracoke, anyway.

Ocracoke as a prospective spot for treasure

After the battle of Ocracoke on November 22, 1718, Lt. Maynard found some gold dust and other "small things of plunder" on Blackbeard's sloop *Adventure*. Maynard remained there for three weeks, searching for anything of value. Other than the sugar and cocoa taken from the *Rose Emelye*, Maynard found nothing. Those facts were mentioned in Captain Gordon's letter to the admiralty.[30] But Maynard didn't search the entire island. Perhaps Blackbeard or even a few of his crewmen buried part of their shares on Ocracoke Island. If they did, William Howard dug it up after he purchased the entire island in 1759.

Chapter Thirty-Six

This leaves us with the most plausible spot for Blackbeard's buried treasure, Bath, North Carolina. Not only was Bath the capital of the province and the center of all business and trade, but it was also where Blackbeard stored part of his sugar for sale at a later date. This was mentioned by Tobias Knight who wrote, "the said sugars was there lodged at the request of the said Thache only til a more Convenient store could be procured."[31] In Chapter Thirty-Five, the legend that Blackbeard hid his treasure on Eaden's tunnel land was thoroughly discussed and rejected.

The Plum Point legend persists today as the strongest contender for the coveted spot of the location where Blackbeard buried his treasure. It's an ideal spot for Blackbeard to keep a stash of money. Plum Point sits literally across the creek from Tobias Knight's home, where Blackbeard used to conduct most of his transactions. It also was located less than two miles from the center of Bath Town, the mercantile center of the province. Although Plum Point was right in the center of everything, it was still remote. The property was unoccupied and wasn't easily accessible from the land. The terrain consisted of an almost impenetrable swamp covered with dense trees and tangled underbrush. A place like that was perfect.

As discussed in Chapter Twenty, Plum Point is also the place where it is widely believed that Blackbeard careened his sloop. A round brick oven, known as Teache's Kettle, was on the property and was supposedly used by Blackbeard to boil tar.[32] While waiting for the tar to dry, there would have been plenty of time for Blackbeard to secretly construct some sort of brick vault to conceal his treasure. If such a vault had been built, wouldn't it have been found by now?

Without a doubt, the most fascinating article concerning Blackbeard's treasure ever to be published was the one that appeared in the February 3, 1929 edition of *The News and Observer*, written by Ben Dixon MacNeil and titled, *Blackbeard's Buried Treasure Found at Last But Mystery of Pirate Gold Not Yet Solved*.[33] Within the article, MacNeil recounts the history of treasure hunting on Plum Point, writing that the digging on Plum Point probably began the day after Blackbeard was killed. "Heaps of earth have been thrown up and deep pits have been sunk into the ground. The place looks like a section of shell-torn battlefield." That part of the article is just background. None of those treasure hunters ever found anything. The main focus of the article is on the ones who did.

The story begins on Christmas 1928, with two men who were wandering through the brush inspecting duck blinds. Apparently, Plum Point was a terrific spot for duck hunting. Unexpectedly, they accidentally came upon many bricks scattered about in the undergrowth. This sparked their

The News & Observer is a news publication in Raleigh, North Carolina, that traces its roots to *The Sentinel*, which was founded in 1865.

curiosity. With no nearby brick structures, they wondered where these bricks came from. A few steps later, they discovered the source of the bricks and their curiosity instantly turned into astonishment. The two men had come to the top of an eight-foot-deep hole that had recently been dug through the sandy topsoil. At the bottom of this hole was a brick vault, which had been broken into. It was immediately obvious that the bricks they had first seen scattered around the area had been the vault's roof. From the fresh dirt around the hole and the untethered condition of the bricks and vault, they deduced that this only could have been unearthed within the last two days.

The vault on Blackbeard's Plum Point location

The top of the vault was about five feet below the level ground. Lowering themselves to the vault itself, they examined this unusual phenomenon closely. The walls of the vault were one foot thick and the inside dimensions were three feet by one and a half feet. One end and the top of the little vault had been broken away entirely, and the bricks had been tossed up out of the excavation. The bricks themselves appeared to be very old. They were similar to the ones used to construct the old church in Bath. The most fascinating part of the vault was the floor. They saw an imprint in the brick floor which appeared to have been made by a wooden chest with iron hoops studded with bolts and placed in the vault while the mortal was still wet and pliable. The imprints of the hoops and bolts were plainly marked, and they saw a lot of rust, which had accumulated over the years.

Once the shock of finding the vault wore off, they began inspecting the surface of the ground near the top of the hole. They saw marks and footprints in the sand and deduced that a heavy object had been dragged to the water's edge and loaded into a boat. In the days that followed, locals recalled seeing three men in a boat about the time the chest would have been taken. Nobody recognized them. Since many duck hunters come to Plum Point, the presence of these three men wasn't considered unusual. The most surprising aspect of this story is that there was only one hole, meaning that these men knew exactly where to dig. This was decades before the invention of metal detectors. MacNeil closed his article with a quote from Charlie Waters, one of the two men who found the broken vault. Charlie said, "They must have known where it was before they came here. It don't look like they had to hunt any for it."

St. Thomas Episcopal Church in historic Bath was built in 1734.

Figure 86: *Illustration of the Vault on Plum Point*

Chapter Thirty-Six

In conclusion, I believe this story is real, and that this was indeed one of the spots where Blackbeard concealed part of his treasure. This vault on Plum Point fits the description of the ideal hiding place faultlessly. All the details align too well for it to be anything else. The location is perfect, Blackbeard was there and had the opportunity to construct the vault, and it is precisely the type of hiding place that Blackbeard would have designed. It was protected from the weather and soil erosion, and it wasn't visible to anyone who didn't know where to look. Unfortunately, the treasure was already found in 1928. This brings us back to the three mysterious men in a rowboat. How did they know where to look? The only logical explanation is that they must have uncovered some sort of document that described the vault's location.

It's ironic that the story of Blackbeard's treasure would end this way. Hundreds of the legends of Blackbeard's treasure have been examined, and some of the more popular ones have been debunked. Philadelphia and Burlington both seem probable, but due to development over the centuries, no traces remain. The one story that seems to be the most plausible, the Plum Point story, ends with a most unexpected plot twist. If the three men who recovered his treasure were using a document of sorts, it certainly could be classified as a treasure map. So, it would appear that the most improbably Hollywood-type ending to the search for Blackbeard's lost treasure might just have been the way it happened. It was found with a treasure map!

The End

APPENDIX

Summary of Pirate Activity and Vessels taken

November–December 1715: Florida Coast, Wreck Site
Benjamin Hornigold: **Sloop *Mary***
- Took Spanish Sloop – November 1715 – Brought it to Nassau – Taken away from him by Henry Jennings
- Took Spanish Sloop – December 1715 – Kept it and renamed it *Benjamin*, sending the Sloop *Mary* back to its original owner on Jamaica

December 1715–January 1716: Florida Coast
Henry Jennings: Sloop *Bersheba* | John Will: Sloop *Eagle*
- Raided Spanish Salvage Camp – December 27, 1715 – Took 120,000 pieces of eight
- Took Spanish Ship near Havana – Looted and released

January–February 1716: Bay of Honduras
Samuel Bellamy & Palsgrave Williams
- Arrived at Florida Wrecks – January 1716
- Took Dutch Ship – Capt. Cornelison – Forced Peter Cornelius Hoof to join – Looted and released
- Took English Sloop – Capt. Young – Kept it and headed north to west end of Cuba

February–March 1716: Cuba to Bahia Honda
Jennings's Fleet:
- Henry Jennings on Sloop *Bersheba*
- Leigh Ashworth on Sloop *Mary*
- Samuel Liddell on *Sloop Cocoa Nut*

- ☠ James Carnegie on Sloop *Discovery* was expected but hadn't arrived yet
- ☠ Took English Sloop – Capt. Young – Bellamy and pirates onboard – They escaped in a periauger – Jennings kept the sloop

Bellamy: using a small Periauger
- ☠ Took French Ship *Mary of Rochell* – Capt. D'Escoubes – Taken by Bellamy in his periauger while Jennings' men debated attacking and watched Bellamy
- ☠ *Mary of Rochell* held prize with Bellamy and his men onboard
- ☠ James Carnegie on Sloop *Discovery* who had just arrived, left to take the French Sloop *Marianne*
- ☠ Samuel Liddell on Sloop *Cocoa Nut* left in disgust

Hornigold: Sloop *Benjamin*
- ☠ Took French Sloop *Marianne* – Ensign le Gardew – Kept it
- ☠ *Marianne* & *Benjamin* chased by Jennings but he couldn't catch them
- ☠ Bellamy took all the treasure from the *Mary of Rochell* and escaped in his periauger
- ☠ Carnegie kept *Mary of Rochell* and gave his Sloop *Discovery* to the French crew
- ☠ Capt. Young's English Sloop was burned
- ☠ Hornigold joined with Bellamy and gave him command of the Sloop *Marianne*

March–November 1716: Waters Around Cuba

Hornigold: Sloop *Benjamin* | Bellamy: Sloop *Marianne*
- ☠ Took English Pink near Havana – Looted and released
- ☠ Joined with the French pirate, Olivier Levasseur (La Buse) on Sloop *Postillion*

Hornigold: Sloop *Benjamin* | Bellamy: Sloop *Marianne* | La Buse: Sloop *Postillion*
- ☠ Took Unidentified Vessels along south coast of Cuba
- ☠ Took English Ship near Havana – Looted and released
- ☠ Took two Spanish Brigantines off Cabo Corrientes – Burned when the Spanish refused to pay ransom – Crew put ashore
- ☠ Took three or four English Sloops at Isle of Pines – Used to careen then released
- ☠ Hornigold sold Sloop *Benjamin* to Capt. Perrin – No information on what replaced it
- ☠ Took unknown number of vessels
- ☠ On request of the others, Hornigold and 26 men left on a small prize sloop
- ☠ Bellamy assumed command of Hornigold's vessel
- ☠ Palsgrave Williams given command of *Marianne*

November–December 1716: North of Hispaniola

Hornigold: Sloop *Delight* | Blackbeard: Unnamed Sloop

- ☠ Took Ship from Bristol – Capt. John Quarry – November 30, 1716 – Looted and released
- ☠ Took Spanish Ship of 40 Guns – December 6, 1716 – Looted and released
- ☠ Took Brigantine *Lamb* eight leagues off of Cape Donna – December 13, 1716 – Capt. Henry Timberlake – Looted and released

March–June 1717: Florida Cape & Caribbean

Hornigold: Sloop *Adventure* (William Howard as QM)

- ☠ Took a Snow from Jamaica – Capt. Blake - March 1717 – Forced Dr. Howell to stay
- ☠ Joined with Captain Napping

Hornigold: Sloop *Adventure* (William Howard as QM) | Napping on unidentified vessel

- ☠ Took Sloop *Bennet* near Friends Islands (Portobelo) – Capt. Hickinbottom – April 1, 1717 – Found chest with 400,000 pieces of eight – Hornigold kept it and gave the *Adventure* to the Bennet's crew

Hornigold: Sloop *Bennet* (William Howard QM) | Napping on unidentified vessel

- ☠ Took Dutch Ship – No further Information
- ☠ Took Dutch Ship – No further Information
- ☠ Took Sloop *Revenge* – April 7, 1717

June–August 1717: Florida Cape & Virginia

Hornigold: Sloop *Bennet* | Blackbeard: Unidentified Sloop

- ☠ Took Sloop from Havana – Capt. Billop – Looted and released
- ☠ Took Sloop from Bermuda – Capt. Thurbar – Looted and released
- ☠ Took Vessel from St. Lucie – Capt. Pritchard – Looted and released
- ☠ Took Ship from London – Kept sails and rigging – Released
- ☠ Careened at Accomack

August 1717: Charles Town & Florida

Bonnet: Sloop *Revenge* (beginning his career as a pirate)

- ☠ Took Brigantine at Charles Town – Capt. Thomas Porter – August 26, 1717 – Vessel taken to Cape Fear – Stripped of rigging, cables, sails, and anchor – Released with passengers from the other sloop
- ☠ Took Sloop at Charles Town – Capt. Joseph Palmer – August 26, 1717 – Crew released on brigantine – used to careen *Revenge* then burned
- ☠ Attacked by Spanish Man-O-War along Florida – *Revenge* damaged – Many killed – Bonnet wounded

September–November 1717: Virginia, Cape May, New York

Blackbeard: Sloop *Revenge* (Bonnet onboard with no command)
- ☠ Took Sloop *Betty* near Cape Charles – Kept pipes of Madera wine—Sunk afterward
- ☠ Took Ship *Good Intent* at Cape May – Capt. Codd – 150 Passengers & 1,000 pounds of cargo – Cargo thrown overboard & Ship released
- ☠ Took Sloop *Robert* Out of Philadelphia at Cape May – Looted and released
- ☠ Took Snow *Sea Nymph* at Cape May – Capt. Budger – Loaded with wheat – Cargo thrown overboard – Kept as a pirate vessel for a short time – Crew released on Spofford's snow
- ☠ Took Snow at Cape May – Capt. Spofford – Loaden with 1,000 Staves – Cargo thrown overboard – Released with crew from *Sea Nymph*

Blackbeard: Sloop *Revenge* | Consort Snow *Sea Nymph*
- ☠ Took Sloop at Cape May – Capt. Peter Peters – 27 Pipes of Wine – Cut his masts at the deck and drove ashore with crew stranded onboard
- ☠ Took Sloop at Cape May – Capt. Grigg – Above 30 Servants – Looted and cut away masts and left it at anchor with the crew onboard
- ☠ Took Sloop from Madera to VA at Cape May – Two Pipes of Wine – Sunk
- ☠ Took Sloop from Antigua at Cape May – Put servants and prisoners from other sloop on board and released
- ☠ Took Ship from Philadelphia at Cape May – Capt. Tover – Looted and released
- ☠ Took Sloop at New York – Capt. Farmer – Took the mast, anchors, cables, and money and put some captured servants on board then set adrift at Sandy Hook
- ☠ Took Great Sloop at New York – Capt. Sipkins – Kept it as a pirate vessel – Mounted 12 guns

Blackbeard: Sloop *Revenge* with consorts *Sea Nymph* & *Great Sloop*
- ☠ Took Large Sloop from Jamaica at New York – Capt. Rolland – Looted and released
- ☠ Took Unidentified Vessel from New England at New York
- ☠ Took Sloop at New York – Capt. Goelet – Half loaden with Cocoa – Cargo thrown overboard and kept as a pirate vessel – Crew allowed to go in *Sea Nymph*
- ☠ *Sea Nymph* released with crew from Goelet's sloop

Blackbeard: Sloop *Revenge* with consorts *Great Sloop* & *Goelet's Sloop*
- ☠ Took Ship at New York – No additional information known
- ☠ Took Brigantine or snow at New York – No additional information known
- ☠ Goelet's sloop discarded after leaving New York

November–December 1717: Antilles

Blackbeard: Sloop *Revenge* with consort *Great Sloop*
- ☠ Took *La Concorde* east of Martinique – Capt. Dosset – November 17, 1717 – 455 slaves & gold dust – taken to Bequia Island to refit – Kept and named *Queen Anne's Revenge*

Bonnet: Sloop *Revenge* (Captain once again) operating while ship is being refitted
- ☠ Took unnamed English vessel at Layou, St. Vincent – November 23 – Burned
- ☠ Took *Dauphin* at Layou, St. Vincent – Capt. Oliver – November 23 – Burned
- ☠ Took French vessel – Capt. Charles St. Amour – November 24 – Held for ransom and released
- ☠ Chased French sloop – November 25 – Escaped
- ☠ *Great Sloop* released with French crew and slaves

Blackbeard: Ship *Queen Anne's Revenge* | Bonnet: Sloop *Revenge*
- ☠ Left Bequia – November 26, 1717
- ☠ Chased French sloop – November 26 – Escaped
- ☠ Took Ship *Great Allen* near Layou, St. Vincent – Capt. Christopher Taylor – November 26 – Much silver and silverware – Ship burned and crew given longboat
- ☠ Attacked Guadeloupe Harbor – November 29 – Burned half the town
- ☠ Took Ship *La Ville de Nantes* at Guadeloupe Harbor – Capt. Sieur Lemaître – November 29 – Kept it temporarily

Blackbeard: Ship *Queen Anne's Revenge* | Bonnet: Sloop *Revenge* | *La Ville de Nantes*
- ☠ Stopped the *Montserrat Merchant* near Guadeloupe – Capt. Hobhouse – November 29 – Ship escaped – Ship's officer Thomas Knight captured and detained
- ☠ Saw an unidentified ship near Nevis Island – Pirates believed it to be HMS *Seaford* – Blackbeard talked them out of an attack
- ☠ Took *New Division* near Nevis Island – Capt. Richard Joy – November 30 – Vessel released along with Thomas Knight
- ☠ Attacked Sandy Point Harbor –December 1 – Used *La Ville de Nantes* as a fire ship
- ☠ Took unidentified vessels at anchor at Sandy Point – Looted and burned
- ☠ Took unidentified vessels at anchor at Sandy Point – Looted and burned
- ☠ Possibly Took unidentified vessels at anchor at Sandy Point – Looted and burned

Blackbeard: Ship *Queen Anne's Revenge* | Bonnet: Sloop *Revenge*
- ☠ Sighted HMS *Seaford* near St. Thomas – December 3 – Ship was avoided
- ☠ Took *Antigua Sloop* near St. Croix – Owner Robert McGill – December 4 – Looted of supplies and guns then burned – crew released on Danish sloop
- ☠ Took Danish Sloop near St. Croix – December 4 – Looted and released with crew from Antigua sloop

- ☠ Took Sloop *Margaret* near Crab Key east of Puerto Rico – Henry Bostock – December 5 – Looted cattle and hogs, gunpowder, five small arms, two cutlasses, books, navigational instruments, linen – Edward Salter forced to join – Sloop and crew released
- ☠ Took *La Volante* at Boca Chica – Capt. Jean Bleu Nesbayes, owner Jean Morange – December 1717 – Crew put ashore and vessel kept
- ☠ Careened and refitted Samana Bay, Hispaniola

January–May 1718: Hispaniola to the Gulf of Honduras

Blackbeard: Ship *Queen Anne's Revenge* | Bonnet: Sloop *Revenge* | *La Volante*
- ☠ Took *Le Roy de Guillawme* near Petit-Goave – Temporarily kept – Fate unknown

Blackbeard: Ship *Queen Anne's Revenge* | Bonnet: Sloop *Revenge* | *La Volante* or *Le Roy de Guillawme*
- ☠ Took unidentified English vessel – Kept for a while
- ☠ Sighted in Gulf of Mexico – "Four sloops and a ship of 42-guns" reported in February 1718

Bonnet: Sloop *Revenge* (sailing on his own to south Gulf of Honduras)
- ☠ Failed attempted capture of Ship of 26 guns *Protestant Caesar* – Capt. William Wyer – March 28, 1718 – Bonnet relieved of command and replaced by Richards

Blackbeard: Ship *Queen Anne's Revenge* | Richards: Sloop *Revenge* | one sloop called the Prize (Possibly *Le Roy de Guillawme* or the English sloop he took)
- ☠ Took Sloop *Adventure* at Turneffe Islands – Captain David Herriot – April 4, 1718 – Kept as a pirate vessel – David Herriot remained with pirates and was tried in November 1718
- ☠ Took Sloop *Land of Promise* taken at Turneffe Islands – Captain Thomas Newton – April 5, 1718 – looted and temporarily kept
- ☠ Pirate fleet headed south to Honduras Bay – Prize sloop abandoned

Blackbeard: Ship *Queen Anne's Revenge* | Richards: Sloop *Revenge* | Hands: Sloop *Adventure* | Sloop *Land of Promise*
- ☠ Took Sloop *Dolphin* between Turneffe and Honduras Bay – Captain James Burchett – April 5 or 6, 1718 – Temporarily kept, then released

Blackbeard's Fleet: *Queen Anne's Revenge*, *Revenge*, *Adventure*, *Dolphin*, and *Land of Promise* taking vessels at Honduras Bay
- ☠ Took Ship of 26 guns *Protestant Caesar* – Capt. William Wyer – April 8, 1718 – Looted and burned three days later – Captain and crew held prisoner
- ☠ Took *Sloop of Jamaica* – Owner and Capt. Jonathan Bernard – April 10, 1718 – Looted and released
- ☠ Took *Sloop William and Samuel* – Captain William Wade – April 10, 1718 – Looted and released
- ☠ Took *Sloop of Jamaica* – Capt. James – April 10, 1718 – Looted and burned

- ☠ Richards on pirate Sloop *Revenge* chased merchant Sloop *Revenge* – Captain William Megerre – Merchant Sloop *Revenge* outran the pirate and escaped
- ☠ Released Sloop *Land of Promise* with Capt. Wyer and the crew of *Protestant Caesar*
- ☠ Released Sloop *Dolphin*
- ☠ Sailed to Turks Islands then to Cayman Islands

Blackbeard: Ship *Queen Anne's Revenge* | Richards: Sloop *Revenge* | Hands: Sloop *Adventure*
- ☠ Took Small Turtler – looted and released
- ☠ Took Spanish Sloop – Kept it – Will become Blackbeard's sloop *Adventure*

May–June 1718: Charles Town to Topsail Inlet

Blackbeard: Ship *Queen Anne's Revenge* | Richards: Sloop *Revenge* | Hands: Sloop *Adventure* | Small Spanish Sloop
- ☠ Arrived outside the bar on May 22, 1718
- ☠ Took Ship *Crowley* from London – Capt. Robert Clarke – Mr. Wraggs and Mr. Marks taken prisoner along with others – Held for ransom for a Chest of Medicine – Looted and Released
- ☠ Took Pilot Boat
- ☠ Took Ship *Ruby* from Charles Town – Capt. James Craigh – Looted and Released
- ☠ Took Ship *William* from Weymouth – Capt. Hewes – Looted and Released
- ☠ Took Ship *Artimesia* from London – Capt. John Dornford – Looted and Released
- ☠ Took Ship *William* from Philadelphia – Capt. Thomas Hurst – Looted and Released
- ☠ Took Brigantine *Princess* from Bristol on way to Topsail – Capt. John Redford – Arrived from Angola with a cargo of 86 slaves – Looted, kept 14 slaves, and released
- ☠ Richards Took Ship *William* from Boston on way to Topsail – Capt. Nathaniel Mason – Looted and released – Blackbeard upset with Richards that he didn't burn it

June 1718: Topsail Inlet
- ☠ *Queen Anne's Revenge* – Capt. Blackbeard – Sank
- ☠ *Adventure* – Capt. Howard & Hands – Sank
- ☠ Blackbeard claims Spanish sloop and names it the *Adventure*
- ☠ Stede Bonnet leaves with *Revenge* and renames it the *Royal James*

July–August 1718: Virginia Capes to Delaware Capes

Bonnet: Sloop *Royal James*
- ☠ Took Pink off Nags Head – looted ten or twelve barrels of pork, and about four hundred weight of bread – Released
- ☠ Took Sloop off Cape Henry bound for Bermuda - Released
- ☠ Took another Sloop off Cape Henry bound for Bermuda - Released

- ☠ Took Sloop off Cape Henry – Looted for provisions and released
- ☠ Took Sloop off Cape Henry – Captured at night – Released
- ☠ Took Ship off Cape Henry – Released
- ☠ Took Snow off Cape Henry – Released
- ☠ Took another Snow off Cape Henry – Released
- ☠ Took Ship bound for Glasgow off Cape Henry – Looted for provisions and tobacco and released
- ☠ Took another Ship bound for Glasgow off Cape Henry – Looted for provisions and released
- ☠ Took Sloop bound from Virginia to Bermudas – Eight pirates put on board, afterwards, they escaped with it
- ☠ Took a third Ship bound from Virginia to Glasgow
- ☠ Took Schooner bound from North Carolina to Boston on way to Cape May, Lat. 38 off Assateague Island – July 28, 1718 – Kept for three days
- ☠ Took Snow at Delaware Bay, 39 degrees – July 29, 1718 – Released
- ☠ Took second Snow at Delaware Bay, 39 degrees – July 29, 1718 – Released
- ☠ Took Sloop bound from Philadelphia to Barbados at Delaware Bay, 39 degrees – July 29, 1718 – Released
- ☠ Took Sloop *Fortune* bound from Philadelphia to Barbados – Captain Thomas Read – July 30, 1718 – Kept and was taken to Cape Fear along with the captain and crew
- ☠ Took Sloop *Francis* bound from Antigua to Philadelphia at Cape Henlopen – Captain Peter Manwareing – July 30, 1718 – Kept and taken to Cape Fear along with the captain and crew
- ☠ Looted the Schoner of pitch and tar – Abandoned
- ☠ Left Cape Henlopen for Cape Fear taking *Fortune* & *Francis* with them on August 1, 1718
- ☠ Arrived at Cape Fear - Began Careening on August 12, 1718

August–September 1718: Pennsylvania, Bermuda, Ocracoke
Blackbeard: Sloop *Adventure*
- ☠ Sailed to Philadelphia July–August 1718
- ☠ Took French Ship *La Toison d'Or* – August 23, 1718 – Captain Eslye Wansbabel
- ☠ Took French Ship *La Rose Emelye* – August 23, 1718 – Captain Jan Geropil
- ☠ Both crews released on *La Toison d'Or*
- ☠ *La Rose Emelye* taken to Ocracoke, looted and burned

September 29, 1718 – Battle of Cape Fear – Bonnet and crew captured

Bonnet: Sloop *Royal James* with prize sloops *Fortune* & *Francis*
- Colonel William Rhett's sloops:
    - Sloop *Henry*, 8 guns – Capt. John Masters (Rhett on board)
    - Sloop *Sea Nymph*, 8 guns – Capt. Fayer Hall
- *Royal James, Fortune*, & *Francis* taken to Charles Town – arrived October 3, 1718

Blackbeard: Sloop *Adventure* Joined by Charles Vane
- October – Party on Ocracoke between the crews of Blackbeard & Vane

Bonnet's Trial in Charles Town
- October 24, 1718 – Bonnet & Herriot escape
- October 28, 1718 – Trial for Bonnet's men began
- November 5, 1718 – Bonnet caught and Herriot killed on Sullivan's Island
- November 11, 1718 – Bonnet found guilty

Blackbeard: Sloop *Adventure* at Ocracoke attacked by Lt. Maynard – November 22, 1718
- Lt. Robert Maynard's sloops:
    - Sloop *Jane*, no guns – Lt. Maynard
    - Sloop *Ranger*, no guns – Midshipman Hyde
- Battle of Ocracoke – Blackbeard killed – *Adventure* taken to Virginia and sold at auction

END NOTES

Chapter 1 The Treasure Hunt Begins Notes:
1. Johnson, *A General History of the Pyrates,* First Edition, 1724, 100-1.

Chapter 2. By Any Other Name Notes:
1. Johnson, *A General History of the Pyrates,* First Edition, 1724, 86.

2. Johnson, *A General History of the Pyrates,* Second Edition, 1724, 70.

3. Timberlake, *Deposition, 17 Dec 1716.*

4. McIlwaine, *Executive Journals of the Colonial Council,* March 11, 1719, 495-6.

5. Duffus, *The Last Days of Black Beard the Pirate,* 169.

6. Brand, *Account of taking Blackbeard,* February 6, 1719.

7. Spotswood, *Capt. Tach's Quartermaster,* Dec. 22, 1718.

8. *Order to Charge William Howard with Piracy,* Oct. 29, 1719.

9. Headlam, *Calendar 1716-1717,* 338.

10. Cracherode, *Report on petitions.*

11. Gordon to Admiralty September 14, 1721, *Report.*

12. *The Boston News-Letter,* Monday, October 28 to Monday, November 4, 1717.

13. Bialuschewski, *Blackbeard off Philadelphia,* 168.

14. Bialuschewski, *Blackbeard off Philadelphia,* 167-8.

15. Ducoin, *Research on the Nantes Ship La Concord,* 5.

16. Brooks, *Quest for Blackbeard,* 148.

17. Pearse, *Account of Phenix at New York,* 10.

18. Brooks, *Quest for Blackbeard,* 128.

19. *Minutes of NC Governor's Council,* May 27, 1719, 180.

Chapter 3. Jacobite Pirates Notes:
1. Bonnet, *The Tryals of Major Stede Bonnet and Other Pirates,* 9.

2. Bonnet, *The Tryals of Major Stede Bonnet and Other Pirates,* 13.

3. Woodard, *The Republic of Pirates*, 196.
4. Cook, *Kings and Queens of England*, 74.
5. Cook, *Kings and Queens of England*, 86.
6. Cook, *Kings and Queens of England*, 93.
7. Cook, *Kings and Queens of England*, 94.
8. Cook, *Kings and Queens of England*, 95.
9. Ibid.
10. *Jacobite Timeline*.
11. Cook, *Kings and Queens of England*, 97.
12. Cook, *Kings and Queens of England*, 104.
13. Barthorp, *The Jacobite Rebellions 1689-1745*, 6.
14. O Ciardha, *Cammock*.
15. *Jacobite Rebellion of 1715*.
16. Barthorp, *The Jacobite Rebellions 1689-1745*, 7.
17. O Ciardha, *Cammock*.
18. Barthorp, *The Jacobite Rebellions 1689-1745*, 7.
19. Lewis, *George Camocke's 1718 Proposal of a Jacobite*.
20. Craton, *A History of the Bahamas*, 100.
21. Ibid.
22. Lewis, *George Camocke's 1718 Proposal of a Jacobite*.

Chapter 4. Charles Johnson and the Beginning of Legends Notes:

1. Bialuschewski, *Daniel Defoe, Nathaniel Mist*, 35.
2. Johnson, *A General History of the Pyrates*, Second Edition 1724, 71.
3. Woodard, *The Republic of Pirates*, 222.
4. Johnson, *A General History of the Pyrates*, Second Edition 1724, 75–76.
5. Brand, *Account of taking Blackbeard*, February 6, 1719.
6. Johnson, *A General History of the Pyrates*, Second Edition 1724, 85.
7. Spotswood, *Capt. Tach's Quartermaster*, Dec. 22, 1718.
8. Bialuschewski, *Daniel Defoe, Nathaniel Mist*, 35.
9. Johnson, *A General History of the Pyrates*, First Edition 1724, 60–100.
10. Johnson, *A General History of the Pyrates*, Second Edition 1724, 70–112.
11. Johnson, *A General History of the Pyrates*, First Edition 1724, 99.
12. Bialuschewski, *Daniel Defoe, Nathaniel Mist*, 21–22.
13. Bialuschewski, *Daniel Defoe, Nathaniel Mist*, 22–25.
14. Markham, *Nathaniel Mist's Journalistic Methods*, 7.

15. Markham, *Nathaniel Mist's Journalistic Methods*, 11.

16. Markham, *Nathaniel Mist's Journalistic Methods*, 42–43.

17. Bialuschewski, *Daniel Defoe, Nathaniel Mist*, 31–32.

18. Johnson, *A General History of the Pyrates*, First Edition 1724, A-1.

19. Bialuschewski, *Daniel Defoe, Nathaniel Mist*, 33.

20. Bialuschewski, *Daniel Defoe, Nathaniel Mist*, 26.

21. Johnson, *A General History of the Pyrates*, Second Edition 1724, A-1.

22. Reiche, *Digitizing the Pyrates*, 7.

23. Ibid.

24. Johnson, *A General History of the Pyrates*, Fourth Edition 1726, A-1.

25. Bialuschewski, *Daniel Defoe, Nathaniel Mist*, 24.

26. Reiche, *Digitizing the Pyrates*, 8.

27. Johnson, *A General History of the Lives and Adventures*, 1734, 203.

28. Johnson, *A General History of the Lives and Adventures*, 1736.

29. Bialuschewski, *Daniel Defoe, Nathaniel Mist*, 37.

Chapter 5. Wealth From the Sea: The 1715 Treasure Fleet Notes:

1. Burgess, *Florida's Golden Galleons*, Appendix D.

2. De Bry, *A Concise History of the 1715 Spanish Plate Fleet*.

3. Ibid.

4. Burgess, *Florida's Golden Galleons*, 40–41.

5. Combs, *Lost Ships of the 1715 Spanish Treasure Fleet*.

6. De Bry, *A Concise History of the 1715 Spanish Plate Fleet*.

7. Ibid.

8. Ibid.

9. Ibid.

10. Burgess, *Florida's Golden Galleons*, 65.

11. De Bry, *A Concise History of the 1715 Spanish Plate Fleet*.

12. *The Boston News-Letter*, Monday, December 26, 1715 to Monday, January 2, 1716.

13. Headlam, *Calendar 1716–1717*, 80.

14. Woodard, *The Republic of Pirates*, 111.

15. Headlam, *Calendar 1716–1717*, 79.

16 Hamilton, *Articles exhibited against Lord Archibald Hamilton*, 15.

17. Headlam, *Calendar 1716–1717*, 79.

18. Hamilton, *Articles exhibited against Lord Archibald Hamilton*, 3.

19. Headlam, *Calendar 1716–1717*, 78–80.

20. Hamilton, *Articles Exhibited against Lord Archibald Hamilton*, V.
21. Hamilton, *Articles Exhibited against Lord Archibald Hamilton*, 3.
22. Ibid.
23. Hamilton, *Articles Exhibited against Lord Archibald Hamilton*, 19.
24. Woodard, *The Republic of Pirates*, 123.
25. Clifford, *Real Pirates*, 42.
26. Woodard, *The Republic of Pirates*, 124.
27. Woodard, *The Republic of Pirates*, 125.
28. *The Trials of eight persons indited for piracy*, 24.
29. Headlam, *Calendar 1716–1717*, 213.
30. Woodard, *The Republic of Pirates*, 130.
31. Headlam, *Calendar 1716–1717*, 165–6.
32. *Deposition of Samuel Liddell*.
33. Headlam, *Calendar 1716–1717*, 212–3.
34. *Deposition of Allen Bernard*.
35. Headlam, *Calendar 1716–1717*, 353–4.
36. Headlam, *Calendar 1716–1717*, 139.
37. Nelson, *The Whydah Pirates Speak*, 30.
38. Headlam, *Calendar 1716–1717*, 213.
39. *Deposition of Samuel Liddell*.
40. *Deposition of Allen Bernard*.

Chapter 6. Ye Grand Pirate Captain Benjamin Hornigold Notes:

1. *The Boston News-Letter*, Monday, May 21 to Monday, May 28, 1716.
2. Headlam, *Calendar 1716–1717*, 382, 176.
3. Brooks, *Quest for Blackbeard*, 202.
4. Brooks, *Quest for Blackbeard*, 57–58.
5. Headlam, *Calendar 1714–1715*, 276, 119.
6. Headlam, *Calendar 1716–1717*, 240. i., 140–1.
7. Nelson, *The Whydah Pirates Speak*, 30.
8. Woodard, *The Republic of Pirates*, 87.
9. Johnson, *A General History of the Pyrates*, Second Edition 1724, 70–71.
10. Headlam, *Calendar 1716–1717*, 240. i., 140–1.
11. *The Trials of eight persons indited for piracy*, 25–26.
12. Headlam, *Calendar 1717–1718*, 296, 148.
13. Headlam, *Calendar 1716–1717*, 425. iii., 231.
14. *The Trials of eight persons indited for piracy*, 25.
15. Headlam, *Calendar 1716–1717*, 267, 149.

16. *The Boston News-Letter*, Monday, May 21 to Monday, May 28, 1716.
17. Headlam, *Calendar 1716–1717*, 690, 363.
18. *The Trials of eight persons indited for piracy*, 25.
19. *The Trials of eight persons indited for piracy*, 25–26.
20. Nelson, *The Whydah Pirates Speak*, 30.
21. *The Boston News-Letter*, Monday, June 17 to Monday, June 24, 1717.

Chapter 7. Blackbeard Begins Notes:

1. Ashmead, *The History of Delaware County*, 458.
2. Headlam, *Calendar 1717–1718*, 298. iii., 150–1.
3. Cooke, *The Virginia Magazine of History and Biography*, 304–7.
4. Timberlake, *Deposition, 17 Dec 1716*.
5. *The Trials of eight persons indited for piracy*, 25.
6. Johnson, *A General History of the Pyrates,* Second Edition, 1724, 70.
7. Headlam, *Calendar 1716–1717*, 328., 176.
8. Headlam, *Calendar 1716–1717*, 635., 338.
9. *Trial of Dr. John Howell*.
10. Ibid.
11. Ibid.
12. Headlam, *Calendar 1716–1717*, 677., 360.
13. *Trial of Dr. John Howell*.
14. Candler, *Letter of Capt. Bartholomew Candler of HMS Winchelsea, 19 Jul 1717*.
15. Headlam, *Calendar 1716–1717*, 677., 360.
16. *Trial of Dr. John Howell*.
17. Johnson, *A General History of the Pyrates,* Second Edition 1724, 70.
18. *The Boston News-Letter*, Monday, July 29 to Monday, August 5, 1717.
19. Brooks, March 15, 1718 issue of *The Weekly Journal* or *Saturday's Post, Quest for Blackbeard*, 333.
20. *The Boston News-Letter*, Monday, November 18 to Monday, November 25, 1717.
21. *The Boston News-Letter*, Monday, December 30 to Monday, January 6, 1718.
22. Ducoin, *Report of Research on the Nantes Ship La Concorde*, XVI.
23. Woodard, *The Republic of Pirates*, 210.
24. Johnson, *A General History of the Pyrates,* Second Edition 1724, 70.
25. Dampier, *Memoirs of a Buccaneer*, 54–56.
26. Upshur, *Eastern-Shore History*, 95–96.
27. *The Boston News-Letter*, Monday, April 29 to Monday, May 6, 1717.
28. Nelson, *The Whydah Pirates Speak*, 26.
29. *The Trials of eight persons indited for piracy*, 2.

30. *Trial of Dr. John Howell.*

Chapter 8. The Gentleman Pirate Notes:

1. Bonnet, *The Tryals of Major Stede Bonnet*, 9.
2. Woodard, *The Republic of Pirates*, 197.
3. Brooks, *Quest for Blackbeard*, 347–48.
4. Brooks, *Quest for Blackbeard*, 349.
5. Candler, *Letter of Capt. Bartholomew Candler of HMS Winchelsea - 19 Jul 1717.*
6. Bonnet, *The Tryals of Major Stede Bonnet*, iii.
7. Johnson, *A General History of the Pyrates*, Second Edition, 1724, 91.
8. Bonnet, *The Tryals of Major Stede Bonnet*, 9.
9. Bonnet, *The Tryals of Major Stede Bonnet*, 13.
10. Johnson, *A General History of the Pyrates*, Second Edition, 1724, 91–92.
11. Headlam, *Calendar 1716–1717*, 595. i, 317.
12. Bonnet, *The Tryals of Major Stede Bonnet*, iii.
13. *The Boston News-Letter*, Monday, October 21 to Monday, October 28, 1717.
14. *The Boston News-Letter*, Monday, October 14 to Monday, October 21, 1717.
15. *The Boston News-Letter*, Monday, November 4 to Monday, November 11, 1717.

Chapter 9. The Unlikely Partnership Notes:

1. Johnson, *A General History of the Pyrates*, First Edition, 1724, 86–87.
2. Johnson, *A General History of the Pyrates*, Second Edition, 1724, 71.
3. Johnson, *A General History of the Pyrates*, Second Edition, 1724, 92–93.
4. Woodard, *The Republic of Pirates*, 202.
5. *The Boston News-Letter*, Monday, November 4 to Monday, November 11, 1717.
6. Brand, *Lyme, December 4, 1717.*
7. Zupko, *A Dictionary of Weights and Measures for the British Isles*, 304.
8. *Order to Charge William Howard with Piracy*, Oct. 29, 1719.
9. Bialuschewski, *Blackbeard off Philadelphia*, 172.
10. *The Boston News-Letter*, Monday, October 28 to Monday, November 4, 1717.
11. *The Boston News-Letter*, Monday, November 4 to Monday, November 11, 1717.
12. Ibid.
13. Bialuschewski, *Blackbeard off Philadelphia*, 168.
14. *Order to Charge William Howard with Piracy*, Oct. 29, 1719.
15. *The Boston News-Letter*, Monday, November 4 to Monday, November 11, 1717.
16. Ibid.
17. Headlam, *Calendar 1717–1718*, 298. iii., 150-1.
18. Dickinson, *Pennsylvania House of Representatives*: House Speaker Biographies.

19. Bialuschewski, *Blackbeard off Philadelphia*, 167.
20. *The Boston News-Letter*, Monday, October 28 to Monday, November 4, 1717.
21. *The Boston News-Letter*, Monday, November 4 to Monday, November 11, 1717.
22. Ibid.
23. Ibid.
24. Bialuschewski, *Blackbeard off Philadelphia*, 167–8.
25. Konstam, *Blackbeard: America's Most Notorious Pirate*, 154.
26. *The Boston News-Letter*, Monday, November 4 to Monday, November 11, 1717.
27. Bialuschewski, *Blackbeard off Philadelphia*, 168.
28. *The Boston News-Letter*, Monday, November 18 to Monday, November 25, 1717.
29. *The Boston News-Letter*, Monday, December 30 to Monday, January 6, 1718.
30. Ducoin, *Report of Research on the Nantes Ship La Concorde*, XVI.
31. Woodard, *The Republic of Pirates*, 210.
32. Ibid.

Chapter 10. Getting the Dates Right Notes:
1. Yotov, *Old Style-New Style Side-By-Side Reference*.
2. Ibid.
3. Lords Proprietors to Knight, *Appointment as Secretary*.
4. Yotov, *Old Style-New Style Side-By-Side Reference*.

Chapter 11. The Queen Anne's Revenge Notes:
1. Ducoin, *Report of Research on the Nantes Ship La Concorde*, 20.
2. Ibid.
3. Ducoin, *Report of Research on the Nantes Ship La Concorde*, 19.
4. Ducoin, *Report of Research on the Nantes Ship La Concorde*, 28.
5. Ducoin, *Report of Research on the Nantes Ship La Concorde*, 1.
6. Ducoin, *Report of Research on the Nantes Ship La Concorde*, 29.
7. *The Boston News-Letter*, Monday, November 4 to Monday, November 11, 1717.
8. Ducoin, *Report of Research on the Nantes Ship La Concorde*, 27.
9. Johnson, *A General History of the Pyrates*, Second Edition 1724, 70–71.
10. Lee, *Blackbeard the Pirate*, 14-15.
11. Headlam, *Calendar 1717–1718, Deposition of Henry Bostock*, 298. iii., 150–1.
12. Ducoin, *Report of Research on the Nantes Ship La Concorde*, 18.
13. Ducoin, *Report of Research on the Nantes Ship La Concorde*, 13.
14. Ducoin, *Report of Research on the Nantes Ship La Concorde*, 16.
15. Ducoin, *Report of Research on the Nantes Ship La Concorde*, 18.
16. Ducoin, *Report of Research on the Nantes Ship La Concorde*, 29.

17. Ducoin, *Report of Research on the Nantes Ship La Concorde*, 27.
18. Dessales, *Annals of the Sovereign Council of Martinique*, 2–3.
19. Ducoin, *Report of Research on the Nantes Ship La Concorde*, 21.
20. Ducoin, *Report of Research on the Nantes Ship La Concorde*, 20.
21. Ducoin, *Report of Research on the Nantes Ship La Concorde*, 21.
22. Ducoin, *Report of Research on the Nantes Ship La Concorde*, 20.
23. Ibid.
24. Ducoin, *Report of Research on the Nantes Ship La Concorde*, 22.
25. Ibid.
26. Ducoin, *Report of Research on the Nantes Ship La Concorde*, 29.
27. Ducoin, *Report of Research on the Nantes Ship La Concorde*, 22.
28. Ducoin, *Report of Research on the Nantes Ship La Concorde*, 29.
29. Ducoin, *Report of Research on the Nantes Ship La Concorde*, 20.
30. Ducoin, *Report of Research on the Nantes Ship La Concorde*, 27–28.
31. Ducoin, *Report of Research on the Nantes Ship La Concorde*, 40.
32. Ducoin, *Report of Research on the Nantes Ship La Concorde*, 28.
33. Ducoin, *Report of Research on the Nantes Ship La Concorde*, 20.
34. Ducoin, *Report of Research on the Nantes Ship La Concorde*, 28.
35. Ducoin, *Report of Research on the Nantes Ship La Concorde*, 19.
36. Ducoin, *Report of Research on the Nantes Ship La Concorde*, 38.
37. Ducoin, *Report of Research on the Nantes Ship La Concorde*, 27.
38. Ducoin, *Report of Research on the Nantes Ship La Concorde*, 13.
39. Taylor, *Declaration*, 3.

Chapter 12. Cruising the Lesser Antilles Notes:

1. Ducoin, *Report of Research on the Nantes Ship La Concorde*, 20.
2. Taylor, *Declaration*, 1.
3. Ibid.
4. Selig, *Eighteenth-Century Currencies*.
5. Headlam, *Calendar 1717–1718*, 298. iii., 151.
6. Taylor, *Declaration*, 1–3.
7. Taylor, *Declaration*, 1.
8. *The Boston News-Letter*, Monday, March 3 to Monday, March 10, 1718.
9. Ducoin, *Report of Research on the Nantes Ship La Concorde*, 22.
10. Ibid.
11. Ibid.
12. Headlam, *Calendar 1717–1718*, 298. ii., 150.
13. Ibid.

14. Woodard, *The Republic of Pirates*, 217.
15. Woodard, *The Republic of Pirates*, 217-8.
16. Headlam, *Calendar 1717-1718*, 298. ii., 150.
17. Woodard, *The Republic of Pirates*, 219.
18. Headlam, *Calendar 1717-1718*, 298. i., 150.
19. Woodard, *The Republic of Pirates*, 219.
20. Headlam, *Calendar 1717-1718*, 298. i, ii., 150.
21. Woodard, *The Republic of Pirates*, 220.
22. Ducoin, *Report of Research on the Nantes Ship La Concorde*, 22.
23. Woodard, *The Republic of Pirates*, 220.
24. Woodard, *The Republic of Pirates*, 221.
25. Bostock, *Depositions of Henry Bostock, 19 Dec 1717*.
26. Headlam, *Calendar 1717-1718*, 298. iii., 150-1.
27. Bostock, *Depositions of Henry Bostock, 19 Dec 1717*.
28. Ibid.
29. Lee, *Blackbeard the Pirate*, 30.
30. Morange, *Testimony of Mr. Jean Morange*.
31. Ibid.
32. Ibid.

Chapter 13. Death's Head Notes:

1. *The Boston News-Letter*, Monday, June 9 to Monday, June 16, 1718.
2. Pringle, *Jolly Roger*, 124.
3. Pringle, *Jolly Roger*, 113.
4. Gosse, *The Pirates' Who's Who*, 251.
5. Quelch, *Arraignment, Tryal, and Condemnation*.
6. *The Boston News-Letter*, Monday, August 12 to Monday, August 19, 1717.
7. Ducoin, *Report on Research on the Nantes Ship La Concorde*, 9.
8. Ibid.
9. Ibid.
10. Ibid.
11. Breverton, *Black Bart Roberts*, 90.

Chapter 14. Cruising the Gulf of Honduras Notes:

1. Ducoin, *Report of Research on the Nantes Ship La Concorde*, 23.
2. Woodard, *The Republic of Pirates*, from the March 28, 1718 edition of the *Jamaican Dispatch*, 240.

3. Woodard, *The Republic of Pirates*, from the June 1718 edition of the *Weekley Journal or British Gazette*, 240.
4. Morange, *Testimony of Mr. Jean Morange*.
5. *The Boston News-Letter*, Monday, June 9 to Monday, June 16, 1718.
6. Bonnet, *The Tryals of Major Stede Bonnet*, 46.
7. *Rony's Tours, Roatan History*.
8. *The Boston News-Letter*, Monday, June 9 to Monday, June 16, 1718.
9. Woodard, *The Republic of Pirates*, 240–1.
10. Johnson, *A General History of the Pyrates*, Second Edition 1724, 71.
11. Bonnet, *The Tryals of Major Stede Bonnet*, 44–45.
12. Ibid.
13. Bonnet, *The Tryals of Major Stede Bonnet*, 45.
14. *The Boston News-Letter*, Monday, June 9 to Monday, June 16, 1718.
15. Jacob, *Letter of Thomas Jacob, HMS Diamond, Statement of Martin Preston*.
16. Bonnet, *The Tryals of Major Stede Bonnet*, 45.
17. Jacob, *Letter of Thomas Jacob, HMS Diamond, Statement of Martin Preston*.
18. *The Boston News-Letter*, Monday, June 9 to Monday, June 16, 1718.
19. Bonnet, *The Tryals of Major Stede Bonnet*, 45.
20. Ibid.
21. Jacob, *Letter of Thomas Jacob, HMS Diamond, Statement of William Wade*.
22. Jacob, *Letter of Thomas Jacob, HMS Diamond, Statement of William Megerre*.
23. *The Boston News-Letter*, Monday, October 13 to Monday, October 20, 1718.
24. Woodard, *The Republic of Pirates*, 243.
25. Bonnet, *The Tryals of Major Stede Bonnet*, 45.
26. *The Boston News-Letter*, Monday, June 9 to Monday, June 16, 1718.
27. Bonnet, *The Tryals of Major Stede Bonnet*, 45.
28. Jacob, *Letter of Thomas Jacob, HMS Diamond, Statement of William Wade*.
29. Jacob, *Letter of Thomas Jacob, HMS Diamond, Statement of Martin Preston*.
30. Bonnet, *The Tryals of Major Stede Bonnet*, 45.
31. Duffus, *The Last Days of Black Beard the Pirate*, 49.
32. Bonnet, *The Tryals of Major Stede Bonnet*, iii.
33. Bonnet, *The Tryals of Major Stede Bonnet*, 46.

Chapter 15. Nassau and the King's Proclamation Notes:

1. Bonnet, *The Tryals of Major Stede Bonnet*, 11.
2. *The Boston News-Letter*, Monday, December 2 to Monday, December 9, 1717.
3. Bell, *British Royal Proclamations*, 1717, September 5, 176.
4. Headlam, *Calendar 1717–1718*, 345, 170–1.

5. Pearse, *Account of Phenix at New York*, 2.

6. Johnson, *A General History of the Pyrates,* Second Edition 1724, 34-35.

7. Woodard, *The Republic of Pirates*, Extract from a letter from South Carolina dated February 2, 1718, printed in the *London Weekley Journal, or British Gazette,* 3 May 1718, 229.

8. Woodard, *The Republic of Pirates*, Extract from a letter from South Carolina dated February 2, 1718, printed in the *London Weekley Journal, or British Gazette,* 3 May 1718, 229–30.

9. Woodard, *The Republic of Pirates*, 232.

10. Pearse, *Account of Phenix at New York*, 10.

11. Woodard, *The Republic of Pirates*, 229.

12. Woodard, *The Republic of Pirates,* ADM 51/690, Pearse to the Admiralty, 3 June 1718, 232.

13. Woodard, *The Republic of Pirates*, 233–4.

14. Pearse, *Account of Phenix at New York*, 2.

15. Pearse, *Account of Phenix at New York*, 1.

16. Woodard, *The Republic of Pirates*, 233–4.

17. *The Boston News-Letter*, Monday, December 16 to Monday, December 23, 1717.

18. Woodard, *The Republic of Pirates*, 233.

19. Pearse, *Account of Phenix at New York*, 1-2.

20. Pearse, *Logbook of HMS Phenix, February to April 1718.*

21. Ibid.

22. Ibid.

23. Pearse, *Account of Phenix at New York*, 1.

24. Pearse, *Account of Phenix at New York*, 1-2.

25. Headlam, *Calendar 1716–1717,* 338.

26. Headlam, *Calendar 1717–1718,* 261.

27. Headlam, *Calendar 1717–1718,* 298. iii., 150–1.

28. Fictum, *The Firsts of Blackbeard.*

29. Woodard, *The Republic of Pirates,* 246.

30. Woodard, *The Republic of Pirates,* 196.

31. Jacob, *Letter of Thomas Jacob, HMS Diamond,* Statement of William Wade.

32. Brand, *Lyme July 12.*

33. Pearse, *Account of Phenix at New York,* 8.

34. Cracherode, *Report on petitions,* 7.

35. Duffus, *The Last Days of Black Beard the Pirate,* 167.

36. Norris, *Beaufort County, North Carolina Deed Book I,* 165.

Chapter 16. Siege of Charles Town Notes:

1. Bonnet, *The Tryals of Major Stede Bonnet,* 45.

2. Johnson. *Letter Dated June 18, 1718 to the Lords Proprietors.*

3. Headlam, *Calendar 1717–1718*, 660., 336–7.
4. Bonnet, *The Tryals of Major Stede Bonnet*, iii.
5. Johnson, *Letter Dated June 18, 1718 to the Lords Proprietors.*
6. Bonnet, *The Tryals of Major Stede Bonnet*, 45.
7. Headlam, *Calendar 1717–1718*, 660., 336–7.
8. Johnson, *Letter Dated June 18, 1718 to the Lords Proprietors.*
9. Bonnet, *The Tryals of Major Stede Bonnet*, 8.
10. Headlam, *Calendar 1717–1718*, 660., 337.
11. Bonnet, *The Tryals of Major Stede Bonnet*, iii.
12. Bonnet, *The Tryals of Major Stede Bonnet*, 45.
13. Headlam, *Calendar 1717–1718*, 660., 336.
14. *Shipping Lists for South Carolina*, p. 51–61.
15. Bonnet, *The Tryals of Major Stede Bonnet*, iii.
16. Bonnet, *The Tryals of Major Stede Bonnet*, 45.
17. Headlam, *Calendar 1717–1718*, 660, 337.
18. *Shipping Lists for South Carolina*, p. 51–61.
19. Ibid.
20. Johnson, *Letter Dated June 18, 1718 to the Lords Proprietors.*
21. Headlam, *Calendar 1717–1718*, 660, 337.
22. Bonnet, *The Tryals of Major Stede Bonnet*, 45.
23. *The Boston News-Letter*, Monday, June 30 to Monday, July 7, 1718.
24. Headlam, *Calendar 1717–1718*, 660, 337.
25. *Shipping Lists for South Carolina*, p. 51–61.
26. *Order to Charge William Howard with Piracy*, Oct. 29, 1719.
27. Bonnet, *The Tryals of Major Stede Bonnet*, 15.
28. Bonnet, *The Tryals of Major Stede Bonnet*, 48.
29. *Shipping Lists for South Carolina*, p. 51–61.
30. Bonnet, *The Tryals of Major Stede Bonnet*, 45.
31. Johnson, *Letter Dated June 18, 1718 to the Lords Proprietors.*
32. Johnson, *A General History of the Pyrates*, Second Edition 1724, 72.
33. Woodard, *The Republic of Pirates*, 249–50.
34. *Shipping Lists for South Carolina*, p. 51–61.
35. Hughson, *Blackbeard & the Carolina Pirates*, 71.
36. Johnson, ed. Hayward, *Key Writings on Subcultures, A General History of the Robberies and Murders*, 1927, 62–63.
37. Pringle, *Jolly Roger*, 194.
38. Lee, *Blackbeard the Pirate*, 43–44.

39. Woodard, *The Republic of Pirates*, 252.
40. *Drug Therapy in Colonial and Revolutionary America*.
41. Woodard, *The Republic of Pirates*, 249.
42. Duffus, *The Last Days of Black Beard the Pirate*, 104.

Chapter 17. Topsail Inlet Notes:

1. *The "Hammock House" c. 1700.*
2. Bonnet, *The Tryals of Major Stede Bonnet*, 45.
3. Wilde-Ramsing and Carnes-McNaughton, *Blackbeard's Sunken Prize*, 47.
4. Bonnet, *The Tryals of Major Stede Bonnet*, iv.
5. Bonnet, *The Tryals of Major Stede Bonnet*, 46.
6. Bonnet, *The Tryals of Major Stede Bonnet*, 11.
7. Bonnet, *The Tryals of Major Stede Bonnet*, 14.
8. Bonnet, *The Tryals of Major Stede Bonnet*, 48.
9. Bonnet, *The Tryals of Major Stede Bonnet*, 19.
10. Bonnet, *The Tryals of Major Stede Bonnet*, 17.
11. Bonnet, *The Tryals of Major Stede Bonnet*, 14–20.
12. Bonnet, *The Tryals of Major Stede Bonnet*, 19.
13. Bonnet, *The Tryals of Major Stede Bonnet*, 46.
14. Bonnet, *The Tryals of Major Stede Bonnet*, 16.
15. Bonnet, *The Tryals of Major Stede Bonnet*, 17.
16. Bonnet, *The Tryals of Major Stede Bonnet*, 11.
17. Bonnet, *The Tryals of Major Stede Bonnet*, 19.
18. Bonnet, *The Tryals of Major Stede Bonnet*, 38.
19. Bonnet, *The Tryals of Major Stede Bonnet*, 45.
20. Brand, *Lyme July 12*.
21. Bonnet, *The Tryals of Major Stede Bonnet*, 46.
22. Bonnet, *The Tryals of Major Stede Bonnet*, iv.
23. Bonnet, *The Tryals of Major Stede Bonnet*, 46.
24. Johnson, *A General History of the Pyrates*, Second Edition 1724, 94.
25. Pearse, *Account of Phenix at New York*, 9.
26. Brand, *Lyme July 12*.
27. Brooks, *Quest for Blackbeard*, 430.
28. Johnson, *A General History of the Pyrates*, Second Edition, 1724, 89.
29. Johnson, *A General History of the Pyrates*, Second Edition 1724, 74–75.
30. Wilde-Ramsing and Carnes-McNaughton, *Blackbeard's Sunken Prize*, 1.
31. Wilde-Ramsing and Carnes-McNaughton, *Blackbeard's Sunken Prize*, 41.
32. Wilde-Ramsing and Carnes-McNaughton, *Blackbeard's Sunken Prize*, 46.

33. Wilde-Ramsing and Carnes-McNaughton, *Blackbeard's Sunken Prize,* 91.

34. Wilde-Ramsing and Carnes-McNaughton, *Blackbeard's Sunken Prize,* 86.

35. Brooks, *Quest for Blackbeard,* 427.

36. Wilde-Ramsing and Carnes-McNaughton, *Blackbeard's Sunken Prize,* 90.

Chapter 18. Life in Bath Notes:

1. Norris, *Beaufort County, North Carolina Deed Book I,* 174.

2. Watson, *Bath: The First Town in North Carolina,* 14.

3. *Minutes of NC Governor's Council, May 27, 1719,* 180, 184.

4. *Lords Proprietors to Knight, Appointment as Secretary.*

5. Reed, *Beaufort County, Two Centuries of Its History,* 49.

6. Watson, *Bath: The First Town in North Carolina,* 16.

7. Watson, *Bath: The First Town in North Carolina,* 14–15.

8. Watson, *Bath: The First Town in North Carolina,* 15.

9. Watson, *Bath: The First Town in North Carolina,* 16.

10. Watson, *Bath: The First Town in North Carolina,* 18.

11. Watson, *Bath: The First Town in North Carolina,* 16.

12. Ibid.

13. Ibid.

14. Watson, *Bath: The First Town in North Carolina,* 16–17.

15. Norris, *Beaufort County, North Carolina Deed Book I,* 174

16. *Lords Proprietors to Knight, Appointment as Secretary.*

17. Watson, *Bath: The First Town in North Carolina,* 17.

18. Watson, *Bath: The First Town in North Carolina,* 21.

19. Watson, *Bath: The First Town in North Carolina,* 17.

20. Norris, *Beaufort County, North Carolina Deed Book I,* 140.

21. Watson, *Bath: The First Town in North Carolina,* 82.

22. Norris, *Beaufort County, North Carolina Deed Book I,* 140.

23. Norris, *Beaufort County, North Carolina Deed Book I,* 149, 170.

24. Norris, *Beaufort County, North Carolina Deed Book I,* 149.

25. Norris, *Beaufort County, North Carolina Deed Book I,* 137.

26. Reed, *Beaufort County, Two Centuries of Its History,* 49.

27. *Minutes of NC Governor's Council, May 27, 1719,* 185.

28. Lee, *Blackbeard the Pirate,* 155.

29. Duffus, *The Last Days of Blackbeard the Pirate,* 183.

30. Norris, *Beaufort County, North Carolina Deed Book I,* 127.

31. Norris, *Beaufort County, North Carolina Deed Book I,* 139.

32. Norris, *Beaufort County, North Carolina Deed Book I,* 140.

33. Ibid.
34. *7 Places Blackbeard's Gold Could've Been Stashed.*
35. Headlam, *Calendar 1717–1718*, 657, 333.
36. Johnson, *A General History of the Pyrates*, Second Edition 1724, 75–76.
37. Johnson, *A General History of the Pyrates*, Second Edition 1724, 77.
38. Ibid.
39. Brand, *Account of taking Blackbeard*, February 6, 1719.
40. Ibid.
41. Norris, *Beaufort County, North Carolina Deed Book I*, 137.
42. Norris, *Beaufort County, North Carolina Deed Book I*, 165.
43. Duffus, *The Last Days of Black Beard the Pirate*, 136–7.

Chapter 19. Philadelphia Notes:

1. *The Boston News-Letter*, Monday, November 4 to Monday, November 11, 1717.
2. Bialuschewski, *Blackbeard off Philadelphia*, 168.
3. Bialuschewski, *Blackbeard off Philadelphia*, 167–8.
4. Ashmead, *History of Delaware County*, 6.
5. Ashmead, *History of Delaware County*, 8.
6. Ashmead, *History of Delaware County*, 456.
7. Mervine, *Pirates and Privateers in the Delaware Bay and River*, 460.
8. Watson, *Annals of Philadelphia*, 1830, 463.
9. Ashmead, *History of Delaware County*, 457.
10. Ashmead, *History of Delaware County*, 458.
11. Watson, *Annals of Philadelphia*, 1844, 225.
12. Watson, *Annals of Philadelphia*, 1830, 463.
13. Watson, *Annals of Philadelphia*, 1844, 223–4.
14. Bialuschewski, *Blackbeard off Philadelphia*, 175.
15. Watson, *Annals of Philadelphia*, cover.
16. Upshur, *Eastern-Shore History*, 95–96.
17. Watson, *Annals of Philadelphia*, 1830, 463.
18. Watson, *Annals of Philadelphia*, 1830, 464.
19. Watson, *Annals of Philadelphia*, 1844, 223.
20. Watson, *Annals of Philadelphia*, 1830, 464.

Chapter 20. I'll Take Two Notes:

1. Boyer, *La Rose Emelye de Nantes*, 93–94.
2. Spotswood, *Capt. Tach's Quartermaster*, Dec. 22, 1718.
3. *Minutes of NC Governor's Council*, May 27, 1719.

4. *The Boston News-Letter*, Monday, November 10 to Monday, November 17, 1718.
5. Brand, *Account of taking Blackbeard*, February 6, 1719.
6. Watson, *Bath: The First Town in North Carolina*, 18.
7. Brand, *Account of taking Blackbeard*, February 6, 1719.
8. Headlam, *Calendar 1717–1718*, 800., 431.
9. *Minutes of NC Governor's Council*, May 27, 1719, 178–80.
10. *Minutes of NC Governor's Council*, May 27, 1719, 185–6.
11. *Minutes of NC Governor's Council*, May 27, 1719, 184.
12. *Minutes of NC Governor's Council*, May 27, 1719, 185.
13. Headlam, *Calendar 1717–1718*, 800., 431–433.
14. Norris, *Beaufort County, North Carolina Deed Book I*, 171.
15. Albertson, *In Ancient Albemarle*, 57.
16. Norris, *Beaufort County, North Carolina Deed Book I*, 171.

Chapter 21. William Howard's Arrest Notes:
1. Lee, *Blackbeard the Pirate*, 99.
2. *The Boston News-Letter*, Monday, December 16 to Monday, December 23, 1717.
3. Headlam, *Calendar 1717–1718*, 333.
4. Spotswood, *Capt. Tach's Quartermaster*, Dec. 22, 1718.
5. Cracherode, *Report on petitions*, 140.
6. *Order to Charge William Howard with Piracy*, Oct. 29, 1719.
7. Cracherode, *Report on petitions*, 140.
8. Brand, *From HMS Lyme*, July 14, 1719.
9. Headlam, *Calendar 1717–1718*, 800., 430.
10. Brand, *From HMS Lyme*, July 14, 1719.
11. Spotswood, *Capt. Tach's Quartermaster*, Dec. 22, 1718.
12. Spotswood, *The official letters of Alexander Spotswood*, 352–3.
13. Ibid.
14. Lee, *Blackbeard the Pirate*, 101.
15. Spotswood, *The official letters of Alexander Spotswood*, 352–3.
16. Ibid.
17. Lee, *Blackbeard the Pirate*, 104.
18. *The Boston News-Letter*, Monday, February 9 to Monday, February 16, 1719.
19. Spotswood, *Letter to Eden, Alexander Spotswood: Biographical Sketch*, 1718 Nov. 7.

Chapter 22. Bonnet on His Own Notes:
1. Bonnet, *The Tryals of Major Stede Bonnet*, 17.
2. Bonnet, *The Tryals of Major Stede Bonnet*, 11.

3. Bonnet, *The Tryals of Major Stede Bonnet*, 14.
4. Bonnet, *The Tryals of Major Stede Bonnet*, 19.
5. Bonnet, *The Tryals of Major Stede Bonnet*, 6.
6. Bonnet, *The Tryals of Major Stede Bonnet*, 38.
7. Bonnet, *The Tryals of Major Stede Bonnet*, 45.
8. Bonnet, *The Tryals of Major Stede Bonnet*, iv.
9. Ibid.
10. Bonnet, *The Tryals of Major Stede Bonnet*, 46.
11. Bonnet, *The Tryals of Major Stede Bonnet*, 42.
12. Bonnet, *The Tryals of Major Stede Bonnet*, 47.
13. Ibid.
14. Bonnet, *The Tryals of Major Stede Bonnet*, 13.
15. Bonnet, *The Tryals of Major Stede Bonnet*, 49.
16. Bonnet, *The Tryals of Major Stede Bonnet*, 13.
17. Bonnet, *The Tryals of Major Stede Bonnet*, 50.
18. Bonnet, *The Tryals of Major Stede Bonnet*, 13.
19. Bonnet, *The Tryals of Major Stede Bonnet*, 47.
20. Bonnet, *The Tryals of Major Stede Bonnet*, iv–vi.
21. Bonnet, *The Tryals of Major Stede Bonnet*, 1.
22. Bonnet, *The Tryals of Major Stede Bonnet*, v.
23. Bonnet, *The Tryals of Major Stede Bonnet*, vi.
24. Bonnet, *The Tryals of Major Stede Bonnet*, 1.
25. Bonnet, *The Tryals of Major Stede Bonnet*, 6.
26. Bonnet, *The Tryals of Major Stede Bonnet*, 15.
27. Bonnet, *The Tryals of Major Stede Bonnet*, vi.
28. Bonnet, *The Tryals of Major Stede Bonnet*, 37.
29. Bonnet, *The Tryals of Major Stede Bonnet*, 40.
30. Bonnet, *The Tryals of Major Stede Bonnet*, 36.
31. Bonnet, *The Tryals of Major Stede Bonnet*, 41.
32. Ramsay, *David. History of South Carolina*, 116.
33. Hughson, *Blackbeard & the Carolina Pirates*, 109.
34. Ramsay, *History of South Carolina*, 116.
35. Ibid.
36. Ramsey, *History of South Carolina*, 119.
37. Ramsey, *History of South Carolina*, 110–1.
38. Bonnet, *The Tryals of Major Stede Bonnet*, 43.

Chapter 23. Charles Vane's Visit Notes:

1. Woodard, Extract from a letter from South Carolina dated February 2, 1718 printed in the *London Weekley Journal, or British Gazette*, 3 May 1718, *The Republic of Pirates*, 229–30.
2. Woodard, *The Republic of Pirates*, 234.
3. Pearse, *Account of Phenix at New York*, 6.
4. Pearse, *Account of Phenix at New York*, 2.
5. Woodard, *The Republic of Pirates*, 246.
6. Woodard, *The Republic of Pirates*, 196.
7. Harrison, *George Cammock*.
8. Lewis, *George Camocke's 1718 Proposal of a Jacobite*.
9. Craton, *A History of the Bahamas*, 100.
10. Lewis, *George Camocke's 1718 Proposal of a Jacobite*.
11. Duffus, *The Last Days of Black Beard the Pirate*, 122.
12. Headlam, *Calendar 1717–1718*, 551. i., 263.
13. Headlam, *Calendar 1717–1718*, 737., 372.
14. Bonnet, *The Tryals of Major Stede Bonnet*, iv.
15. *The Boston News-Letter*, Monday, October 13 to Monday, October 20, 1718.
16. Bonnet, *The Tryals of Major Stede Bonnet*, iv.
17. Johnson, *A General History of the Pyrates*, Second Edition 1724, 145.
18. Ellms, *The Pirates Own Book*, 352.
19. Ellms, *The Pirates Own Book*, 339.
20. Duffus, *The Last Days of Black Beard the Pirate*, 126.
21. Ibid.
22. Duffus, *The Last Days of Black Beard the Pirate*, 127.
23. Spotswood, *Capt. Tach's Quartermaster*, Dec. 22, 1718.

Chapter 24. Brand and Maynard's Invasion of a Province Notes:

1. Brand, *Account of taking Blackbeard*, February 6, 1719.
2. Norris, *Beaufort County, North Carolina Deed Book I*, 137.
3. Spotswood, *Capt. Tach's Quartermaster*, Dec. 22, 1718.
4. Headlam, *Calendar 1717–1718*, 333.
5. *Minutes of NC Governor's Council*, May 27, 1719, 184.
6. *Minutes of NC Governor's Council*, May 27, 1719, 185.
7. *Minutes of NC Governor's Council*, May 27, 1719, 183.
8. *Minutes of NC Governor's Council*, May 27, 1719, 179.
9. *Minutes of NC Governor's Council*, May 27, 1719, 183.
10. Watson, *Bath: The First Town in North Carolina*, 81.
11. Gordon to Admiralty September 14, 1721, *Report*.

12. Bell, *British Royal Proclamations*, 1717, September 5, 176.
13. *Order to Charge William Howard with Piracy,* Oct. 29, 1719.
14. Headlam, *Calendar 1717-1718,* 800., 430-1.
15. Spotswood, *Capt. Tach's Quartermaster,* Dec. 22, 1718.
16. *The Boston News-Letter,* Monday, December 16 to Monday, December 23, 1717.
17. Brand, *Account of taking Blackbeard,* February 6, 1719.
18. Spotswood, *Capt. Tach's Quartermaster,* Dec. 22, 1718.
19. Palmer, *Calendar of Virginia 1652-1781,* 190.
20. Spotswood, *Capt. Tach's Quartermaster,* Dec. 22, 1718.
21. Ibid.
22. Gordon to Admiralty September 14, 1721, *Report.*
23. Cooke, *The Virginia Magazine of History and Biography,* 304-7.
24. Ibid.
25. Johnson, *A General History of the Pyrates,* Second Edition 1724, 82.
26. Rankin, *The Pirates of North Carolina,* 54.
27. Lee, *Blackbeard the Pirate,* 115.
28. Shomette, *Pirates on the Chesapeake,* 327.
29. Lyme Log, *Lyme Kiquotan Rd.,* Table.
30. Hughson, *Blackbeard & the Carolina Pirates,* 77.
31. Pringle, *Jolly Roger,* 203.
32. Rankin, *The Pirates of North Carolina,* 55.
33. Lee, *Blackbeard the Pirate,* 109.
34. Woodard, *The Republic of Pirates,* 289.
35. Konstam, *Black Beard: America's Most Notorious Pirate,* 242.
36. Gordon to Admiralty September 14, 1721, *Report.*
37. Johnson, *A General History of the Pyrates,* Second Edition 1724, 88-89.

Chapter 25. Battle of Ocracoke Notes:

1. Brand, *Account of taking Blackbeard,* February 6, 1719.
2. Cooke, *The Virginia Magazine of History and Biography,* pp. 304-7.
3. Gordon to Admiralty September 14, 1721, *Report.*
4. Johnson, *A General History of the Pyrates,* Second Edition 1724, 80.
5. Headlam, *Calendar 1717-1718,* 800., 431.
6. Ibid.
7. *The Boston News-Letter,* Monday, February 23 to Monday, March 2, 1719
8. Johnson, *A General History of the Pyrates,* Second Edition 1724, 81-82.
9. *The Boston News-Letter,* Monday, February 23 to Monday, March 2, 1719.
10. Ibid.

11. Johnson, *A General History of the Pyrates*, Second Edition 1724, 82–83.

12. Spotswood, *Capt. Tach's Quartermaster*, Dec. 22, 1718.

13. Johnson, *A General History of the Pyrates*, Second Edition, 1724, 85.

14. *The Boston News-Letter*, Monday, February 23 to Monday, March 2, 1719.

15. Johnson, *A General History of the Pyrates*, Second Edition 1724, 83–84.

Chapter 26. Aftermath in Ocracoke and Bath Notes:

1. Brand, *Account of taking Blackbeard*, February 6, 1719.

2. Cooke, *The Virginia Magazine of History and Biography*, pp. 304–7.

3. Ibid.

4. Brand, *Account of taking Blackbeard*, February 6, 1719.

5. Gordon to Admiralty September 14, 1721, *Report*.

6. Cooke, *The Virginia Magazine of History and Biography*, pp. 304–7.

7. Brand, *Account of taking Blackbeard*, February 6, 1719.

8. Ibid.

9. Gordon to Admiralty September 14, 1721, *Report*.

10. Brooks, *Quest for Blackbeard*, 368.

11. *Minutes of NC Governor's Council*, May 27, 1719.

12. Norris, *Beaufort County*, North Carolina Deed Book I, 137.

13. Duffus, *The Last Days of Black Beard the Pirate*, 147.

14. Brand, *Account of taking Blackbeard*, February 6, 1719.

15. Spotswood, *Letter to Eden*, Nov. 7, 1718.

16. Brand, *Account of taking Blackbeard*, February 6, 1719.

Chapter 27. Politics and Legalities Notes:

1. *Minutes of NC Governor's Council, May 27, 1719*, 184.

2. Lee, *Blackbeard the Pirate*, 155.

3. Duffus, *The Last Days of Blackbeard the Pirate*, 183.

4. *Minutes of NC Governor's Council, May 27, 1719.* 185.

5. *Minutes of NC Governor's Council, May 27, 1719*, 181.

6. *Minutes of NC Governor's Council, May 27, 1719*, 183.

7. Johnson, *A General History of the Pyrates*, Second Edition, 1724, 76.

8. Brand, *Account of taking Blackbeard*, February 6, 1719.

9. Duffus, *The Last Days of Blackbeard the Pirate*, 161–2.

10. Gordon to Admiralty September 14, 1721, *Report*.

11. *Documenting the American South*, Letter from Thomas Pollock to Charles Eden, December 8, 1718, 319–20.

12. Ibid.

13. Spotswood, *The official letters of Alexander Spotswood*, 352–3.
14. Lee, *Blackbeard the Pirate*, 110–11.
15. Lee, *Blackbeard the Pirate*, 133.
16. Spotswood, *The official letters of Alexander Spotswood*, 541.
17. McCarty, *Some Remarkable Proceedings*, Dec. 1, 1718.
18. Lee, *Blackbeard the Pirate*, 129–130.
19. Spotswood, *Capt. Tach's Quartermaster*, Dec. 22, 1718.
20. Lee, *Blackbeard the Pirate*, 130–131.
21. Bell, *British Royal Proclamations*, 1701, March 6, 155.
22. Ibid.
23. *Order to Charge William Howard with Piracy*, Oct. 29, 1719.
24. Headlam, *Calendar 1717–1718*, xviii.
25. Brand, *Lyme Virginia March 12, 1719*.
26. *Documenting the American South*, Minutes of the North Carolina Council, December 30, 1718–December 31, 1718, 321–4.
27. *Documenting the American South*, Minutes of the North Carolina Governor's Council, July 28, 1719 - August 01, 1719, 357–64.
28. Brand, *Account of taking Blackbeard*, February 6, 1719.
29. Brand, *Lyme Virginia March 12, 1719*.
30. Ellms, *The Pirates Own Book*, 345.
31. Duffus, HMS *Pearl's* logbook, *The Last Days of Blackbeard the Pirate*, 169.

Chapter 28. The Answers Lie with Cracherode Notes:
1. Johnson, *A General History of the Pyrates*, Second Edition, 1724, 90.
2. Brand, *Account of taking Blackbeard*, February 6, 1719.
3. Johnson, *A General History of the Pyrates*, Second Edition, 1724, 86.
4. Cracherode, *Report on petitions*, 136.
5. Cracherode, *Report on petitions*, 139.
6. Cracherode, *Report on petitions*, 139–40.
7. Cracherode, *Report on petitions*, 138–9.
8. Cracherode, *Report on petitions*, 139.
9. Spotswood, *Capt. Tach's Quartermaster*, Dec. 22, 1718.
10. Johnson, *A General History of the Pyrates*, Second Edition, 1724, 85.
11. Johnson, *A General History of the Pyrates*, Second Edition, 1724, 86.
12. Norris, *Beaufort County, North Carolina Deed Book I*, 149.
13. Norris, *Beaufort County, North Carolina Deed Book I*, 170.
14. Norris, *Beaufort County, North Carolina Deed Book I*, 165.
15. Norris, *Beaufort County, North Carolina Deed Book I*, 149.

Chapter 29. Chained in Williamsburg Notes:
1. Duffus, *The Last Days of Black Beard the Pirate*, 169.
2. Spotswood, *Capt. Tach's Quartermaster*, Dec. 22, 1718.
3. Ibid.
4. Brand, *Account of taking Blackbeard*, February 6, 1719.
5. McIlwaine, *Executive Journals of the Colonial Council*, 495–6.
6. *Minutes of NC Governor's Council*, May 27, 1719.
7. Duffus, HMS *Pearl's* Logbook, *The Last Days of Black Beard the Pirate*, 174.
8. Bell, *British Royal Proclamations*, 1718, December 21, 178.
9. Duffus, *The Last Days of Black Beard the Pirate*, 172.
10. Brand, *From HMS Lyme*, July 14, 1719.
11. Spotswood, *Capt. Tach's Quartermaster*, Dec. 22, 1718.
12. McIlwaine, *Executive Journals of the Colonial Council*, 497.
13. *Minutes of NC Governor's Council*, May 27, 1719.
14. Lee, *Blackbeard the Pirate*, 108.
15. *Minutes of NC Governor's Council*, May 27, 1719.
16. Irwin, *Petition to Council for two horses*, Dec. 23, 1720.
17. *Minutes of NC Governor's Council*, May 27, 1719.
18. Spotswood, *The Official Letters of Alexander Spotswood*, 316–7.
19. Johnson, *A General History of the Pyrates*, Second Edition, 1724, 86.
20. Watson, *Bath: The First Town in North Carolina*, 12.
21. Norris, *Beaufort County, North Carolina Deed Book I*, 165.
22. Ibid.
23. Norris, *Beaufort County, North Carolina Deed Book I*, 149.
24. Norris, *Beaufort County, North Carolina Deed Book I*, 164.
25. Hughson, *Blackbeard & the Carolina Pirates*, 78.
26. Spotswood, *The Official Letters of Alexander Spotswood*, 316–7.
27. Gordon to Secretary to Admiralty Oct. 2, 1719, *Disposal*, 2.
28. Gordon to Admiralty September 14, 1721, *Report*.

Chapter 30. Tobias Knight's Ordeal Notes:
1. *Minutes of NC Governor's Council*, May 27, 1719.
2. Ibid.
3. Brand, *From HMS Lyme*, July 14, 1719.
4. Johnson, *A General History of the Pyrates* Second Edition, 1724, 76.
5. Duffus, *The Last Days of Black Beard the Pirate*, 113–4.
6. Duffus, *The Last Days of Black Beard the Pirate*, 116.

7. *Minutes of NC Governor's Council*, May 27, 1719.

8. Lee, *Blackbeard the Pirate*, 151.

9. Lee, *Blackbeard the Pirate*, 153.

10. *Minutes of NC Governor's Council*, May 27, 1719.

11. Lee, *Blackbeard the Pirate*, 155.

12. *The Boston News-Letter*, Monday, March 3 to Monday, March 10, 1718.

13. Headlam, *Calendar 1717–1718*, 151.

14. Lee, *Blackbeard the Pirate*, 64–65.

15. Headlam, *Calendar of State Papers, 1722-1723*, 91., 36.

16. Johnson, *A General History of the Pyrates*, Second Edition 1724, 75.

Chapter 31. Will the Real Israel Hands Please Stand Up? Notes:

1. Johnson, *A General History of the Pyrates*, Second Edition 1724, 72.

2. Johnson, *A General History of the Pyrates*, Second Edition 1724, 74.

3. Bonnet, *The Tryals of Major Stede Bonnet*, 45.

4. Johnson, *A General History of the Pyrates*, Second Edition 1724, 86–87.

5. *Minutes of NC Governor's Council*, May 27, 1719, 178.

6. Johnson, *A General History of the Pyrates*, Second Edition 1724, 87.

7. Brand, *From HMS Lyme*, July 14, 1719.

8. Johnson, *A General History of the Pyrates*, Second Edition 1724, 87.

9. Cracherode, *Report on petitions*, 139.

10. Cracherode, *Report on petitions*, 136.

11. *Minutes of NC Governor's Council*, May 27, 1719, 178.

12. Bonnet, *The Tryals of Major Stede Bonnet*, 45.

Chapter 32. Meet Edward Thache—A Bristol Man Born Notes:

1. Johnson, *A General History of the Pyrates*, Second Edition 1724, 70.

2. Johnson, *A General History of the Pyrates*, First Edition 1724, 86.

3. Watson, *Annals of Philadelphia*, 1844, 220.

4. Rankin, *The Pirates of North Carolina*, 43.

5. Lee, *Blackbeard the Pirate*, 177.

6. Lee, *Blackbeard the Pirate*, 177–8.

7. Lee, *Blackbeard the Pirate*, 178.

8. Konstam, *The History of Pirates*, 12.

9. Brooks, *Quest for Blackbeard*, 154.

10. Brooks, *Quest for Blackbeard*, 156.

11. Brooks, *Quest for Blackbeard*, 154.

12. Brooks, *Quest for Blackbeard*, 165.

13. Brooks, *Quest for Blackbeard*, 166.
14. Brooks, *Quest for Blackbeard*, 144.
15. Brooks, *Quest for Blackbeard*, 148.
16. Brooks, *Quest for Blackbeard*, 167.
17. Brooks, *Quest for Blackbeard*, 186–7.
18. *The Boston News-Letter Boston*, Monday, September 30, to Monday, October 7, 1706.
19. Brooks, *Quest for Blackbeard*, 195–6.
20. Brooks, *Quest for Blackbeard*, 151.
21. Brooks, *Quest for Blackbeard*, 149.
22. Brooks, *Quest for Blackbeard*, 143.
23. Brooks, *Quest for Blackbeard*, 113–4.

Chapter 33. Meet Black Beard Notes:

1. Duffus, *The Last Days of Black Beard the Pirate*, 81–82.
2. Norris, *Beaufort County, North Carolina Deed Book I*, 127–88.
3. Norris, *Beaufort County, North Carolina Deed Book I*, 176–7.
4. Norris, *Beaufort County, North Carolina Deed Book I*, 144–5.
5. *Origins: Are the Bairds a Clan?*
6. *Anglo-Spanish Hostility in Early South Carolina.*
7. Norris, *Beaufort County, North Carolina Deed Book I*, 154.
8. Norris, *Beaufort County, North Carolina Deed Book I*, 144.
9. Norris, *Beaufort County, North Carolina Deed Book I*, 130.
10. Norris, *Beaufort County, North Carolina Deed Book I*, 143.
11. Norris, *Beaufort County, North Carolina Deed Book I*, 139.
12. Ibid.
13. Norris, *Beaufort County, North Carolina Deed Book I*, 138–9.
14. Norris, *Beaufort County, North Carolina Deed Book I*, 152.
15. Watson, *Bath: The First Town in North Carolina*, 12.
16. Norris, *Beaufort County, North Carolina Deed Book I*, 165.
17. Willis, *Beaufort County, N.C. Deeds Vol. 1*, 42.
18. Norris, *Beaufort County, North Carolina Deed Book I*, 165.
19. Norris, *Beaufort County, North Carolina Deed Book I*, 144.
20. Norris, *Beaufort County, North Carolina Deed Book I*, 165.
21. Duffus, *The Last Days of Black Beard the Pirate*, 136–7.
22. Duffus, *The Last Days of Black Beard the Pirate*, 167.
23. Norris, *Beaufort County, North Carolina Deed Book I*, 139.
24. *Lords Proprietors to Knight, Appointment as Secretary.*
25. Norris, *Beaufort County, North Carolina Deed Book I*, 145.

26. *Minutes of NC Governor's Council*, May 27, 1719, 180–1.
27. *Minutes of NC Governor's Council*, May 27, 1719, 184.
28. Duffus, *The Last Days of Black Beard the Pirate*, 185–7.

Chapter 34. Meet Mrs. Blackbeard Notes:

1. Johnson, *A General History of the Pyrates*, Second Edition 1724, 75–76.
2. Brand, *Account of taking Blackbeard*, February 6, 1719.
3. Headlam, *Calendar 1717–1718*, 149.
4. Lee, *Blackbeard the Pirate*, 74.
5. Lee, *Blackbeard the Pirate*, 74–75.
6. Lee, *Blackbeard the Pirate*, 198.
7. Brooks, *Quest for Blackbeard*, 471–9.
8. Brooks, *Quest for Blackbeard*, 472–7.
9. Norris, *Beaufort County, North Carolina Deed Book I*, 159–60.

Chapter 35. Twisted Tales Notes:

1. Watson, *Annals of Philadelphia and Pennsylvania*, 1844, 221.
2. Pringle, *Jolly Roger*, 207.
3. Robinson, *Pirates on the Piscataqua*.
4. Ibid.
5. Duffus, *The Last Days of Black Beard the Pirate*, 202.
6. *History of Blackbeard Island*.
7. *Blackbeard Island National Wildlife Refuge*.
8. *History of Sapelo Island*.
9. Plunkett, *Blackbeard's Tower*.
10. *Blackbeard's Castle*.
11. *The Colorful Pirate History of Amelia Island*.
12. Albertson, *In Ancient Albemarle*, 55.
13. Albertson, *In Ancient Albemarle*, 56–57.
14. Albertson, *In Ancient Albemarle*, 58.
15. Albertson, *In Ancient Albemarle*, 55.
16. Duffus, *The Last Days of Black Beard the Pirate*, 208.
17. Albertson, *In Ancient Albemarle*, 55.
18. Norris, *Beaufort County, North Carolina Deed Book I*, 138.
19. Norris, *Beaufort County, North Carolina Deed Book I*, 139–40.
20. Johnson, *A General History of the Pyrates*, Second Edition 1724, 85–86.
21. *7 Places Blackbeard's Gold Could've Been Stashed*.
22. Norris, *Beaufort County, North Carolina Deed Book I*, 140.

23. Ibid.

24. Albertson, Catherine. *In Ancient Albemarle*, 58.

25. Norris, *Beaufort County, North Carolina Deed Book I*, 141.

26. Norris, *Beaufort County, North Carolina Deed Book I*, 143.

27. Norris, *Beaufort County, North Carolina Deed Book I*, 149.

28. Spotswood, Capt. Tach's Quartermaster, Dec. 22, 1718.

29. Norris, *Beaufort County, North Carolina Deed Book I*, 149.

30. *The Lost Treasure of Edward Teach.*

Chapter 36. Blackbeard's Treasure Notes:

1. Bonnet, *The Tryals of Major Stede Bonnet*, 45.

2. Cracherode, *Report on petitions*, 7.

3. *Order to Charge William Howard with Piracy*, Oct. 29, 1719.

4. Headlam, *Calendar 1717–1718*, 660., 336–7.

5. Bonnet, *The Tryals of Major Stede Bonnet*, iv.

6. Bonnet, *The Tryals of Major Stede Bonnet*, 45.

7. Wilde-Ramsing and Carnes-McNaughton, *Blackbeard's Sunken Prize*, 1.

8. Wilde-Ramsing and Carnes-McNaughton, *Blackbeard's Sunken Prize*, 150.

9. Wilde-Ramsing and Carnes-McNaughton, *Blackbeard's Sunken Prize*, 34.

10. Ducoin, *Report of Research in French Archives on the Nantes Ship La Concorde*, 27–29.

11. Ducoin, *Report of Research in French Archives on the Nantes Ship La Concorde*, 21.

12. Taylor, *Declaration*, 1.

13. Selig, *Eighteenth-Century Currencies*.

14. Bostock, *Depositions of Henry Bostock, 19 Dec 1717*.

15. Woodard, *The Republic of Pirates*, 240.

16. *The Boston News-Letter*, Monday, June 30 to Monday, July 7, 1718.

17. Bonnet, *The Tryals of Major Stede Bonnet*, 8.

18. Spotswood, *Capt. Tach's Quartermaster*, Dec. 22, 1718.

19. Duffus, *The Last Days of Black Beard the Pirate*, 172.

20. Norris, *Beaufort County, North Carolina Deed Book I*, 165.

21. Norris, *Beaufort County, North Carolina Deed Book I*, 147–9.

22. Norris, *Beaufort County, North Carolina Deed Book I*, 150.

23. Watson, *Annals of Philadelphia*, 1850, 223–4.

24. Watson, *Annals of Philadelphia*, 1830, 463–4.

25. Watson, *Annals of Philadelphia*, 1850, 32.

26. Watson, *Annals of Philadelphia*, 1830, 466.

27. Watson, *Annals of Philadelphia*, 1850, 223.

28. "Digs for Pirate's Hoard," *New York Times*, Friday, October 8, 1926, 25.

29. Finding of Human Skull Spurs Hunt For Pirate Blackbeard's Buried Treasure, *New York Times*, Saturday, October 9, 1926, 1.
30. Gordon to Admiralty September 14, 1721, *Report*.
31. *Minutes of NC Governor's Council*, May 27, 1719, 183.
32. Albertson, *In Ancient Albemarle*, 57.
33. MacNiel, Ben Dixon, *The News and Observer*, 9.

GLOSSARY

A

Act of Settlement
an English Act passed by Parliament in 1701 that forbade a Catholic from becoming a monarch in England.

Admiral
naval rank of a commander of a fleet or an officer of very high rank.

Admiralty Council (Board of Admiralty)
an assembly of admirals who are appointed to manage the day-to-day operations of the English Royal Navy.

Aft
nautical term for the rear of a boat, ship, or any other type of vessel.

B

Back Staff
late sixteenth-century navigational device used at sea. It was an improved version of the cross staff and works on the same principle except the user looks away from the sun to get a sighting.

Barrel
a medium-sized oval-shaped wooden container that holds 31.5 gallons of liquid or dry goods.

Baldric
belt for a sword or other piece of equipment worn over one shoulder and reaching down to the opposite hip.

Below Decks
nautical term meaning the inside space of a ship or vessel which is literally below the main deck.

Bermuda Sloop
seventeenth- and eighteenth-century single-masted vessel with a triangular sail rigged fore and aft with a square sail above that was designed for trade, measured approximately 70 feet in length, and had a large cargo hold.

Blunderbuss
short-barreled large-bored gun with a flared muzzle.

Boarding Action
when the crew of one vessel jumps over onto the second vessel during an attack.

Boatswain
senior member of the ship's crew who is in charge of operations on the deck.

Bow
front of a ship, boat, or vessel.

Brigantine
two-masted vessel with a square-rigged foremast and a fore-and-aft-rigged mainmast.

Broadside
the nearly simultaneous firing of all the guns from one side of a warship.

C

Cagg
a small cask, usually ten gallons or less.

Canoe
an eighteenth-century term for a small rowboat.

Calendar of State Papers
a very detailed index and collection of a wide range of documents relating to the American and West Indian colonies between 1574 and 1739.

Careening
the cleaning and repairing of the hull of a ship or vessel below the water line which requires taking it out of the water.

Chip Log
late sixteenth-century navigational device used at sea to measure the speed of the vessel. It is a quarter circle of wood attached to a long line on a reel with knots tied every 47 feet 3 inches.

Coasting Sloop
a smaller sloop designed for local trade rather than trade requiring travel across the ocean.

Colonel
military rank below a general for an officer who commands a regiment or a similar sized group of soldiers.

Colors
military term meaning a national flag. To raise the flag is to raise the colors. Colors flying is synonymous with flags waving.

Commissions
instruction, command, or authority given to individuals in writing.

Council of Trade and Plantations (Board of Trade and Plantations)
a committee of the Privy Council of the British colonial office that oversaw colonial affairs.

Consort Vessel
small vessel used to carry supplies in support of other vessels.

Cross Staff
early sixteenth-century navigational instrument used to measure the sun's angle and determine the vessel's latitude.

D

Death's Head
the depiction of a human skull often accompanied by a set of crossed bones that symbolizes death.

Deposition
a sworn statement made verbally or in writing that is recorded for legal use.

Dory
a small rowboat pointed at both ends.

Dutch Flute
sixteenth- through eighteenth-century Dutch merchant ship with a rounded hull designed to carry large amounts of cargo.

F

Firing in Volley
military term meaning for all the soldiers in a designated group or line to fire their weapons at one time.

First Mate
senior member of the ship's crew who second in command to the quartermaster.

Florida Straights
water passage between Florida and Cuba.

Fore Mast
mast in the front of the vessel usually in front of the mainmast.

Forecastle
structure on the front of a ship or vessel usually containing all the ropes that are not in use.

Frigate
three-masted ship designed for war but is smaller in size and armament than a ship-o-the-line.

Fully Rigged
nautical term for a ship with three masts all of with have square sails.

G

Gaff Rigged
sail configuration plan for a four-sided sail rigged fore and aft and fastened at all four points with to a large spar connected to the center of the mast and hoisted up with lines connected to the top of the mast.

Galley
large vessel with up to three banks of oars that uses oars as its primary propulsion but also has sails.

Gaol
seventeenth- and eighteenth-century term for a jail.

Glorious Revolution
revolution by the English Parliament in 1688 to overthrow King James II of England and place William III, Prince of Orange, who was James' nephew and son-in-law, and Mary II, who was James' daughter, on the throne.

Gregorian Calendar
calendar named after Pope Gregory XIII, who introduced it in October 1582 as a refinement of the Julian calendar, although it was not adopted in England until 1752.

Gulf Stream
a warm and swift Atlantic Ocean current that originates in the Gulf of Mexico and flows through the Florida Straight and up the eastern coast.

Gunner
naval rank for a member of the crew who is an expert with naval ordinance and who is responsible for aiming and firing one or more of the ship's guns during battle.

H

Harbor Pilot

person who is thoroughly familiar with a harbor and guides large vessels into port.

HMS

English nautical abbreviation meaning His/Her Majesty's Ships and precedes the name of all vessels officially listed as part of the Royal Navy.

Heave To

a nautical term meaning to stop a vessel, normally at a port or alongside of another vessel.

Hogshead

a large oval-shaped wooden container holding 63 gallons of liquid or dry goods.

Hull

the body of a vessel.

J

Jacobite

a supporter of James Edward Stuart's claim to the throne of Great Britan which derives from "Jacob," the Latin word for James.

Jacobite Revolution

a series of revolutions primarily conducted by the Scottish and Irish people in 1688, 1715, and 1745 to restore James II or his descendants to the throne of England.

Jamaica Sloop

seventeenth- and eighteenth-century single-masted vessel with a triangular sail rigged fore and aft with a square sail above that was designed for trade, measured approximately 70 feet in length, but had a narrower construction than other sloops in increase speed.

Jolly Roger

a pirate flag that came from the French term "Jolie Rouge" which means "Happy Red".

Julian Calendar

calendar named after Julius Caesar, who introduced it in 46 BC as a reform of the Roman calendar.

K

Kangxi Porcelain

a type of fine blue and white porcelain made in China between 1662 and 1722.

L

Larboard
the port, or left side of a ship, boat, or vessel.

Leeward
the side of a vessel that is opposite to the direction of the wind.

Letters of Marquee
document issued by an agent of a government giving the holder permission to attack all vessels of nations listed in the document which are deemed as enemies to the issuing government.

Line
nautical term meaning a working rope that is connected to a sail, anchor, or any part of a vessel used to sail it (a rope becomes a line as soon as it is attached to a part of the vessel).

Lords Proprietors
the governing body of a colony established through a royal charter.

M

Main Mast
tallest and most important mast on a ship or vessel.

Main Sail
on vessels rigged fore and aft, it is the largest sail and on vessels that are square rigged, it is the largest and lowest sail on the mast.

Manila Galleon
Spanish merchant vessel that routinely traveled between the Philippines and the west coast of Mexico, primarily between Manila and Acapulco.

Man-O-War
class of warships in the 17th and 18th centuries that are the largest in size and armament.

Marooned
the act of being set ashore and left alone on a deserted island.

Master
another term for a ship's captain usually applied to captains of small vessels.

Match
a length of hemp or flax cord that had been chemically treated to make it burn slowly and consistently for an extended period, used in the eighteenth-century to fire matchlock muskets or to fire artillery and naval guns.

Mate
member of a ship's crew.

Merchantmen
seventeenth- and eighteenth-century term for any merchant ship or vessel.

Midshipman
naval rank for an officer cadet or someone who is in training to become a naval officer.

Militia
military force that is raised from the civilian population to supplement a regular military force.

Mizzen Mast
mast directly behind the main mast.

O

Ordinance
military and naval term for artillery or guns onboard a ship.

P

Pattararas
another term for a swivel gun, a small caliber gun that is mounted with a swivel device on the wall of a fort or on the rail of a vessel that can be easily aimed and fired at close range.

Periauger
a small shallow draft vessel with either one or two masts that can either be sailed or rowed with oars.

Pieces of Eight
term for a Spanish coin valued at eight reals but could also apply to other Spanish real coins of lesser value.

Pink
three-masted, square-rigged sailing vessel, typically with a narrow, overhanging stern.

Pilot
a mariner who has specific navigational knowledge of dangerous or congested waterways, such as harbors or river mouths.

Pipe
a large oval-shaped wooden container holding 126 gallons of liquid or day goods.

Pirate
those who rob or commit illegal violence at sea or on the shores of the sea.

Pirate Banyan
gathering of pirate crews to socialize and form alliances, usually involving large quantities of food, drink, and general camaraderie.

Port
left side of a ship, boat, or vessel.

Pounds Sterling
denomination of English currency equal to 20 schillings.

Pretender or Pretender Across the Sea
eighteenth-century term for James Edward Stuart, the claimant to the throne of England who was often referred to as the "pretender" and who lived across the sea in France.

Privateers
private person or ship that engages in maritime warfare under a commission also known as letters of marque.

Prize Vessel
any vessel that has been captured by another vessel.

Pull Alongside
nautical term meaning to bring two vessels side by side.

Q

Quadrant
ancient navigational instrument consisting of a graduated quarter circle and a sighting mechanism used for taking angular measurements in order to determine latitude.

Quakers
member of the Religious Society of Friends, a Christian movement founded and devoted to peaceful principles, with a belief in the doctrine of "inner light" and rejection of formal ministry and all set forms of worship.

Quarter
military term meaning mercy shown toward an enemy or opponent, generally used as "No Quarter" or "Quarter Given".

Quarterdeck
raised deck behind the main or mizzen mast of a ship or vessel where navigation is done or where the captain commands the vessel.

Quartermaster
within a naval context of the 17th and 18th centuries, it was the rank of the senior member of the crew (not including the officers) who was responsible for the care and discipline of the rest of the crew.

R

Rail of a Ship
the railing around the outside of a vessel that prevents someone from falling overboard.

Rating of Ships
classification of warships which indicated size and armament.

Re-provision
when used as a nautical term, it means to take on all manner of supplies on board a vessel.

Refit
to supply or to reconfigure a vessel to prepare for a voyage.

Rigged Fore and Aft
configuration of sails on a vessel where triangular sails are set along the keel.

Rigging
the system of ropes, cables, or chains to support a ship's masts (standard rigging) and to control or set the yards and sails (running rigging); and the action of providing a sailing ship with rigging.

S

Sailing Master
nautical rank for the officer responsible for setting and adjusting the sails of a ship or vessel.

Schooner
two-masted vessel with sails that are rigged fore and aft.

Scuttle
intentionally sinking a vessel.

Ship
three-masted vessel that is fully rigged with square sails.

Ship of the Line
largest and most powerful naval warships with either a first, second, or third rating usually carrying at least 80 guns.

Shot
military term for any type of projectile fired from a gun.

Sloop
single-masted vessel with a triangular sail rigged fore and aft.

Snow
large, two-masted merchant vessel that is rigged with square sails and can also be constructed with oars (galley-built snot).

Spanish Galleon
large Spanish merchant ship commonly used between the 16th and 18th centuries with three or four masts with the foremast and the mainmast rigged with square sails and the mizzenmast and aft mast rigged with lateen sails and a small square sail on a high-rising bowsprit.

Square Rigged
vessel with traditional square sails that are generally perpendicular to the keel of the vessel.

Starboard
right side of a ship, boat, or vessel.

Stern
outer back rear of a ship, boat, or vessel.

Superstructure
any structure or cabin built on or above the main deck of a ship or vessel.

Sweeps
nautical slang term for the long oars used on any vessel.

Swivel Gun
small caliber gun that is mounted with a swivel device on the wall of a fort or on the rail of a vessel that can be easily aimed and fired at close range.

T

Teredo Worms
a wormlike marine mollusk that bores into the wooden hull of a vessel and can cause substantial damage.

Top Sail
square sail immediately above the mainsail on a square-rigged vessel.

Tory
an English political party opposed King George.

Touch Hole
small hole in the side or top of a gun barrel through which the gunpowder inside the barrel is ignited.

Treaty of Utrecht
treaty signed in 1713 that ended the War of the Spanish Succession.

U

Under Full Sail
nautical term meaning that all the sails are raised in order to sail as fast as possible.

Union Act of 1707
an act passed under Queen Anne's rule that officially joined England and Scotland as one nation with one ruler.

V

Vice Admiral
rank for a senior naval officer below the rank of admiral.

W

Warship
a well-armed vessel designed for battle.

War of the Spanish Succession
war that resulted from the disputed succession to the throne of Spain between France and Spain on one side and England, the Dutch Republic, and the Holy Roman Empire on the other, that lasted from 1701 to 1714.

Weather Gauge
nautical term that means your vessel is upwind of the other vessel, which gives you far more maneuverability that the vessel downwind.

West Indies
islands in the Caribbean that include the Greater Antilles, the Lesser Antilles, and the Lucayan Archipelago.

Whigs
an English political party that supported King George.

BIBLIOGRAPHY

7 Places Blackbeard's Gold Could've Been Stashed. 2015. Retrieved from https://www.mentalfloss.com/article/68887/7-places-blackbeards-gold-couldve-been-stashed on June 10, 2023.

About John Fripp, I. Retrieved from https://www.geni.com/people/John-Fripp-I/6000000013496023354 on May 3, 2021.

Albertson, Catherine. *In Ancient Albemarle*. Raleigh: Published by the North Carolina Society Daughter of the Revolution, 1914. Retrieved from the East Carolina University Library https://digital.lib.ecu.edu/16894 on March 28, 2024.

Anglo-Spanish Hostility in Early South Carolina, 1670–1748. Copyright 2022, Charleston County Public Library. Retrieved from https://www.ccpl.org/charleston-time-machine/anglo-spanish-hostility-early-south-carolina-1670–1748 on March 12, 2024.

Arsenault, Kathleen Hardee. *Looking for treasure on Amelia Island*. 2021. Retrieved from https://fernandinaobserver.com/featured-story/looking-for-treasure-on-amelia-island/ on October 21, 2022.

Ashmead, Henry G. *History of Delaware County, Pennsylvania*. L.H. Everts & Co. Philadelphia, 1884.

Barthorp, Michael. *The Jacobite Rebellions 1689–1745*. London: Osprey Publishers, 1995.

Bell, Colin, John Campbell, and Joseph Cooper, Producers. *British Royal Proclamations Relating to America 1603–1783*. Project Gutenberg. Retrieved from https://www.gutenberg.org/ebooks/46167 on January 25, 2024.

Bialuschewski, Arne. *Blackbeard off Philadelphia: Documents Pertaining to the Campaign against the Pirates in 1717 and 1718*. Pennsylvania Magazine of History and Biography, Vol. CXXXIV, No 2, April 2010.

Bialuschewski, Arne. *Daniel Defoe, Nathaniel Mist, and the Golden Age of Piracy*. The Papers of the Bibliographical Society of America, 2004.

Blackbeard Island National Wildlife Refuge. Retrieved from https://gosouthsavannah.com/tybee-island-and-coast/blackbeard-island.html on December 2, 2023.

Blackbeard Legends. Retrieved from https://www.ncmuseumofhistory.org/legends-nc/blackbeard-legends on November 15, 2021.

Blackbeard's Castle. St. Thomas, U.S. Virgin Islands. Retrieved from https://www.atlasobscura.com/places/blackbeard-s-castle on March 21, 2024.

Blackbeard's Treasure on Plum Point NC. Fleming's Bond Website, 2013. Retrieved from https://flemingsbond.com/blackbeards-treasure-on-plum-point-nc/on May 17, 2022.

Bonnet, Stede. *The Tryals of Major Stede Bonnet and Other Pirates.* London: Printed for Benj. Cowse at the Rose and Crown in St. Paul's Church, 1719. Retrieved from https://www.loc.gov/item/33008758/ on November 5, 2023.

Bostock, Henry. *Depositions of Henry Bostock, 19 Dec 1717.* Retrieved from http://baylusbrooks.com/index_files/Page3912.htm on January 12, 2024.

Boston News-Letter. Boston Mass.: B. Green, 1704–1726.

Boyer, Pierre. *La Rose Emelye de Nantes.* Rapports des capitaines à l'Amirauté de Nantes, December 27, 1718. Les Archives départementales. Translation by bryna@charteroak-genealogy.com. Retrieved from https://archives-numerisees.loire-atlantique.fr/v2/ad44/visualiseur/amiraute.html?id=440323905 on March 1, 2024.

Brand. *Lyme Elizabeth River, Virginia, December 4, 1717.* ADM 1/1472 subsection 11. [166] Library of Virginia, Richmond.

Brand. *Lyme July 12, 1718 Cape Henry bearing West ½ N:°50 Leagues.* ADM 1/1472. State Archives Building, Raleigh, NC.

Brand. *Lyme Virginia March 12, 1718/19.* Raleigh, NC.: State Archives Building, ADM 1/1472.

Brand to Lordships. *From HMS Lyme in Galleons Reach.* 1719 July 14. Richmond, VA.: Library of Virginia, ADM 1/1472. [reel 166 subsection 11].

Brand to Secretary of Admiralty. *Account of taking Blackbeard (detailed).* 1718/19 Feb. 6. Raleigh, NC.: State Archives Building, PRO-ADM 1/1472. [72.992.1–4].

Brigham, Clarence S. ed. *British Royal Proclamations Relating to America 1603–1783.* Worcester, Massachusetts: Published by the American Antiquarian Society, 1911. Retrieved from https://archive.org/details/royalproclamations12brigrich/page/178/mode/2up?q=%22suppressing+pirates%22 on January 16, 2024.

British Fourth Rate ship of the line 'Windsor' (1695). Retrieved from https://threedecks.org/index.php?display_type=show_ship&id=165 on May 4, 2021.

Brooks, Baylus C. *Lt. Robert Maynard: Rewards of Killing a Pirate!* Retrieved from https://bcbrooks.blogspot.com/2018/07/lt-robert-maynard-rewards-of-killing.html on October 23, 2023.

Brooks, Baylus C. *Quest for Blackbeard: The True Story of Edward Thache and His World.* Lake City, Florida: Lulu Press, Inc., 2016.

Breverton, Terry. *Black Bart Roberts, The Greatest Pirate of Them All.* St, Gretna, Louisiana: Pelican Publishing Company, Inc., 2004.

Burgess, Robert T. and Carl J. Clausen. *Florida's Golden Galleons; The Search for the 1715 Spanish Treasure Fleet.* Port Salerno, Florida: Florida Classics Library, 1982.

Cabell, Craig, Graham A. Thomas, and Allan Richards. *The Hunt for Blackbeard: The World's Most Notorious Pirate.* Barnsley, Great Britain: Pen and Sword Books, Ltd., 2012.

Candler, Bartholomew. *Letter of Capt. Bartholomew Candler of HMS Winchelsea - 19 Jul 1717.* ADM 1/1597 Winchelsea off Crooked Island in the Windward Passage. Retrieved from http://baylusbrooks.com/index_files/Page34546.htm on February 17, 2024.

Caribbean Archaeology, 2019. Retrieved from https://www.academia.edu/40249657/Archaeology_of_Piracy_between_Caribbean_Sea_and_the_North_American_Coast_of_17th_and_18th_Centuries_Shipwrecks_Material_Culture_and_Terrestrial_Perspectives on September 5, 2021.

Clifford, Barry and Kenneth J. Kinkor with Sharon Simpson. *Real Pirates: The Untold Story of The Whydah from Slave Ship to Pirate Ship.* Washington D. C.: National Geographic Society, 2007.

Combs, Joe C. *Lost Ships of the 1715 Spanish Treasure Fleet.* Retrieved from https://joeccombs2nd.com/2014/08/03/lost-ships-of-the-1715-spanish-treasure-fleet-lost-treasures-part-4/ on February 3, 2024.

Cooke, Arthur L. "British Newspaper Accounts of Blackbeard's Death." *The Virginia Magazine of History and Biography* 61, No.3 (1953): 304–307, Virginia Historical Society, 1953. Retrieved from https://www.jstor.org/stable/4245947 on January 3, 2024.

Cook, D. V. *Kings and Queens of England.* New York: Weathervane Books, 1978.

Cordingly, David. *Under the Black Flag.* New York: Random House, Inc., 2006.

Cracherode to Treasury Board. *Report on petitions for rewards for capturing pirates.* 18 March 1720. Richmond: Library of Virginia, PRO-T 1/227. Treasury Papers–In Letters, [SR 01227.1248 117.ff.134–141].

Craton, Mihael. *A History of the Bahamas.* London: Collins, 1962.

Dampier, William. *Memoirs of a Buccaneer, Dampier's New Voyage Round the World, 1697.* Mineola, New York: Dover Publications, Inc. 1968.

Day, Jean. *Blackbeard and the Queen Anne's Revenge.* New Bern, North Carolina: Griffin & Tilghman, 2007.

De Bry, John. *A Concise History of the 1715 Spanish Plate Fleet.* 1715 Fleet Society, Inc, 2024. Retrieved from https://1715fleetsociety.com/history/ on November 28, 2023.

De Bry, John. *Archaeology of Piracy Between Caribbean Sea and the North American Coast of 17th and 18th Centuries: Shipwrecks, Material Culture and Terrestrial Perspectives.* Journal of Ducoin, Jaques. *Report of Research in French Archives on the Nantes Ship La Concorde Captured by Pirates in 1717.* Raleigh: North Carolina Department of Cultural Resources, 2001. Retrieved from https://www.qaronline.org/compte-rendu-de-recherches-dans-les-archives-francaises-sur-le-navire-nantais-la-concorde-capture on September 14, 2023.

Dessales. *Annals of the Sovereign Council of Martinique or Historical table of the government of this colony from its first establishment to the present day to which we have attached the reasoned analysis of the laws which were published there with reflections on the usefulness or the insufficiency of each of these laws in particular.* Bergerac: JB Puynesge, 1786. Retrieved from https://recherche-anom.culture.gouv.fr/archive/resultats/complexe/lineaire/FRANOM_00010/n:110?RECH_S=+COL+C8+A+23+F%C2%B0+39%2C+COL+C8+A+23+F%C2%B0+39&RECH_TYP=and&RECH_unitid=COL+C8+A+23+F%C2%B0+39%2C+COL+C8+A+23+F%C2%B0+39&type=complexe&RECH_Field=RECH_lieux-&RECH_Letter=M on December 28, 2023.

Documenting the American South, Colonial and State Records of North Carolina, Volume 2, Retrieved from https://docsouth.unc.edu/index.html on March 4, 2024.

"Digs for Pirate's Hoard," *New York Times*, Friday, October 8, 1926.

Dolin, Eric Jay. *Black Flags, Blue Waters.* New York, London: Liverlight Publishing Corporation, 2018.

Downing, Clement. *A Compendious History of the Indian Wars; with An Account of the Rise, Progress, Strength, and Forces of Angria the Pyrate.* London: T. Cooper, at the Globe in Pater-noster Row. 1737. Retrieved from https://books.google.com/books?id=AKs-BAAAAQAAJ&printsec=frontcover&source=gbs_ge_summary_r&cad=0#v=onepage&q&f=false on April 1, 2024.

Drug Therapy in Colonial and Revolutionary America. Johns Hopkins Hospital Pharmacy, NCBI Literature Resources, 1976. Retrieved from Https://pubmed.ncbi.nlm.nih.gov/782235/#:~:text=Therapy%20in%20the%2017th%20and,the%20most%20widely%20used%20drugs on March 30, 2024.

Duffus, Kevin P. *The Last Days of Black Beard the Pirate*. Raleigh: Looking Glass Productions, Inc., 2008.

Ellms, Charles. *The Pirates Own Book: Authentic Narratives of the Most Celebrated Sea Robbers*. New York: Dover Publications, Inc., an unabridged reproduction of a work original published by the Marine Research Society, Salem, Massachusetts in 1924, 1993.

Fictum, David. *The Firsts of Blackbeard: Exploring Edward Thatch's Early Days as a Pirate*, 2015. Retrieved in February 2023 from https://csphistorical.com/2015/10/18/the-firsts-of-blackbeard-exploring-edward-thatchs-early-days-as-a-pirate/.

"Finding of Human Skull Spurs Hunt For Pirate Blackbeard's Buried Treasure," *New York Times*, Saturday, October 9, 1926

Goodall, Jamie L. H. *Pirates of the Chesapeake Bay*. Charleston, South Carolina: The History Press, 2020.

Gordon to Secretary to Admiralty. *Disposal of Blackbeard's effects*. October 2, 1719. PRO-ADM 1/1826. [72.1110.1–3] State Archives Building, Raleigh, NC.

Gordon to Admiralty September 14, 1721. *Report Relating to the Destroying the Pirate Blackbeard or Thatch*. PRO-ADM 1/1826. [72.1115.1–4] State Archives Building, Raleigh, NC.

Gosse, Philip. *The Pirates' Who's Who, Giving Particulars of the Lives & Deaths of the Pirates & Buccaneers*. Charles E. Lauriat Company, Boston, 1924.

Hamilton, Archibald, Lord, -1754. *Articles exhibited against Lord Archibald Hamilton, Late Governour of Jamaica: With Sundry Depositions and Proofs Relating to the Same*. London: S.N., Printed in the Year, 1717 Pdf. Retrieved from https://www.loc.gov/item/2005575719/ on December 29, 2023.

Harrison, Cy. *George Cammock (1665–1732)*. Three Decks' Forum, 2023. Retrieved from https://threedecks.org/index.php?display_type=show_crewman&id=4311on November 22, 2023.

Headlam, Cecil, ed. *Calendar of State Papers, Colonial Series, America and West Indies, January, 1719–February, 1720*. London: His Majesty's Stationery Office, 1933. Digitized from IA15135066–01. https://archive.org/details/sim_great-britain-public-record-america-and-the-west-indies_january-1710-february-1720/page/92/mode/2up on December 15, 2023.

Headlam, Cecil, ed. *Calendar of State Papers, Colonial Series, America and West Indies, August 1714–Dec. 1715*. London: His Majesty's Stationery Office, 1928. Digitized from IA1513506–01. Retrieved from https://archive.org/details/sim_great-britain-public-record-america-and-the-west-indies_august-1714-december-1715/mode/2up on December 15, 2023.

Headlam, Cecil, ed. *Calendar of State Papers, Colonial Series, America and West Indies, Jan. 1716–July, 1717*. London: His Majesty's Stationery Office, 1930. Digitized from IA1513506–01.

Retrieved from https://archive.org/details/sim_great-britain-public-record-america-and-the-west-indies_1716–1717_index/page/366/mode/2up on December 15, 2023.

Headlam, Cecil, ed. *Calendar of State Papers, Colonial Series, America and West Indies, August 1717–Dec. 1718.* London: His Majesty's Stationery Office, 1930. Digitized from IA1513506–01. Retrieved from https://archive.org/details/sim_great-britain-public-record-america-and-the-west-indies_august-1717-december-1718/mode/2up on December 15, 2023.

Headlam, Cecil, ed. *Calendar of State Papers, Colonial Series, America and West Indies, January, 1719 to February, 1720.* London: His Majesty's Stationery Office, 1933. Digitized from IA1513506–01. Retrieved from https://archive.org/details/sim_great-britain-public-record-america-and-the-west-indies_january-1719-february-1720/mode/2up on December 15, 2023.

Headlam, Cecil, ed. *Calendar of State Papers, Colonial Series, America and West Indies, 1722 to 1723.* London: His Majesty's Stationery Office, 1934. Digitized from IA1513506–01. Retrieved from https://archive.org/details/sim_great-britain-public-record-america-and-the-west-indies_1722–1723/mode/2up on December 15, 2023.

History of Blackbeard Island. Retrieved from https://www.fws.gov/refuge/blackbeard_island/about/history.html on January 21, 2021.

History of Sapelo Island. University of Georgia Marine Institute. Retrieved from http://coastgis.marsci.uga.edu/summit/saphistory.htm on March 20, 2024.

Hughson, Shirley Carter. *Blackbeard & the Carolina Pirates.* Hampton, Virginia: Reproduced from the 1894 edition by Port Hampton Press, 1894.

Hurd, G. Winfield. "Blackbeard Isle Lures Vain Hunters for Gold." *Evening Star, Washington, D.C.* July 18, 1937. (Accessed from https://www.newspapers.com/image/618941608/?terms=Blackbeard&match=1 on May 16, 2023).

Irwin, Henry. *Petition to Council for two horses.* 23 Dec. 1720. Calendar of Virginia State Papers p. 200. Original manuscript in Folder 30, Box 45, Id 36138, Colonial Papers. Library of Virginia, Richmond. Retrieved from Rosetta.virginiamemory.com:1801/delivery/DeliveryManagerServlet?dps_pid=IE3448582 on January 15, 2024.

Jacob, Thomas. *Letter of Thomas Jacob, HMS Diamond, Statement of Martin Preston.* ADM 1/1982. Retrieved from http://baylusbrooks.com/index_files/Page39237.htm on February 18, 2024.

Jacob, Thomas. *Letter of Thomas Jacob, HMS Diamond, Statement of William Megerre.* ADM 1/1982. Retrieved from http://baylusbrooks.com/index_files/Page39143.htm on February 18, 2024.

Jacob, Thomas. *Letter of Thomas Jacob, HMS Diamond, Statement of William Wade.* ADM 1/1982. Retrieved from http://baylusbrooks.com/index_files/Page39050.htm on February 18, 2024.

Jacobite Rebellion of 1715. National Archives of the United Kingdon, (SP 35/33 f.28). Retrieved from https://www.nationalarchives.gov.uk/education/resources/jacobite-1715/ on October 14, 2023.

Jacobite Timeline. National Archives of the United Kingdon. Retrieved from https://www.nationalarchives.gov.uk/wp-content/uploads/2014/06/jacobite-timeline.jpg on October 18, 2023.

Johnson, Charles. *A General and True History of the Lives and Actions of the most Famous Highwaymen, Murderers, Street-Robbers, &c. To which is added, A Genuine Account of the* Johnson, Charles. *A General History of the Robberies and Murders Of the Most Notorious PYRATES, and also Their Policies, Discipline and Government, From their first Rise and Settlement in the Island of Providence, in 1717, to the present Year 1724.* London: Printed for Ch. Rivington at the Bible and Crown in St. Paul's Church-Yard, F. Lacy at the Ship near the Temple-Gate, and J. Stone next the Crow Coffee-house the back of Greys-Inn, 1724.

Johnson, Charles. *A General History of the Lives and Adventures of The Most Famous Highwaymen, Murderers, Street-Robbers, &C. To which is added, A Genuine Account of the Voyages and Plunders of the most notorious Pyrates.* London: Printed for and Sold by J. Janeway, in White-Fryers, 1734. Retrieved from https://tenpound.com/bookmans-log/book/a-general-history-of-the-lives-and-adventures-of-the-most-famous-highwaymen-murderers-street-robbers-c-to-which-is-added-a-genuine-account-of-the-voyages-and-plunders-of-the-most-notorious-pyrate-4 on September 10, 2023.

Johnson, Charles. *A General History of the Lives and Adventures of The Most Famous Highwaymen, Murderers, Street-Robbers, &C. To which is added, A Genuine Account of the Voyages and Plunders of the most notorious Pyrates. Interspersed with diverse Tales, and pleasant Songs. And Adorned with Six and Twenty Large Copper Plates, Engraved by the book Makers.* London: Printed for and Sold by Olive Payne, at Horace's Head, in Round-Court in the Strand, over-against York Buildings, 1736. Retrieved from https://www.rct.uk/collection/1052025/a-general-history-of-the-lives-and-adventures-of-the-most-famous-highwaymen on September 10, 2023.

Johnson, Charles. *A General History of the Pyrates from Their Rise and Settlement in the Island of Providence, to the present Time. With the remarkable Actions and Adventures of the two Female Pyrates Mary Read and Anne Bonny.* London: T. Warner, 1724. https://archive.org/details/generalhistoryof00defo/page/n3/mode/2up

Johnson, Charles. *A General History of the Pyrates from Their Rise and Settlement in the Island of Providence, to the present Time. With the remarkable Actions and Adventures of the two*

Female Pyrates Mary Read and Anne Bonny. London: T. Woodard, at the Half-Moon, over against St. Dunstan's Church, Fleet-Street, 1726.

Johnson, Charles. Ed. Arthur L. Hayward. *Key Writings on Subcultures 1535–1727 Classics from the Underworld Volume IV, A General History of the Robberies and Murders of the Most Notorious Pirates, from their First Rise and Settlement in the Island of Providence to the Present Year*. London and New York: George Routledge & Sons, LTD., 1927. Retrieved from https://archive.org/details/a-general-history-of-the-robberies-and-murders-of-the-most-notorious-pirates-fro/mode/2up.

Johnson, Robert. *Letter Dated June 18, 1718 to the Lords Proprietors*. CSP 556. GR 033 (BMP D449) Charleston County Library.

Karraker, Cyrus H. *Piracy was a Business*. New Hampshire: Richard R. Smith Publisher, Inc., 1953. Retrieved from https://archive.org/details/piracywasabusiness_202309/mode/2up.

Konstam, Angus. *Black Beard: America's Most Notorious Pirate*. Hoboken, New Jersey: John Wiley & Sons, Inc., 2006.

Konstam, Angus. *The History of Pirates*. Guilford, CT: The Lyons Press, 1999.

Lee, Robert E. *Blackbeard the Pirate*. Winston-Salem, North Carolina: Published by John F. Blair, 1974.

Lewis, Harry M. *George Camocke's 1718 Proposal of a Jacobite–Pirate Alliance, The Mariner's Mirror*. 2021. Retrieved from https://doi.org/10.1080/00253359.2021.1940524 on August 24, 2022.

Lords Proprietors of the Province of Carolina to Tobias Knight. *Appointment as Secretary of Northern parts of Carolina*. January 24, 1712. Archives of North Carolina.

Lyme Log November 8–29, 1718. *Lyme Kiquotan Rd., Virginia*. Pro-Adm. 51/4250 [72.2278.1–2]. State Archives Building, Raleigh, NC.

MacNiel, Ben Dixon. "Blackbeard's Buried Treasure Found at Last But Mystery of Pirate Gold Not Yet Solved." *The News and Observer*. February 3, 1929. https://www.newspapers.com/image/?clipping_id=15249733&fcfToken=eyJhbG-ciOiJIUzI1NiIsInR5cCI6IkpXVCJ9.eyJmcmVlLXZpZXctaWQiOjI3MTIzMT-g5LCJpYXQiOjE2MDQ3NTYxMjksImV4cCI6MTYwNDg0MjUyOX0.yqTsdYCMjCCZHvD-0OFGHSVTLDU9cx7ECjTepEc4ypI (Accessed on May 14, 2021).

Markham, James Walter. *Nathaniel Mist's Journalistic Methods: An Analysis of The Weekly Journal, 1717–1719*. Austin: University of Texas, 1940. Retrieved from https://repositories.lib.utexas.edu/handle/2152/45646 on September 14, 2023.

Marquis, Samuel. *Blackbeard: The Birth of America*. Mount Sopris Publishing, 2018.

McIlwaine, Henry R. ed. *Executive Journals of the Colonial Council of Virginia* Vol. III. Richmond: Davis Bottom, Superintendent of Public Printing, 1928. Virginia State Library, Richmond. Retrieved from https://archive.org/details/executivejournal_c03virg/page/497/mode/1up on January 12, 2024.

McCarty, Daniel to the King. *Some Remarkable Proceedings in the Assembly of Virginia. Anno 1718.* 1718 Dec. 1. C.O.5/1318, Reel 41, f. 312. Library of Virginia, Richmond.

Mervine, William M. *Pirates and Privateers in the Delaware Bay and River*. The Pennsylvania Magazine of History and Biography, 1908, Vol. 32, No. 4 (1908), pp. 459–470 Published by: University of Pennsylvania Press. Retrieved from https://www.jstor.org/stable/20085447 on November 29, 2023.

Minutes of the North Carolina Governor's Council, May 27, 1719, including a deposition, a remonstrance, and correspondence concerning Tobias Knight's business with Edward Teach. North Carolina Council. May 27, 1719. North Carolina State Archives.

Morange, Jean. *Testimony of Mr. Jean Morange, owner of the boat La Volante, from Saint-Pierre, regarding the capture of this vessel by the bandits and the mistreatment suffered by the crew at the hands of the Spanish from Puerto Rico.* Secretary of State for the Navy - Correspondence on the arrival of Martinique (1635–1815). 1718. French National Archives. Retrieved from https://recherche-anom.culture.gouv.fr/ark:/61561/zn401kekiv on December 15, 2023.

Moseley to Thomas Cary, Governor, North Carolina, August 28, 1707. Colonial Court: Court of Chancery, 1689–1775. Colonial Court Records. North Carolina Digital Collections. Retrieved from https://digital.ncdcr.gov/Documents/Detail/colonial-court-of-chancery-1689-1775/409438?item=409585.

Nelson, Laura. *The Whydah Pirates Speak*. Columbia, South Carolina: Postillion LLC, 2020.

Norris, Allen Hart. *Beaufort County, North Carolina Deed Book I, 1696–1729: Records of Bath County, North Carolina*. Washington: The Beaufort County Genealogical Society, 2003.

O Ciardha, Eamonn. *Cammock (Camocke, Camock), George*. 2009. Retrieved from https://www.dib.ie/biography/cammock-camocke-camock-george-a1408 on September 14, 2023.

Order to Charge William Howard with piracy, 1718 Oct. 29. Colonial Papers. Library of Virginia. Retrieved from https://lva.primo.exlibrisgroup.com/permalink/01LVA_INST/1cgm05i/alma990015765170205756 on January 3, 2024.

Origins: Are the Bairds a Clan? Kincardineshire, Scotland: Clan Baird Society Worldwide. Retrieved from https://www.clanbairdsocietyworldwide.co.uk/news on December 2, 2023.

Palmer, William. P., ed. *Calendar of Virginia State Papers and Other Manuscripts, 1652-1781, Preserved in the Capitol at Richmond*. Richmond: R.F. Walker, Superintendent of Public Printing, 1875. Digitized by Google from the library of Oxford University. Retrieved from https://books.google.com/books?id=pKwFAAAAQAAJ&oe=UTF-8 on December 15, 2023.

Pearse, Vincent. *Account of Phenix at New York 1717–1718*. ADM 1.2282 North Carolina Archives.

Pearse, Vincent. *Logbook of HMS Phenix, February to April 1718*. ADM 51/690. Retrieved from http://baylusbrooks.com/index_files/Page16923.htm on March 1, 2024.

Pyrates Mary Read and Anne Bonny. London: T. Warner, 1724. https://archive.org/details/generalhistoryof00defo/page/n3/mode/2up

Plunkett, Dennis. *Blackbeard's Tower*. 2014. Retrieved from https://www.bahamastourcenter.com/blackbeards- tower/#:~:text=Blackbeard's%20Tower%20is%20one%20of,of%20the%20Pirates'%20Golden%20Era on January 6, 2021.

Pringle, Patrick. *Jolly Roger: The Story of the Great Age of Piracy*. Mineola, New York: Dover Publications, Inc., 2001, is an unabridged republication of the original edition published by W. W. Norton and Company Inc., New York, 1953

Quelch, John. *The arraignment, tryal, and condemnation, of Capt. John Quelch, for sundry piracies, robberies, and murder, committed on the coast of Brasil, who were found guilty, in Boston on the thirteenth of June, 1704. To which are also added, some papers that were produc'd at the tryal above said*. Printed for Benn. Bragg in Avmary-Lane, London, 1704. Retrieved from: https://archive.org/details/bim_eighteenth-century_the-arraignment-tryal-_quelch-john_1704 on March 10, 2024.

Rankin, Hugh. *The Pirates of North Carolina*. Raleigh: North Carolina Division of Archives and History, 1960, reprinted 2001.

Reed, Wingate C. *Beaufort County: Two Centuries of Its History*. Raleigh: The Edwards & Broughton Co., 1962.

Reiche, Ingrid. *Digitizing the Pyrates: Making a Digital Critical Edition of Captain Charles Johnson's A General History of the Pyrates* (1724–1726). Master's thesis, Carleton University, 2016. https://repository.library.carleton.ca/concern/etds/cj82k810d.

Roberts, Nancy. *Blackbeard and Other Pirates of the Atlantic Coast*. Durham, NC. Blair Publishing, 1993.

Robinson, Dennis J. *Pirates on the Piscataqua*. 2006. Retrieved from https://www.seacoastnh.com/seeking-blackbeards-pirate-treasure/?showall=1 on March 30, 2024.

Rony's Tours, Roatan History. 2023. Retrieved from https://ronystours.com/roatan-info/.

Selig, Robert. *Eighteenth-Century Currencies*. Published in *The Brigade Dispatch, Vol. XLIII No. 3, Autumn 2013*. Retrieved from chrome-extension://efaidnbmnnnibpcajpcglclefindmkaj/http://w3r-archive.org/history/library/seligreptria6.pdf on January 4, 2024.

Shipping Lists for South Carolina 1717–1721, CO5 / 508–509, Charleston Public Library, p. 51–61.

Shomette, Donald G. *Pirates on the Chesapeake.* Atglen, Pennsylvania: Tidewater Publishers, 1985.

Spotswood to Commissioners for Trade and Plantations. *Capt. Tach's Quartermaster apprehension and trial.* 1718 Dec. 22. CO5/1318. [41.ff.291–298]. Library of Virginia, Richmond.

Spotswood, Alexander, Brock, ed. *The Official Letters of Alexander Spotswood, Lieutenant-Governor of the Colony of Virginia, 1710–1722.* Richmond: Virginia Historical Society, 1882. Retrieved from https://archive.org/details/officialletterso12virg/page/n1057/mode/2up on January 11, 2024.

Spotswood Letter to Eden. *Alexander Spotswood: Biographical Sketch.* 1718 Nov. 7. F222.V81 M68 W 41. [1930 p. 150–151]. Virginia Museum of History & Culture, Richmond.

State of Merchant ships leaving Martinique for French ports during the last six months of 1717. Table. Secretary of State for the Navy—Correspondence on the arrival of Martinique (1635–1815). French National Archives. Retrieved From https://recherche-anom.culture.gouv.fr/ark:/61561/zn401ywtssf on December 15, 2023.

Taylor, Christophe. *Déclaration du Sieur Christophle Taylor, Capitaine ou navire Anglois le Alleyne, entue à Saint-Vincent et brûlé par les forbans January 12, 1718.* Secretary of State for the Navy - Correspondence on the arrival of Martinique (1635–1815). 1718. French National Archives. Retrieved from: https://recherche-anom.culture.gouv.fr/ark:/61561/zn401xv-suxj/daogrp/0/1on December 15, 2023.

The Colorful Pirate History of Amelia Island. Retrieved from https://www.ameliavacations.com/the-colorful-pirate-history-of-amelia-island/ on October 21, 2022.

The Deposition of Allen Bernard Captaine's Quarter Master of ye Sloop Bersheba Captn. Henry Jennings Comander. Retrieved from https://petercorneliushoof.blogspot.com/2014/08/a-story-of-henry-jennings-part-one.html on February 13, 2024.

The Deposition of Samuel Liddell– Jamaica. August 7, 1716. Jamaican Council Minutes ff.49–50. Retrieved from https://petercorneliushoof.blogspot.com/2015/03/a-further-story-of-henry-jennings.html on February 13, 2024.

The "Hammock House" c. 1700. Retrieved from https://blackbeardthepirate.com/history.htm on November 30, 2023.

The History and Lives Of all the most Notorious PIRATES and their CREWS, From Capt. Avery, who first settled at Madagascar, to Capt. John Gow, and James Williams, his Lieutenant, &c. who were hang'd at Execution Dock, June 11, 1725, for Piracy and Murther; and afterwards hang'd in Chains between Blackwall and Depford. London: Edward Midwinter, at the Looking-Glass on London-Bridge, 1725. Retrieved from https://www.loc.gov/item/10031667 on September 14, 2023.

The Lost Treasure of Edward Teach. Virginia Beach Live the Life. Retrieved from https://www.visitvirginiabeach.com/blog/post/vabeach-haunts-the-lost-treasure-of-edward-teach/ on January 15, 2021.

Trial of Dr. John Howell, 18–29 December 1721. National Archives, London, CO 23/1 (1717–1725), 42 iii. Retrieved from http://baylusbrooks.com/index_files/Page2224.htm on December 14, 2023.

The Trials of eight persons indited for piracy &c. Of whom two were acquitted, and the rest found guilty. Boston: Printed by B. Green, for John Edwards, and Sold at his Shop in King's Street, 1718. Retrieved from https://quod.lib.umich.edu/cgi/t/text/pageviewer-idx-?c=evans;idno=n01688.0001.001;cc=evans;node=N01688.0001.001:3;seq=28;page=-root;view=text;size=100 on February 13, 2024.

Timberlake, Henry. *Deposition, 17 Dec 1716.* Jamaica. Retrieved from http://baylusbrooks.com/index_files/Page2378.htm on December 14, 2023.

Tyler, Leon G. *Tyler's Quarterly Historical and Genealogical Magazine.* Vol. 1 No. 1. July 1919. Digitized by Google from the library of the University of Virginia and uploaded to the Internet Archive. Retrieved from https://archive.org/details/tylersquarterly00tylegoog/page/n48/mode/2up?q=Howard on December 24, 2023.

Smith, Austin, *The Gentleman Pirate: Stede Bonnet's Leadership and Legacy.* Retrieved from https://41380642.weebly.com/ on February 2, 2023.

Upshur, Thomas T. *Eastern-Shore History.* Virginia Historical Society, The Virginia Magazine of History and Biography, 1901.

Vallar, Cindy. *Benjamin Hornigold The Pirates' Pirate.* Retrieved from http://www.cindyvallar.com/hornigold.html on January 20, 2021.

Watson, Alan D. *Bath: The First Town in North Carolina.* Raleigh: North Carolina Department of Cultural Resourceless, 2005.

Watson, John F. *Annals of Philadelphia and Pennsylvania, Being a Collection of Memoirs, Anecdote, and Incidents of the City and Its Inhabitants from the Days of the Pilgrim Founders.* Philadelphia: Uriah Hunt, No. 147 Market Street, 1830. Retrieved from https://digital.library.pitt.edu/islandora/object/pitt:31735056288065 on January 15, 2024.

Watson, John F. *Annals of Philadelphia and Pennsylvania, in the Olden Time; Being a Collection of Memoirs, Anecdote, and Incidents of the City and Its Inhabitants, and of the Earliest Settlements of the Inland Part of Pennsylvania, from the Days of the Fonders, Vol II.* Philadelphia: J. Penington, U. Hunt – New York, Baker & Crane, 1844. Retrieved from archive.org/details/annalsofphiladel02watsuoft/mode/2up on October 23, 2024.

Wilde-Ramsing, Mark U., and Linda F. Carnes-McNaughton. *Blackbeard's Sunken Prize: The 300-Year Voyage of Queen Anne's Revenge.* Chapel Hill: The University of North Carolina Press, 2018.

Wills, Laura. *Bath County, North Carolina Early Wills 1696–1739.* Signal Mountain: Mountain Press, Reprinted 2019.

Willis, Laura. *Beaufort County, N.C. Deeds Vol. 1 (Oct. 1700–July 1709).* Melber, KY: Simmons Historical Publications, 2000.

Wilson, David. *Pirates, Merchants, and Imperial Authority in the British Atlantic, 1716–1726.* Glasgow: University of Strathclyde, 2017.

Virginia. Lieutenant-Governor, Alexander Spotswood, and R. A Brock. *The official letters of Alexander Spotswood, Lieutenant-Governor of the colony of Virginia, -1722, now first printed from the manuscript in the collections of the Virginia Historical Society.* Richmond, Va., Virginia Historical -85, 1882. Pdf. https://www.loc.gov/item/01005713/.

Woodard, Colin. *The Republic of Pirates.* New York: Harper Collins Publishers, 2007.

Yotov, Petko. *Old Style-New Style Side-By-Side Reference.* Retrieved from http://5ko.free.fr/en/jul.php?y=1717 on November 23, 2003.

Zupko, Ronald E. *A Dictionary of Weights and Measures for the British Isles: The Middle Ages to the Twentieth Century.* Philadelphia: American Philosophical Society, 1985.

Index

Symbols

1715 treasure fleet v, vii, 21, 23, 24, 241, 380, 429

A

Accomack 64, 65, 66

Adventure (sloop)

    Blackbeard 5, 7, 14, 173, 178–9, 187, 198–204, 207–8, 210, 216, 253, 255–6, 258–9, 261–3, 265, 268, 272, 275, 279–81, 288,-9, 296, 299–300, 302–4, 309, 314, 347, 356, 359, 365

    Herriot 123, 133–6, 139–140, 153, 156, 158, 165, 167–9, 176, 226, 311–2, 356, 359

    Horingold 57–59, 370–1

Albemarle 66, 208, 251, 269–70, 279, 345, 347

Allein, Richard 68, 157, 361

Antigua 85, 114–6, 118, 120, 188, 220, 221, 234, 236, 334

Arrot, Louis 100, 104

Artimesia (ship) 159

Ashworth, Leigh 27, 35, 143–4, 162

Assateague Island, Virginia 219

Ayscough, John 80

B

Bahamas 5, 21, 25, 39, 56–57, 92, 146, 148, 153, 162, 232

Bahia Honda, Cuba 30–31, 33, 40

Barbados (Barbadoes) 13, 68–74, 77, 79, 87, 99, 118, 191, 219–20, 222, 234–6, 323–4, 329–30, 334

Basse-Terre 110–11

Bath (Bath Town, North Carolina) 13–14, 121–2, 151–2, 170–2, 175, 178–87, 202–9, 216, 239, 241–4, 246, 248, 250–2, 254, 266, 269–70, 272, 274–5, 279–80, 282, 287–92, 294–7, 302, 306, 310, 312, 321–2, 324–9, 332–8, 342, 346–8, 356–8, 361–2, 365–7

Bay Islands, Honduras 28–29, 131

Beard Creek 324–6

Beard, Edward 322, 326, 329, 331, 337

Beard, James 322–9, 331–2, 336–7

Beard, William 323, 329–30

Beaufort, North Carolina 29, 168, 177, 322, 327, 336, 359

Bellamy, Samuel 28–30, 32–35, 40, 42–49, 55, 67, 71, 88, 92, 106, 110, 139, 143–4, 326

Bell, William 205, 207, 243–4, 294–5, 302–3, 308–9

Benjamin (sloop) 34, 41–45, 47–48

Bennet (sloop) 57–59

Bennett, Benjamin 58, 60, 143, 233

Bequia Island 63, 90, 98, 100, 102–5, 107

Bermuda 25, 61–62, 74, 105, 121, 126, 140, 143–4, 198–9, 201, 208, 232–33, 348

Bernard, Allen 30, 36

Bersheba (sloop) 26–27, 30–34, 37, 41, 144

Betty (sloop) 79–80, 84–85, 211, 213

Bicoya 99, 104

Blackbeard

  Teach, Edward 4, 65, 77, 83, 135, 137, 285, 315, 317, 320, 331, 345, 348

    Tach, Tache, Tatch, Teach, Teech, Thach, Thache, Thatch, Theach, Titche 4–7, 13–15, 44, 53–54, 57, 60–63, 65, 77, 79–80, 82–83, 87–91, 96, 99, 110, 113, 117, 133–7, 139–140, 154, 156, 158, 160–1, 163, 169–74, 176, 184–5, 188–91, 194, 202–3, 205, 207–8, 211–4, 217–8, 228, 237, 243, 245–9, 251–2, 256–7, 259–64, 268, 270, 273–6, 278, 281, 284–5, 288–9, 296, 299, 301–4, 307–8, 311–2, 314–20, 322, 329, 331, 333, 335, 344–5, 347–48, 359, 366

Black Caesar 14, 125, 128, 131–9, 287, 292, 297, 299

Boca Chica 122

Bondavias, Jean 60, 144

Bonnet, Stede 8, 11, 15, 18, 67–80, 99–107, 109, 112–4, 119, 122, 130–4, 136, 140, 142, 151, 155, 157, 159, 160–1, 165, 168–78, 191, 216–28, 230, 232, 235–6, 244, 284, 314, 320, 322, 327, 329, 330–2, 351–3, 356–61

Bonny, Anne 12, 16–17, 318

Bostock, Henry 53, 89–99, 108, 116–7, 120–1, 150, 309, 360

Boston, Massachusetts 108, 110, 131, 135–6, 138–9, 160–1

Boyer, Pierre 199–201

Brand, Ellis 5, 13, 79, 95, 151, 173–175, 185, 203–5, 210, 214–5, 241–52, 254–6, 259v60, 262, 267–76, 278–80, 282, 285–7, 289, 291–2, 294–5, 299, 300, 304, 307, 309, 313, 327, 334, 348, 358, 361

Brimstone-Hill 114–5

Bristol, England 4, 15, 52, 54–55, 71, 82, 84–85, 112, 145, 160, 315–7, 320

Brown, John 46, 48, 55

Burgess, Josiah 144, 162, 183, 347

Burlington, New Jersey 363–4, 368

C

Cammock, George (Camocke, Camocke) 10–11, 232, 236

Candler, Bartholomew 59, 69

Cape Charles 71, 79–80, 211, 213

Cape Cod 67, 88

Cape Fear viii, 72, 181, 189, 219, 223–5

Cape Henlopen 219–20

Cape Henry 218–9, 249, 348–50

Cape May 78, 80, 82–83, 86–87, 89, 92, 187, 189, 192, 217, 219

Captain Codd 84, 89

Captain Edwards 69, 72, 74, 112–3, 227. *See also* Bonnet, Stede

Captain Goelet 88, 92

Captain Hickinbottom 58–59

Captain Pritchard 62, 90

Captain Read 219–20, 222–3

Captain Richards 134–5, 138, 140, 156, 161, 234, 311, 353–4

Captain Rolland 82, 87

Captain Young 29–31, 34–35

Caribbean 1, 8, 21–22, 28–29, 45, 51–52, 61, 64, 70, 78, 92, 96, 98, 103, 106–7, 109–10, 113–4, 119, 122, 131, 142, 151, 198, 236, 319, 362

Carnegie, James 28

Carolina 64, 73–74, 154, 161–3, 168, 175, 181, 188–9, 219, 224, 228, 237, 247, 250, 275, 278, 298–9, 316, 356

Carpenter, Joseph 47

Cedar Island, Virginia 66

Chamberlayne, Edmund 206, 242, 272–3, 305–6, 308

Charleston (Charlestown, Charles Town, Charles-Town) xii, 16, 68, 72, 124, 134, 140, 142, 151, 153–65, 169, 216, 223, 225–7, 230, 234–6, 311, 320, 322, 323–5, 329–30, 343, 352–5, 357–8, 360–1

Charnock, Francis 37

Chesapeake Bay 79, 210

Cocoa Nut (sloop) 27, 30, 33

Crab Key 116

Cracherode, Anthony 283–4, 313

Crosby, Joshua 6, 88, 189

Crowley (ship) 155–6, 159, 353, 357

Cuba vii, 28–31, 41, 45–46, 58–59, 140, 212–3

D

Daniell, Robert 47, 57, 181

Dauphin (sloop) 101–2, 107, 109

de Echevers, Don Antonio 22

Delaware vii, viii, 53, 86, 192, 195, 219

Delaware Bay 72, 80, 82, 84, 212–3, 219

Delaware Capes 53, 63, 72, 80, 82–4, 86, 88, 90–2, 113, 187, 191–2, 195, 198, 212–3, 219–20

Delight (vessel) 52–54

del Valle, Don Juan Francisco 26

Diamond (HMS ship) 25, 135, 137–9, 151, 233

Dickinson, Jonathan 6, 86, 88, 189–90

Discovery (sloop) 28, 31, 33–4, 143

Dolphin (sloop) 135–6, 139–40

Dosset, Pierre 6, 96

Dossitt, George 31, 37

Drummond, Edward 315–7

Eagle (sloop) 26–27

Eden, Charles 170, 179, 181, 241, 251, 270, 275, 288, 302, 310

Eels, Joseph 30, 34–35

Eleuthera Island 41

Endeavour (vessel) 71

Ernaud, François 97–98, 103–4

Florida 21–26, 28, 41–42, 47, 58, 74–75, 79, 92, 140, 142, 148, 151, 153, 171, 198, 327, 330, 343–5, 362

Fortune (sloop) 68–69, 219, 222–3, 225

Francis (sloop) 219, 221–3, 225

Friends Islands 58–59

Fripp Island ix, 354–8, 361–2

Gilmor, Robert 71

Godin, Stephen 154–7, 160, 357

Good Intent (ship) 84, 212–3

Goose Creek 323

Gordon, George 210

Governor Johnson 154–6, 159, 161, 163, 227–8, 357

Governor Spotswood 5, 14, 35, 71, 180, 183, 201, 205, 207, 210, 239, 245–6, 255, 270, 273, 294, 302, 309, 361

Great Allen (ship) 107–8, 110, 268, 309, 360

Great Sloop (vessel) 87–88, 92, 97, 100, 105

Grenadines 63, 90, 98–99

Guadeloupe 110, 112, 115

Gulf of Honduras 29, 121, 129–31, 140, 151

Hamilton, Archibald 26–27

Hampton, Virginia 79, 210, 215, 280, 292–3, 298–9, 315–6, 339–41, 348, 361

Hands, Hesikia 202, 302, 314

   Hezekiah 286–87, 292, 295, 297, 313–4, 353–4, 356–9

Hands, Israel 134, 153–4, 156–8, 162, 169–74, 176, 202, 217–9, 221, 286–7, 292, 295, 297, 299, 302, 306, 311–4, 353–4, 356–9

Harbour Island 56–57, 126, 147–8, 231

Havana, Cuba 22, 24–27, 33, 43, 45, 47, 61, 120, 154, 169, 173, 312, 357

Haywood, Peter 27

Henry (sloop) 223–5

Herriot, David 123, 130, 133–4, 139–40, 153, 156, 158–9, 169–71, 173–4, 218, 226, 312, 314, 357, 359

Higgins, Jeremiah 42, 48

Holloway, John 214

Honduras 29–31, 35, 46, 121, 129–31, 135–6, 140, 142, 151, 171

Hoof, Peter Cornelius 29

Hornigold, Benjamin (Hornigold, Hornigole, Hornygold) 28, 34–35, 38–45, 41–42, 44, 47–49, 52–55, 57–64, 67, 69, 77–78, 85, 89–91, 99, 119, 143–6, 150–2, 186, 197, 211, 228, 231–2, 321, 327, 330, 351, 354, 358

Howard, William 5, 58, 60–61, 67, 79–80, 84, 139, 150, 152, 160, 168–9, 176, 186–7, 200–201, 210–3, 215, 242, 244, 278, 290, 293, 298, 312, 321, 327, 353, 356–9, 361, 365

Howell, John 57–58, 60

Hunter, Robert 6, 188

Isle of Pines, Cuba 46–47

J

Jacobite 8–9, 11, 18, 70–71, 106, 144, 150, 192, 217, 231–3, 236–7, 316, 320, 324, 330–1, 351

Jamaica 5–7, 15, 25–30, 33–36, 39, 41–42, 51–55, 57–59, 61–62, 69, 74–75, 77, 82–83, 87, 118, 129, 132–8, 140, 144, 147–8, 188–9, 191, 196–7, 315, 317–20, 322, 331–2

James (sloop) 324

Jane (sloop) 249, 253–6, 259, 261–2, 264–5, 267, 275, 32–3

Jennings, Henry 8, 26, 28–29, 36, 41, 44, 143–4

Johnson, Governor Robert 154–5, 227

Joy, Richard 114

K

Kecoughtan, Virginia (Kikotan) ix, 210–11, 213, 248, 249, 251, 274, 280, 290, 294, 339, 361

Killing, James 8, 220, 221, 222, 223

Kingston 27, 189, 319

Kiquotan, Virginia 210–1

Knight, Thomas 112–4

Knight, Tobias 7, 179, 181–2, 185, 205–8, 242–4, 268–70, 272–4, 276, 279, 289, 294–6, 300–310, 327–9, 332, 346, 366

L

La Buse 46–47, 48, 62, 370. *See also* Levasseur, Olivier

La Concorde (ship) 6, 93–94, 96–103, 105–7, 111, 123, 129–30, 165, 177, 212, 360

Lamb (brigantine) 5, 51–53, 55

Land of Promise (sloop) 134–6, 139

La Rose Emelye (ship) 199–205, 244

La Ville de Nantes (ship) 110, 113, 115

La Volante (vessel) 122, 124, 129

Layou 101, 104, 360

Le Roy de Guillaume (vessel) 129–30

Lesser Antilles 92, 107, 109

Levasseur, Olivier. *See* La Buse

Liddell, Samuel 27, 30, 36

Lieutenant Symonds 145–6, 231, 248, 254, 267

Logan, James 6, 80, 83, 89, 188–9, 196

Long Island, New York 63, 72, 78, 82, 86, 90

Lyme (HMS ship) 79, 174–5, 210, 241, 245, 247–9, 255–6, 259, 274, 299, 341

M

Manila Galleons 22

Manwareing, Peter 221

Marcus Hook xii, 191–4, 196–7

Margaret, (sloop) 116–7, 120–1, 150, 196–7, 348

Marianne, (sloop) 33–35, 42–43, 45–46, 48

Martinique (Martinico) 44, 78, 90, 92–93, 96–99, 101–105, 107–9, 110–11, 124, 130, 199, 201, 299

Martin, John (Martyn) 60–61, 67, 150–2, 186, 287, 292, 286–7, 299, 321, 325, 327, 354–5, 357–8, 361

Mary Anne (sloop) 43, 46, 48

Maryland 63, 86, 90, 174, 278, 341

Mary of Rochell (ship) 31–34

Mary Ormond 335–8

Mary Read 12, 16–17, 318

Mary (sloop) 27, 30, 32–34, 41

Massachusetts 28, 41, 126, 278

Maynard, Robert 121, 248, 252, 266, 349

Mayougany 101–2, 107

Mesnier, Charles 97

Mist, Nathaniel 16–17, 20, 62

Montserrat Merchant (ship) 93, 112–3

Morgan, Henry 29, 131

Moseley, Edward 179–182, 185, 202–203, 241–4, 250–2, 256, 265, 269–70, 273, 279, 294–5, 309, 362

Mr. Marks 156–8, 162, 163–4, 354, 357

Musson, Mathew 5, 47, 56–57

Napping, James (Nappin/Knapping) 58–59, 69, 126

Nassau 5, 8, 11, 21, 35, 39–42, 44–45, 48–49, 55–57, 60–61, 67, 75–76, 78–79, 141–53, 165, 188, 191, 196, 231–3, 236, 322, 330–31, 343–4, 351, 358, 360–2

Nevis Island 13, 113, 323–4

New Division (vessel) 114

New England 1, 54, 72–73, 87–88, 135–6, 138–40, 192, 362

New Jersey 6, 80, 188, 363, 365

Newton, Thomas 134–5, 139

New York 6, 48, 61, 64, 66, 71, 74, 80, 82, 87, 97, 145, 188, 231, 278, 316, 324, 362, 364

Noland, Richard 58, 60

North Carolina xii, 7, 14, 29, 62, 66, 72–74, 90, 97, 121, 161, 165, 167, 169–70, 173–4, 177, 179–81, 183–6, 203–11, 215, 219–20, 222–5, 227, 239, 241–6, 248–51, 255, 260, 267, 270, 274, 276, 278–9, 281, 289, 291–2, 29–6, 299–304, 309, 312–3, 316, 322, 324–7, 331, 334, 336, 345–7, 351, 354, 359, 362, 366

Ocracoke, Oackrycook, Ocacock, Oerecock, Oerecrock, Okercock, Okerecock 2, 13–14, 151, 173, 175, 178–9, 184, 187, 202–3, 208, 216, 218, 237–9, 241–3, 247–8, 250, 252–6, 260, 266, 269, 274–5, 282, 287, 289–93, 295–7, 302–3, 306, 309, 327–8, 333, 339–40, 349, 356, 361–2, 365

Odell, Samuel 253, 257, 287–8, 299

Olivier, Pierre Raimond 101

Ormond, Mary 335

Ormond, Wyriott 336–7

Palmer, Joseph 72–73

Pamlico Sound 173, 178, 208, 218, 239, 246, 252, 362

Pearl (HMS ship) 5, 210–1, 213–5, 242, 245, 247–9, 255, 259–60, 267, 274, 280–1, 290, 292–3, 341

Pearse, Vincent 145–6, 174

Pell, Ignatius 142, 151, 161, 170–2, 226

Pennsylvania 80, 86, 188–9, 191–5, 221, 315–6, 332, 340, 363

Peters, Peter 85–86

Phenix (HMS ship) 7, 145–8, 150–1, 174, 231

Philadelphia 6, 63, 79–80, 82–84, 86, 88–90, 159, 187–97, 213, 219, 221, 315–6, 332, 340, 362–4, 368

Pindar, William 60

Plum Point 187, 208–9, 366–8

Pollock, Thomas 179–82, 275, 302

Porter, Thomas 72–74

Portobelo (Portobello) 45–46, 58

Postillion (sloop) 46–47

Pretender 8–10, 71, 222, 232

Protestant Caesar (ship) 125, 128, 131–9

Providence Island 41

Quarry 52, 55, 370

Queen Anne 8, 10, 106

Queen Anne's Creek 251

Queen Anne's Revenge (ship) 96, 106–8, 110, 113, 115–7, 120–3, 130, 133–6, 138–40, 153, 156, 164–5, 167–9, 173, 175–7, 185, 210, 212, 216, 227, 232, 241, 311–2, 348, 352, 354–7, 359–61

Ranger (sloop) 232–3, 249, 253–6, 259–63, 267, 274, 376

Rattan Island 132, 137

Read, Mary 12

Read, Thomas 219, 221–2

Revenge (sloop) 59, 69–76, 78–80, 83–89, 92, 96–98, 100, 102–4, 106–8, 110, 112–7, 119–23, 129–31, 133–41, 153, 156, 160–1, 164–5, 167–9, 171–3, 175–8, 185, 187–91, 210, 212–3, 216–7, 227, 232, 241, 311–2, 330–1, 348, 352–7, 359–61

Rhett, William 223, 236

Rhode Island 72, 134

Richardson, Joseph 85

Roatán 131, 137

Roberts, John 16, 128, 326

Robins, James (Robbins) 182, 287–9, 297–9, 361

Rogers, Woodes 57, 149–50, 232–3

Royal James (sloop) 8–10, 71, 173, 217–9, 222–5, 232

Ruby (ship) 158–9

#

Saint Christopher 114–5, 117, 121

Saint Croix 116

Saint-Michel de Nantes (vessel) 126

Saint Vincent 100, 268, 360

Salter, Edward 118, 121, 286–7, 292, 297–9, 348, 361–2

Sandy Hook 87, 372

Saturday's Post 16–17, 62

Scarborough (HMS ship) 13

Seaford 13, 113, 115–6

Sea Nymph (snow) 84–8

Simon, Sieur 101, 107, 109

South Carolina xii, 5, 47, 56, 61, 68, 72–74, 154–6, 160, 162, 170, 180, 212–3, 216, 226, 228, 235, 278, 313, 323–4, 326, 330, 353

Spotswood, Alexander 5, 183, 241, 276

St. Amour, Charles 102–4

St. Croix 120

Stuart, James Edward 8–10, 71, 106, 173, 217, 232

St. Vincent 100–105, 107–8, 110, 309

Sullivan's Island 226, 236

T

Taylor, Christopher 105, 107–10, 360

Thache, Cox 7, 331

Thache, Sr., Edward 319

The Boston News-Letter xii, 6, 39–40, 48, 61–63, 67, 72–74, 79–80, 82–87, 89–91, 94, 110, 125–6, 130, 132, 135–6, 139, 142, 155, 159, 188, 190, 203–5, 215, 235, 255, 257, 260–1, 264, 319, 361

The Weekly Journal 16–17, 62

Thurbar (sloop) 61–63

Timberlake, Henry 5, 51, 53, 55

Toison d'Or (ship) 199–201

Turbes (vessel) 71

Turbett, Andrew 71

Turneffe Islands 131–133

V

Vane, Charles (Vean/Veine) 8, 11, 28, 106, 144–146, 148, 150, 223–4, 231–7, 239–40, 351

Vickers, John 42, 45

Virginia xi, xii, 5, 14, 35, 47, 62–67, 71–72, 78–80, 82, 85–86, 89–90, 99, 113, 173–4, 180–1, 183, 185, 187, 198, 201, 205–7, 209–15, 218–20, 239, 241–2, 244–8, 250–1, 255, 260, 263, 267, 270–1, 273, 275–9, 281–2, 295, 298–302, 305, 307–8, 313, 315–6, 324, 334, 340–1, 348–9, 361

W

Walker, Thomas 40–41, 55–57

Whydah (ship) 67, 92, 96–99

William and Samuel (sloop) 137, 139–140

Williamsburg, Virginia xii, 79, 121, 202, 206, 210–1, 213–5, 241, 243–4, 246, 290–1, 293, 295–9, 301, 307, 312, 314, 316, 327, 339–41, 348, 357, 361

William (ship) 159, 161, 165

Williams, Palsgrave 28, 48, 144

Wills, John 26

Winchelsea (HMS ship) 59, 69

Windsor (HMS ship) 55, 250, 319–20

Wragg, Samuel 15–7, 353, 357

Wright, Pearce 59

Wyer, William 131

Winchelsea (HMS ship) 59, 69

Windsor (HMS ship) 55, 250, 319–20

Wragg, Samuel 15–7, 353, 357

Wright, Pearce 59

Wyer, William 131

STAY IN THE KNOW

Be among the first to know about Robert Jacob's appearances, lectures, presentations, and other pirate news.

Also available from Robert Jacob are his award-winning bestsellers

A Pirate's Life in the Golden Age of Piracy

ISBN: 978-1-937801-91-5 (Hard Cover)
ISBN: 978-1-950075-09-6 (Paperback)

Think you know everything about Pirates? The facts are that the historical record on pirates is vague, contradictory, and rarely accurate. Digging deep into the true history of Piracy and those who lived this life, Robert Jacob has unearthed a treasure of information that allows you to see, feel, and experience the true life and motivation of pirates in their Golden Age. A Pirate's Life in the Golden Age of Piracy will fascinate and transport you back in time with rich stories and visual accuracy. You're in for some surprising revelations that will leave you even more intrigued!

Pirates of the Florida Coast: Truths, Legends, and Myths

ISBN: 978-1-950075-59-1 (Hard Cover)

Florida has a long and rich history of pirates plundering its shores, cities, and shipping lanes. Tales of Florida pirates abound from Key West to Pensacola and from Miami to Jacksonville. But how much of it is true? Digging deep into the true history of piracy along the Florida coast, Robert Jacob has unearthed a treasure of information that reveals the truths and identifies the myths about Florida Pirates, from Sir Francis Drake to José Gaspar. Pirates of the Florida Coast will transport you back to old Florida and paint a vivid picture of piracy throughout the years.

www.RobertJacobAuthor.com

www.ingramcontent.com/pod-product-compliance
Lightning Source LLC
Chambersburg PA
CBHW060333010526
44117CB00017B/2815